CARBON CAPTURE AND STORAGE

CO_2 MANAGEMENT TECHNOLOGIES

CARBON CAPTURE AND STORAGE

CO_2 MANAGEMENT TECHNOLOGIES

Edited by

Amitava Bandyopadhyay, PhD

Apple Academic Press

TORONTO NEW JERSEY

Apple Academic Press Inc.
3333 Mistwell Crescent
Oakville, ON L6L 0A2
Canada

Apple Academic Press Inc.
9 Spinnaker Way
Waretown, NJ 08758
USA

First issued in paperback 2021

Exclusive worldwide distribution by CRC Press, a member of Taylor & Francis Group

ISBN 13: 978-1-77463-341-0 (pbk)
ISBN 13: 978-1-77188-021-3 (hbk)

Library of Congress Control Number: 2014932150

Library and Archives Canada Cataloguing in Publication

Carbon capture and storage: CO_2 management technologies/edited by Amitava Bandyopadhyay, Ph.D.

Includes bibliographical references and index.
ISBN 978-1-77188-021-3 (bound)
1. Carbon sequestration--Technological innovations. 2. Geological carbon sequestration-Technological innovations. 3. Carbon dioxide mitigation--Technological innovations.
I. Bandyopadhyay, Amitava (Chemical engineer), author, editor of compilation

TD885.5.C3C27 2014   628.5'32   C2014-900664-0

Apple Academic Press also publishes its books in a variety of electronic formats. Some content that appears in print may not be available in electronic format. For information about Apple Academic Press products, visit our website at **www.appleacademicpress.com** and the CRC Press website at **www.crcpress.com**

ABOUT THE EDITOR

AMITAVA BANDYOPADHYAY, PhD

Dr. Amitava Bandyopadhyay is currently the Visiting Associate Professor at the School of Environment, Resources and Development at the Asian Institute of Technology in Thailand. Before joining as a faculty member in the Department of Chemical Engineering at the University of Calcutta, he served for more than 10 years in the West Bengal Pollution Control Board (WBPCB), the State Environmental Regulatory Agency, dealing with the implementation of environmental rules and regulations. He was a Faculty Member of the World Bank concerning the development of emission standards for Indian petrochemical industries. He has worked in several organizations before joining in the State Environmental Regulatory Agency. Besides his active background in teaching, he focuses his research on waste minimization, advanced separation processes, CO_2 capture, and emission control.

He has published more than 70 articles in peer-reviewed journals and authored two books. He is one of the topical editors of water and environmental engineering in *CLEAN—Soil, Air, Water* (Wiley-Blackwell), has acted as one of the guest editors for *Separation and Purification Technology* (Elsevier) on CO_2 Capture Technology, and also serves on the editorial boards and advisory boards of several international journals including *Environmental Quality Management* (Wiley-Blackwell). He has reviewed projects funded at both international and national levels and is a member of various institutions and governmental departments in India. Dr. Bandyopadhyay has received several laurels for his excellent research works, importantly the Outstanding Paper Award 2012 from Clean Technologies & Environmental Policy for contributing an article—Chapter 1 in this book.

ABOUT THE EDITOR

[illegible]

CONTENTS

PART III: Biological Sequestration of CO_2

PART IV: Current Research Trends in CO_2 Capture using Ionic Liquids

PART V: Air Capture of CO_2

ACKNOWLEDGMENT AND HOW TO CITE

The editor and publisher thank each of the authors who contributed to this book, whether by granting their permission individually or by releasing their research as open source articles. The chapters in this book were previously published in various places in various formats. To cite the work contained in this book and to view the individual permissions, please refer to the citation at the beginning of each chapter. Each chapter was read individually and carefully selected by the editors. The result is a book that provides a nuanced study of the recent advances in the study of the capture and storage of carbon dioxide.

LIST OF CONTRIBUTORS

Per Aagaard
Department of Geosciences, University of Oslo, Pb. 1047, Blindern, Oslo, Norway

Ugur Atikol
Department of Mechanical Engineering, Faculty of Engineering, Eastern Mediterranean University, Famagusta, Via Mersin 10, North Cyprus, Turkey

Amitava Bandyopadhyay
Department of Chemical Engineering, University of Calcutta, 92, A.P.C. Road, Kolkata, 700009, India

Manindra Nath Biswas
Department of Leather Technology, Government College of Engineering & Leather Technology, Salt Lake City, Kolkata 700 098, India

Fitzgerald L. Booker
Plant Science Research Unit, United States Department of Agriculture, Agricultural Research Service, Raleigh, North Carolina, United States of America and Department of Crop Science, North Carolina State University, Raleigh, North Carolina, United States of America

Kent O. Burkey
Plant Science Research Unit, United States Department of Agriculture, Agricultural Research Service, Raleigh, North Carolina, United States of America and Department of Crop Science, North Carolina State University, Raleigh, North Carolina, United States of America

Susan A. Carroll
Lawrence Livermore National Laboratory, 7000 East Avenue, Livermore CA 94550, USA

Lei Cheng
Department of Plant Pathology, North Carolina State University, Raleigh, North Carolina, United States of America

Jared L. Deforest
Department of Environmental and Plant Biology, Ohio University, Athens, Ohio, United States of America

Ilka C. Feller
Smithsonian Environmental Research Center, Edgewater, Maryland, United States of America

Edwin L. Fiscus
Plant Science Research Unit, United States Department of Agriculture, Agricultural Research Service, Raleigh, North Carolina, United States of America and Department of Crop Science, North Carolina State University, Raleigh, North Carolina, United States of America

Patrick Grimes
Grimes Associates, Scotch Plains, NJ 07076

David Gutiérrez-Tauste
Renewable Energies R&D Department, LEITAT Technological Center, Carrer de la Innovació, 2, Terrassa 08225 Barcelona, Spain

Eric T. Harvill
Department of Veterinary and Biomedical Sciences, The Pennsylvania State University, University Park, Pennsylvania, United States of America

Helge Hellevang
Department of Geosciences, University of Oslo, Pb. 1047, Blindern, Oslo, Norway

Sara E. Hester
Department of Veterinary and Biomedical Sciences, The Pennsylvania State University, University Park, Pennsylvania, United States of America and Graduate Program in Biochemistry, Microbiology, and Molecular Biology, The Pennsylvania State University, University Park, Pennsylvania, United States of America

Shuijin Hu
Department of Plant Pathology, North Carolina State University, Raleigh, North Carolina, United States of America

Mehrdad Khamooshi
Department of Mechanical Engineering, Faculty of Engineering, Eastern Mediterranean University, Famagusta, Via Mersin 10, North Cyprus, Turkey

Klaus S. Lackner
Columbia University, New York, NY 10027

Qingping Li
China National Offshore Oil Corporation Research Center, Beijing 100027, China

Yu Liu
Key Laboratory of Ocean Energy Utilization and Energy Conservation of Ministry of Education, Dalian University of Technology, Dalian 116024, China

Catherine E. Lovelock
School of Biological Sciences, The University of Queensland, St Lucia, Queensland, Australia

Minghsun Lui
Department of Microbiology, Immunology, and Molecular Genetics, David Geffen School of Medicine at UCLA, University of California Los Angeles, Los Angeles, California, United States of America

Walt W. McNab
Lawrence Livermore National Laboratory, 7000 East Avenue, Livermore CA 94550, USA

Tracy Nicholson
National Animal Disease Center, Agricultural Research Service, United States Department of Agriculture, Ames, Iowa, United State of America

Daryl Nowacki
Department of Veterinary and Biomedical Sciences, The Pennsylvania State University, University Park, Pennsylvania, United States of America and Graduate Program in Biochemistry, Microbiology, and Molecular Biology, The Pennsylvania State University, University Park, Pennsylvania, United States of America

Kiyan Parham
Department of Mechanical Engineering, Faculty of Engineering, Eastern Mediterranean University, Famagusta, Via Mersin 10, North Cyprus, Turkey

Thi Hai Van Pham
Department of Geosciences, University of Oslo, Pb. 1047, Blindern, Oslo, Norway

Alain C. Pierre
Institut de Recherches sur la Catalyse et L'environnement de Lyon, Université Claude Bernard Lyon 1 CNRS, UMR 5256, 2 Avenue Albert Einstein, 69626 Villeurbanne, France

Xuke Ruan
Key Laboratory of Ocean Energy Utilization and Energy Conservation of Ministry of Education, Dalian University of Technology, Dalian 116024, China

Roger W. Ruess
Institute of Arctic Biology, University of Alaska Fairbanks, Fairbanks, Alaska, United States of America

Thomas W. Rufty
Department of Crop Science, North Carolina State University, Raleigh, North Carolina, United States of America

H. David Shew
Department of Plant Pathology, North Carolina State University, Raleigh, North Carolina, United States of America

James Skinner
Renewable Energies R&D Department, LEITAT Technological Center, Carrer de la Innovació, 2, Terrassa 08225 Barcelona, Spain

Yongchen Song
Key Laboratory of Ocean Energy Utilization and Energy Conservation of Ministry of Education, Dalian University of Technology, Dalian 116024, China

Elena Torralba-Calleja
Renewable Energies R&D Department, LEITAT Technological Center, Carrer de la Innovació, 2, Terrassa 08225 Barcelona, Spain

Sharon C. Torres
Lawrence Livermore National Laboratory, 7000 East Avenue, Livermore CA 94550, USA

Cong Tu
Department of Plant Pathology, North Carolina State University, Raleigh, North Carolina, United States of America

Lucian Wielopolski
Environmental Science Department, Brookhaven National Laboratory, Bldg. 490, Upton, NY 11973, USA

Mingjun Yang
Key Laboratory of Ocean Energy Utilization and Energy Conservation of Ministry of Education, Dalian University of Technology, Dalian 116024, China

Jiafei Zhao
Key Laboratory of Ocean Energy Utilization and Energy Conservation of Ministry of Education, Dalian University of Technology, Dalian 116024, China

Hans-J. Ziock
Los Alamos National Laboratory, Los Alamos, NM 87544

INTRODUCTION

Carbon capture and storage (CCS) refers to a set of methods for the mitigation, remediation, and storage of industrial CO_2 emissions. This current book addresses the technologies currently being applied and developed, and those most in need of further research. The book as a whole discusses methods of carbon capture in industrial settings, while the various sections look at topics such as biological and geological approaches to carbon sequestration, introducing ionic liquids as a method of carbon capture, and new approaches to capturing CO_2 from ambient air.

Carbon dioxide (CO_2), one of the green house gases (GHGs), has been well known for more than a century. Its emission from the combustion of fossil fuels, in addition to other industrial sources, is adversely affecting the climate on earth. Climate change is emerging as a risk all over the world that has generated public concern. Estimates have indicated that power production contributes to the tune of 70% of the total CO_2 released into the atmosphere from fossil fuel combustion worldwide. Capturing and securely storing CO_2 from the global combustion systems thus constitutes an important and achievable target. A legion of researchers have thus far developed absorbents to remove CO_2 from combustion facilities that are currently recognized globally as most effective. The cost of capturing CO_2 can be reduced by finding a low-cost solvent that can minimize energy requirements, equipment size, and corrosion. Monoethanolamine is being used for removing CO_2 from the exhaust streams and is a subject inculcated over a period of about 80 years. A host of such amines are being investigated and put into practice. However, commercializations of such operating plants for capturing CO_2 from power plants in the world are few and far between. On the other hand, aqueous ammonia is the other chemical solvent for capturing CO_2 that has proven experimentally to be more effective than amine-based processes. Chapter 1, by Bandyopadhyay, aims at critically elucidating relative merits and demerits of ammonia and amine-based CO_2 capture options from the exhausts of coal-fired thermal power plants (TPPs). It includes the life cycle CO_2 emissions for both the processes. Finally, it is estimated that a total emission of about 152 Mt CO_2-equivalent could occur after use of 100 Mt ammonium bicarbonate (NH_4HCO_3) as synthetic N-fertilizer that is about 50% of the total CO_2 captured (315 Mt) for producing the fertilizer, NH_4HCO_3. Clearly, this estimate demonstrates that the synthetic N-fertilizer, NH_4HCO_3, produced by NH_3 scrubbing of CO_2 from fossil fuel (e.g., coal) fired TPP could have a significant beneficial environmental impact so far as GHG emission is concerned.

The emission of CO_2 into the atmosphere is causing the majority of the global warming, and thus various end-of-pipe treatment methods have evolved to capture CO_2 from fixed point sources. In Chapter 2, Bandyopadhyay and Biswas deal with CO_2 capture from a simulated gas stream using dilute NaOH solution in a spray column using a two-phase critical flow atomizer capable of producing very fine sprays with high degree of uniformity and moving at very high velocities. Experimentation was carried out to investigate the percentage removal of CO_2 as well as interfacial area as functions of different variables. The maximum percentage removal of CO_2 observed was about 99.96% for a QL/QG ratio of 6.0 m^3/1000 ACM (liquid flow rate of 1.83×10^{-5} m^3/s and gas flow rate of 3.33×10^{-3} m^3/s) and for a CO_2 feed rate of 100 l/h, while the observed values of interfacial area were in the range of 22.62–88.35 m^2/m^3 within the framework of the experimentation. A simple correlation was developed for predicting the interfacial area as functions of various pertinent variables of the system. Experimental data fitted excellently well with the correlation. The comparison of the interfacial area observed between the present system and the existing systems revealed that the present system produced higher values of interfacial area than the existing systems and hence the performance of the system was better than the existing system.

CO_2 hydrate formation and dissociation is crucial for hydrate-based CO_2 capture and storage. Experimental and calculated phase equilibrium conditions of carbon dioxide (CO_2) hydrate in porous medium were investigated in Chapter 3, by Yang and colleagues. Glass beads were used to form the porous medium. The experimental data were generated using a graphical method. The results indicated the decrease of pore size resulted in the increase of the equilibrium pressure of CO_2 hydrate. Magnetic resonance imaging (MRI) was used to investigate the priority formation site of CO_2 hydrate in different porous media, and the results showed that the hydrate form firstly in BZ-02 glass beads under the same pressure and temperature. An improved model was used to predict CO_2 hydrate equilibrium conditions, and the predictions showed good agreement with experimental measurements.

There are two distinct objectives in monitoring geological carbon sequestration (GCS): Deep monitoring of the reservoir's integrity and plume movement and near-surface monitoring (NSM) to ensure public health and the safety of the environment. However, the minimum detection limits of the current instrumentation for NSM is too high for detecting weak signals that are embedded in the background levels of the natural variations, and the data obtained represents point measurements in space and time. In Chapter 4, Wielopolskie introduces a new approach for NSM, based on gamma-ray spectroscopy induced by inelastic neutron scatterings (INS). This technique offers novel and unique characteristics providing the following: (1) High sensitivity with a reducible error of measurement and detection limits, and, (2) temporal- and spatial-integration of carbon in soil that results from underground CO_2 seepage. Preliminary field results validated this approach showing carbon suppression of 14% in the first year and 7% in the second year. In addition the temporal behavior of the error propagation is presented and it is shown that for a signal at the level of the minimum detection level the error asymptotically approaches 47%.

In the past decade, the capture of anthropic carbonic dioxide and its storage or transformation have emerged as major tasks to achieve, in order to control the increasing atmospheric temperature of our planet. One possibility rests on the use of carbonic anhydrase enzymes, which have been long known to accelerate the hydration of neutral aqueous CO_2 molecules to ionic bicarbonate HCO_3- species. In Chapter 5, by Pierre, the principle underlying the use of these enzymes is summarized. Their main characteristics, including their structure and catalysis kinetics, are presented. A special section is next devoted to the main types of CO_2 capture reactors under development, to possibly use these enzymes industrially. Finally, the possible application of carbonic anhydrases to directly store the captured CO_2 as inert solid carbonates deserves a review presented in a final section.

In Chapter 6, Pham and colleagues consider continental flood basalts (CFB) as potential CO_2 storage sites because of their high reactivity and abundant divalent metal ions that can potentially trap carbon for geological timescales. Moreover, laterally extensive CFB are found in many place in the world within reasonable distances from major CO_2 point emission sources. Based on the mineral and glass composition of the Columbia River Basalt (CRB), the authors estimated the potential of CFB to store CO_2 in secondary carbonates. They simulated the system using kinetic dependent dissolution of primary basalt-minerals (pyroxene, feldspar and glass) and the local equilibrium assumption for secondary phases (weathering products). The simulations were divided into closed-system batch simulations at a constant CO_2 pressure of 100 bar with sensitivity studies of temperature and reactive surface area, an evaluation of the reactivity of H_2O in $scCO_2$, and finally 1D reactive diffusion simulations giving reactivity at CO_2 pressures varying from 0 to 100 bar. Although the uncertainty in reactive surface area and corresponding reaction rates are large, the article estimated the potential for CO_2 mineral storage and identified factors that control the maximum extent of carbonation. The simulations showed that formation of carbonates from basalt at 40 C may be limited to the formation of siderite and possibly FeMg carbonates. Calcium was largely consumed by zeolite and oxide instead of forming carbonates. At higher temperatures (60 – 100 C), magnesite is suggested to form together with siderite and ankerite. The maximum potential of CO_2 stored as solid carbonates, if CO_2 is supplied to the reactions unlimited, is shown to depend on the availability of pore space as the hydration and carbonation reactions increase the solid volume and clog the pore space. For systems such as in the $scCO_2$ phase with limited amount of water, the total carbonation potential is limited by the amount of water present for hydration of basalt.

Reactive-transport simulation is a tool that is being used to estimate long-term trapping of CO_2, and wellbore and cap rock integrity for geologic CO_2 storage. In Chapter 7, Carroll and colleagues reacted end member components of a heterolithic sandstone and shale unit that forms the upper section of the In Salah Gas Project carbon storage reservoir in Krechba, Algeria with supercritical CO_2, brine, and with/without cement at reservoir conditions to develop experimentally constrained geochemical models for use in reactive transport simulations. The authors observed marked changes in solution composition when CO_2 reacted with cement, sandstone, and shale components at reservoir conditions. The geochemical model for the reaction of sandstone and shale with CO_2 and brine is a simple one in which albite, chlorite, illite and carbonate minerals partially dissolve and boehmite, smectite, and amorphous silica precipitate. The geochemical model for the wellbore environment is also fairly simple, in which alkaline cements and rock react with CO_2-rich brines to form an Fe containing calcite, amorphous silica, smectite and boehmite or amorphous $Al(OH)_3$. Our research shows that relatively simple geochemical models can describe the dominant reactions that are likely to occur when CO_2 is stored in deep saline aquifers sealed with overlying shale cap rocks, as well as the dominant reactions for cement carbonation at the wellbore interface.

Sensing the environment allows pathogenic bacteria to coordinately regulate gene expression to maximize survival within or outside of a host. In Chapter 8, Hester and colleagues show that *Bordetella* species regulate virulence factor expression in response to carbon dioxide levels that mimic in vivo conditions within the respiratory tract. We found strains of Bordetella bronchiseptica that did not produce adenylate cyclase toxin (ACT) when grown in liquid or solid media with ambient air aeration, but produced ACT and additional antigens when grown in air supplemented to 5% CO_2. Transcriptome analysis and quantitative real time-PCR analysis revealed that strain 761, as well as strain RB50, increased transcription of genes encoding ACT, filamentous hemagglutinin (FHA), pertactin, fimbriae and the type III secretion system in 5% CO2 conditions, relative to ambient air. Furthermore, transcription of cyaA and fhaB in response to 5% CO_2 was increased even in the absence of BvgS. In vitro analysis also revealed increases in cytotoxicity and adherence when strains were grown in 5% CO_2. The human pathogens *B. pertussis* and *B. parapertussis* also increased transcription of several virulence factors when grown in 5% CO_2, indicating that this response is conserved among the classical bordetellae. Together, our data indicate that *Bordetella* species can sense and respond to physiologically relevant changes in CO_2 concentrations by regulating virulence factors important for colonization, persistence and evasion of the host immune response.

CO_2 emissions from cleared mangrove areas may be substantial, increasing the costs of continued losses of these ecosystems, particularly in mangroves that have highly organic soils. In Chapter 9, Lovelock and colleagues measured CO_2 efflux from mangrove soils that had been cleared for up to 20 years on the islands of Twin Cays, Belize. The authors also disturbed these cleared peat soils to assess what disturbance of soils after clearing may have on CO_2 efflux. CO_2 efflux from soils declines from time of clearing from ~10 600 tonnes km^{-2} $year^{-1}$ in the first year to 3000 tonnes km^2 $year^{-1}$ after 20 years since clearing. Disturbing peat leads to short term increases in CO_2 efflux (27 umol m^{-2} s^{-1}), but this had returned to baseline levels within 2 days. Deforesting mangroves that grow on peat soils results in CO_2 emissions that are comparable to rates estimated for peat collapse in other tropical ecosystems. Preventing deforestation presents an opportunity for countries to benefit from carbon payments for preservation of threatened carbon stocks.

Climate change factors such as elevated atmospheric carbon dioxide (CO_2) and ozone (O_3) can exert significant impacts on soil microbes and the ecosystem level processes they mediate. However, the underlying mechanisms by which soil microbes respond to these environmental changes remain poorly understood. The prevailing hypothesis, which states that CO_2- or O_3-induced changes in carbon (C) availability dominate microbial responses, is primarily based on results from nitrogen (N)-limiting forests and grasslands. It remains largely unexplored how soil microbes respond to elevated CO_2 and O_3 in N-rich or N-aggrading systems, which severely hinders our ability to predict the long-term soil C dynamics in agroecosystems. Using a long-term field study conducted in a no-till wheat-soybean rotation system with open-top chambers, Chapter 10, by Cheng and colleagues, showed that elevated CO_2 but not O_3 had a potent influence on soil microbes. Elevated CO_2 (1.5×ambient) significantly increased, while O_3 (1.4×ambient) reduced, aboveground (and presumably belowground) plant residue C and N inputs to soil. However, only elevated CO_2 significantly affected soil microbial biomass, activities (namely heterotrophic respiration) and community composition. The enhancement of microbial biomass and activities by elevated CO_2 largely occurred in the third and fourth years of the experiment and coincided with increased soil N availability, likely due to CO_2-stimulation of symbiotic N_2 fixation in soybean. Fungal biomass and the fungi:bacteria ratio decreased under both ambient and elevated CO_2 by the third year and also coincided with increased soil N availability; but they were significantly higher under elevated than ambient CO_2. These results suggest that more attention should be directed towards assessing the impact of N availability on microbial activities and decomposition in projections of soil organic C balance in N-rich systems under future CO_2 scenarios.

The cycle performance of refrigeration cycles depends not only on their configuration, but also on thermodynamic properties of working pairs regularly composed of refrigerant and absorbent. The commonly used working pairs in absorption cycles are aqueous solutions of either lithium bromide water or ammonia water. However, corrosion, crystallization, high working pressure, and toxicity are their major disadvantages in industrial applications. Therefore, seeking more advantageous working pairs with good thermal stability, with minimum corrosion, and without crystallization has become the research focus in the past two decades. Ionic liquids (ILs) are room-temperature melting salts that can remain in the liquid state at near or below room temperature. ILs have attracted considerable attention due to their unique properties, such as negligible vapor pressure, nonflammability, thermal stability, good solubility, low melting points, and staying in the liquid state over a wide temperature range from room temperature to about 300°C. The previously mentioned highly favorable properties of ILs motivated Khamooshi and colleagues in Chapter 11 to carry out the present research and review the available ILs found in the literature as the working fluids of absorption cycles. Absorption cycles contain absorption heat pumps, absorption chillers, and absorption transformers.

The growing concern of climate change and global warming has in turn given rise to a thriving research field dedicated to finding solutions. One particular area which has received considerable attention is the lowering of carbon dioxide emissions from large-scale sources, that is, fossil fuel power. Capter 12, by Torralba-Calleja and colleagues, focuses on ionic liquids being used as novel media for CO_2 capture. In particular, solubility data and experimental techniques are used at a laboratory scale. Cited CO_2 absorption data for imidazolium-, pyrrolidinium-, pyridinium-, quaternary-ammonium-, and tetra-alkyl-phosphonium-based ionic liquids is reviewed, expressed as mole fractions (X) of CO_2 to ionic liquid. The following experimental techniques are featured: gravimetric analysis, the pressure drop method, and the view-cell method.

The goal of carbon sequestration is to take CO_2 that would otherwise accumulate in the atmosphere and put it in safe and permanent storage. Most proposed methods would capture CO_2 from concentrated sources like power plants. Indeed, on-site capture is the most sensible approach for large sources and initially offers the most cost-effective avenue to sequestration. For distributed, mobile sources like cars, on-board capture at affordable cost would not be feasible. Yet, in order to stabilize atmospheric levels of CO_2, these emissions, too, will need to be curtailed. Chapter 13, by Lackner and colleagues, suggests that extraction of CO_2 from air could provide a viable and cost-effective alternative to changing the transportation infrastructure to non-carbonaceous fuels. Ambient CO_2 in the air could be removed from natural airflow passing over absorber surfaces. The CO_2 captured would compensate for CO_2 emission from power generation two orders of magnitude larger than the power, which could have instead been extracted from the same airflow by a windmill of similar size. The authors outline several approaches, and show that the major cost is in the sorbent recovery and not in the capture process. Air extraction is an appealing concept, because it separates the source from disposal. One could collect CO_2 after the fact and from any source. Air extraction could reduce atmospheric CO_2 levels without making the existing energy or transportation infrastructure obsolete. There would be no need for a network of pipelines shipping CO_2 from its source to its disposal site. The atmosphere would act as a temporary storage and transport system. The authors also discuss the potential impact of such a technology on the climate change debate and outline how such an approach could actually be implemented.

PART I

ABSORPTION, ADSORPTION, AND MEMBRANE BASED SEPARATION PROCESSES FOR CO_2 CAPTURE

CHAPTER 1

AMINE VERSUS AMMONIA ABSORPTION OF CO_2 AS A MEASURE OF REDUCING GHG EMISSION: A CRITICAL ANALYSIS

AMITAVA BANDYOPADHYAY

1.1 INTRODUCTION

The report published by the Inter Governmental Panel on Climate Change (IPCC 1990) clearly demonstrates that human activities result in the generation of four greenhouse gases (GHGs) into the atmosphere. These gases are carbon dioxide (CO_2), methane (CH_4), chlorofluorocarbons (CFCs), and nitrous oxide (N_2O) that are contributing to the global warming phenomena considerably. The global warming caused by the increased levels of these gases is one of the most serious environmental threats to the human race at present (Yeh and Bai 1999). CO_2 emitted into the atmosphere is assumed to cause the greatest adverse impact on the observed green house effect accounting for approximately 55% of the observed global warming (IPCC 1990).

The risks associated with the climate change are increasing. The growing awareness of this fact has brought the interests of researchers for the abatement of CO_2 since 1989 (Diao et al. 2004). Besides CO_2 abatement that is commonly known as the CO_2 capture, the importance of its sequestration is also being gradually addressed all around the globe. The Kyoto Protocol to the United Nations Framework Convention on Climate Change (UNFCC 1997) put forth an embargo on the major contributing nations to reduce CO_2 emissions by 6% below the level as was in 1990. That the protection of the climate system should occur "on the basis of equity and in accordance with Parties" was noted categorically by the UNFCC. The leadership role in combating climate change and its adverse impacts should be taken by the industrialized nations as the majority of the historical cumulative emissions were caused by them. Therefore, they were given specified commitments for the reduction of emissions in the Kyoto Protocol (Harald et al. 2002). Furthermore, the Kyoto Protocol has also elucidated scientific and economic aspects at length (Bolin 1998; Banks 2000). The arithmetic aspect under the scientific regime based on the CO_2 emission intensity levels were estimated by Sun (2003). The emissions of CO_2 from coal were decreased, not only in their relative share but also in their absolute value due to its supply in the total fuel pool was reduced (Sun 2002, 2003). However, attention on researches on the flue gas emitted from the thermal power plants (TPPs) is growing because 30% of the total global fossil fuel is used for power generation that emits considerable amount of CO_2. It is estimated that around 6 billion tons of carbon emission occur globally by burning fossil fuels out of which about 1.8 billion tons is contributed from TPPs alone (Martin and Meyer 1999). It is further estimated from 1995 database (Sun 2002) that the USA is the largest emitter of CO_2 in the world amounting to 23.7% of the total CO_2 emission while China stands second place making up 13.6%. Considering the seriousness and urgency involved in the matter, the reduction of CO_2 emission from burning fossil fuels assumes considerable importance so as to slow down the trend of global warming.

Legions of researches are being carried out over the past few decades to reduce the CO_2 emission into the atmosphere. These studies came up with suggestive strategies for CO_2 emission reduction, for instance, fuel alternative, energy conservation, and improving efficiencies of TPPs (Blok et al. 1993; Huang 1993; Bai and Wei 1996). Implementation of these strategies, however, may have a subtle impact on the CO_2 emission reduction. As a result, various end-of-pipe treatment methods have been given serious attention to capture/reduce and recover CO_2 from the flue gas streams. These methods are chemical (gas–liquid) absorption, physical adsorption, cryogenic separation, membrane separation, biological fixation, and oxy-fuel combustion with CO_2 recycling (Wolsky et al. 1994; Kimura et al. 1995; Nishikawa et al. 1995). The chemical absorption among these methods has been studied extensively for reducing CO_2 emission from fossil fuel-fired TPPs considering it as a reliable and relatively low cost method (Chakma 1995). Ostensibly, very limited technologies are available commercially for CO_2 capture suitable for TPPs. None of these technologies has so far been deployed at typical base loaded plant since these are not cost effective. The major researches that were carried out on amine and on ammonia-based absorption systems for CO_2 capture are briefly discussed here. Yeh and Bai (1999) experimentally investigated on the evaluation of two reagents viz. aqueous NH_3 and monotheanolamine (MEA) for scrubbing of CO_2. The performances of these two solvents were compared in terms of CO_2 removal efficiency and absorption capacity. They showed experimentally that both the CO_2 removal efficiency and absorption capacity using NH_3 as a solvent were better than those of MEA as a solvent under similar operating conditions chosen in their investigation. Ciferno et al. ($_2$005) reported on a technical and economic scooping analysis that compared two different CO_2 absorption processes, viz. NH_3 absorption of CO_2 and MEA-based CO_2 absorption processes. This analysis was based on the research into aqueous NH_3-based CO_2 capture conducted under the aegis of National Energy Technology Laboratory (NETL). An economic scooping study was conducted to quantify the potential benefits of aqueous NH_3-based technology developing heat and material balances for a pulverized coal (PC)-fired TPP. Estimated were also the differences in capital and operating costs relative to the base case amine for CO_2 capture for the PC-fired TPP of same size. Reportedly, NH_3-based system achieved much better performance than the MEA-based system. McLarnon and Duncan (2005) reported that Powerspan developed performances of CO_2 capture processes based on MEA and aqueous NH_3 as solvents. In aqueous NH_3-based CO_2 absorption process, the rich solvent was stripped off to regenerate

NH_3 and to release CO_2. Ammonia was thus not consumed in the scrubbing process, and no separate by-product was created. A 1-MW pilot plant demonstration was scheduled to begin in 2008, which would produce 20 tons (ca.) of sequestration-ready CO_2 per day. The objective of the pilot was to demonstrate on CO_2 capture through integration with the multi-pollutant control process, i.e., CO_2 capture undertaken after SO_2 and fine particulates were captured. Also performance of MEA-based process was compared with that of the NH_3-based process that clearly indicated the superiority of the NH_3-based system over the MEA-based system in terms of performance. Dave et al. (2009) presented results of ASPEN simulations of a CO_2 removal and recovery plant that was intended for capturing CO_2 from a 500-MW conventional coal-fired TPP flue gas stream. They investigated into the performance of CO_2 capture process with aqueous solutions of 2-amino-2-methyl-1-propanol (AMP), methyldiethanolamine (MDEA), and NH_3 as solvents and compared the performances of each solvent with the conventional 30% by weight of MEA solution. A laboratory scale wetted wall gas–liquid contactor was further, experimentally investigated by them to generate mass transfer data so as to validate the simulated process condition for CO_2 capture by aqueous NH_3. The ASPEN-derived results further showed that 30% by weight of AMP-based process had the lowest overall energy requirement amongst the solvents chosen in their investigation. However, they did not investigate into aspects like refinement of product CO_2, its compression, transportation, and storage. Aspects of economics derived from the merits and demerits of amine and NH_3-based processes were also not reported. Thus, AMP-based process apparently looked energetically favorable, but intrinsically this study is falling short of a complete process.

Critical appraisal of these comparative studies indicates that CO_2 absorption (capture/removal) using aqueous NH_3 solution classically meets the demands of the GHG mitigation options compared to that of amine-based system under similar hydrodynamic conditions despite both these methods were having their own advantages and limitations. An attempt has, therefore, been made in this article to assess critically the relative merits and demerits of these two chemical absorption systems for CO_2 capture as a measure of reducing the GHG emission. While critically elucidating the aforesaid relative merits and demerits of NH_3-based and MEA-based CO_2

absorption processes, investigations carried out by earlier researchers to compare their performances will be analyzed elaborately. However, in this analysis, the study reported by Dave et al. (2009) will not be considered further due to the shortcomings associated with it as mentioned earlier.

1.2 AMINE-BASED CO_2 ABSORPTION PROCESS

The CO_2 absorption from the mixture of sulfur-containing acidic gases had been a subject in the history of chemical engineering since early part of the last century. The concept of separating CO_2 from flue gas streams was commenced in the 1970s as an economic source of CO_2 especially for enhanced oil recovery (EOR) operations rather than as an option for the GHG emission reduction. In the USA, several CO_2 capture plants were constructed for commercial applications in the late 1970s as well as in early 1980s (Kaplan 1982; Pauley et al. 1984). All these CO_2 capture plants were based on the chemical absorption processes using MEA-based solvent. MEA is a homologue of alkanolamines. The credit for developing alkanolamines as absorbents for acidic gases goes to Bottoms (1930). He was granted a patent in 1930 for such application. The historic development of CO_2 removal using aqueous amines are classically elucidated by Kohl and Nielsen (1997). Thus, this technology was developed over 80 years ago as a general, non-selective solvent to remove acidic gas impurities like, hydrogen sulfide (H_2S), CO_2 from refinery operations as well as sweetening of natural gas streams (Kohl and Nielsen 1997). Some of the initial developers of MEA-based CO_2 capture technology were Fluor Daniel Inc., Dow Chemical Co., Kerr-McGee Chemical Corp. and ABB Lummus Crest Inc. This technology is capable of capturing typically about 75–90% of the CO_2 and producing a nearly pure CO_2 (>99%) in the product stream.

Besides, the absorption of CO_2 in alkali was developed since middle of the last century in studying various aspects of mass transfer in chemical engineering (Lynn et al. 1955). Development of devices for carrying out gas–liquid mass transfer operations is still important to the researchers. Developing such devices means improving the intrinsic mass transfer design parameters like interfacial area of contact and true gas side as well as liquid side mass transfer coefficients. CO_2 absorption in various alkaline solutions had been shown to perform excellently well in determining the intrinsic mass transfer design parameters, for instance, the interfacial area of contact and true liquid side mass transfer coefficient. In these deterministic studies, researchers have classically demonstrated the uses of reactions of CO_2 with various alkaline solutions like generic amines, mixture of amines, hindered amines, sodium hydroxide (NaOH), and sodium carbonate/sodium bicarbonate (Na_2CO_3/$NaHCO_3$) mixture or potassium carbonate/potassium bicarbonate (K_2CO_3/$KHCO_3$) mixture (Danckwerts and Sharma 1966; Astarita 1967; Danckwerts 1970). Since this section is aimed at assessing the applicability of amines as a solvent for CO_2 removal, the discussion will be restricted within the amine-based solvents only.

In the amine-based CO_2 capture plant, the flue gas is contacted with the amine such as MEA, in a packed bed absorption tower. The principal reactions occurring when solutions of MEA are used to absorb CO_2 may be represented (Kohl and Nielsen 1997) by ionization of water (Eq. 1), hydrolysis as well as ionization of dissolved CO_2 (Eq. 2), protonation of alkanolamine (Eq. 3), and carbamate formation (Eq. 4) as given below:

$$H_2O = H^+ + OH^- \qquad (1)$$

$$CO_2 + H_2O = HCO_3^- + H^+ \qquad (2)$$

$$RNH_2 + H^+ = RNH_3^+ \qquad (3)$$

$$RNH_2 + CO_2 = RNHCOO^- + H^+ \qquad (4)$$

All the above reactions account for the principal species present in aqueous alkanolamine treating solutions. Additional reactions, however, may occur, which produce species other than those specified, but these are not considered important in the CO_2 absorption process. The flue gases after absorber are washed to recover any residual MEA and exhausted to the atmosphere. The CO_2-rich solvent is passed through a desorber in which a counter-current steam stripped off CO_2 from the solvent producing a stream of H_2O and CO_2. The H_2O is condensed out leaving a stream of CO_2 (purity >99%) that is ready for compression. The CO_2-lean solvent, on the other hand, is cooled in a condenser and recycled back to the absorber.

Although these reactions relate specifically to primary amines, such as MEA, they can squarely be applied to secondary amines, such as diethanolamine (DEA), by suitably modifying the structural formula of the amine. However, tertiary amine undergoes reactions 1 through 4, but cannot react directly with CO_2 to form carbamates by reaction 4 due to the absence of α–H atom. If the reaction 4 is predominant, as it often occurs with primary amines, then the carbamate ion ties up with an alkanol-ammonium ion produced in reaction 3. As a result, the capacity of the solution for CO_2 is almost limited to 0.5 mol of CO_2/mol of amine. This is true even at relatively high partial pressures of CO_2 in the gas to be treated. This limitation is attributed to the high stability of the carbamate and its relatively low rate of hydrolysis to bicarbonate. In contrast, a ratio of 1 mol of CO_2/mol of amine can theoretically be achieved for the tertiary amines which are unable to form carbamates. In order to overcome this difficulty of operation as in the case of MDEA, it is necessary to add an activator, typically another amine, which increases the rate of hydration of dissolved CO_2. This has, therefore, given birth to the absorption of CO_2 in mixed amines as described later.

Methyldiethanolamine (MDEA) has gained considerable importance as a nonselective solvent for removing high concentration of CO_2 owing to its low energy requirements, high capacity, excellent stability, and other associated factors. Its chief disadvantage is its low rate of reaction with CO_2 that has resulted in the low absorption rate as well. Investigation showed that the addition of primary or secondary amines, for instance MEA and DEA, had increased the rate of CO_2 absorption significantly without diminishing the previously noted advantages of MDEA (Mshewa and Rochelle 1994). Their model calculations over a wide range of temperatures and partial pressures predicted that the overall gas phase mass transfer coefficient for CO_2 absorption in a solution containing 40% MDEA and 10% DEA was 1.7–3.4 times greater than that for CO_2 absorption in a 50% MDEA solution under typical absorption column conditions. The relevant kinetic data of DEA had taken from the available literature for the purpose of the model predictions.

TABLE 1: Commercial anthropogenic CO_2 capture facilities as of June 2009

Commercial CO_2 source	Type of operation	Use	CO_2 capture capacity	
			TPD	MTPA
IMC Global Soda Ash Plant (USA)	Coal fired power plant	Soda ash production	800	0.3
Warrior Run Power Plant (USA)	Coal fired power plant	Food/beverage	330	0.1
Schwarze Pumpe Pilot Plant (Germany)	Coal fired oxyfuel combustion	Various	202	0.1
Shady Point Power Plant (USA)	Coal fired power plant	Food/beverage	200	0.1
Great Plains Synfuels Plant (USA)	Coal gasification	EOR	5,480	2.0
Sumitomo Chemicals Plant (Japan)	Natural gas-fired power plant	Various	200	0.1
Prosint Methanol Production Plant (Brazil)	Methanol	Food/beverage	90	0.0
Enid Fertilizer Plant (USA)	Fertilizer	Urea, EOR	1,850	0.7
Indian Farmers Fertilizer Company (India)	Fertilizer	Urea, NPK, DAP, NP	900	0.3
Ruwais Fertilizer Industries (UAE)	Fertilizer	Urea	400	0.1
Luzhou Natural Gas Chemicals (China)	Fertilizer	Urea	160	0.1
Petronas Fertilizer (Malaysia)	Fertilizer	Urea	160	0.1
Shute Creek Natural Gas Processing Plant (USA)	Natural gas processing	EOR	15,870	5.8
Val Verde Natural Gas Plants (USA)	Natural gas processing	EOR	3,970	1.4
In Salah Natural Gas Production Facility (Algeria)	Natural gas processing	Geologic storage	3,290	1.2
Sleipner West Field (North Sea, Norway)	Natural gas processing	Geologic storage	2,740	1.0
Snohvit LNG Project (Barents Sea, Norway)	Natural gas processing	Geologic storage	1,920	0.7
DTE Turtle Lake Gas Processing Plant (USA)	Natural gas processing	EOR/geologic storage	600	0.2

TABLE 2: Commercial CO_2 scrubbing solvent

Solvent type	Solvent		Process conditions
	Proprietary name	Chemical name	
Physical solvents	Rectisol	Methanol	−10/−70°C, >2 MPa
	Purisol	N-2-methyl-2-pyrolidone	−20/+40°C, >2 MPa
	Selexol	Diemethyl ethers of polyethyleneglycol	−40°C, 2–3 MPa
	Fluor solvent	Propylene carbonate	Below ambient temperatures, 3.1–6.9 MPa
Chemical solvents	Organic (amine based)		
	MEA	2.5N MEA and inhibitors	40°C, ambient-intermediate pressures
	Amine guard	5N MEA and inhibitors	
	Econamine	6N Diglycolamine	80–120°C, 6.3 MPa
	ADIP	2-4N Diisopropanolamine, 2N MDEA	35–40°C, >0.1 MPa
	MDEA	2N MDEA	
	Flexsorb, KS-1, KS-2, KS-3	Hindered amine	
	Inorganic		
	Benfield and versions	Potassium carbonate and catalysts. Lurgi and Catacarb processes with arsenic trioxide	70–120°C, 2.2–7 MPa
Physical/ chemical solvents	Sulfinol-D, Sulfinol-M	Mixture of DIPA or MDEA, water and tetrahydrothiopene or diethylamine	>0.5 MPa
	Amisol	Mixture of methanol and MEA, DEA, diisopropylamine or diethylamine	5/40°C, >1 MP

A different class of absorbents known as the sterically hindered amines has reported in the literature to control the CO_2/amine reaction (Goldstein 1983; Sartori and Savage 1983; Chludzinski et al. 1986). Some of the sterically hindered amines reported for CO_2 absorption are AMP, 1,8-p menthanediamine (MDA), and 2-piperidine ethanol (PE). The CO_2 absorption characteristics of sterically hindered amines are sufficiently similar to those of the alkanolamines although they are not necessarily alkanolamines. The hindered amines are used as promoters in hot K_2CO_3 systems as component of organic solvent and as the principal agent in aqueous solutions for the selective absorption of CO_2. Sterically hindered amine with a specifically designed molecular configuration can yield independent performance based on their individual selectivity toward CO_2 absorption. The pilot and commercial plant data revealed that substantial savings in capital and operating cost could be achieved with the hindered amines.

A comprehensive assessment of the commercially available CO_2 capture technologies as of June 2009 was reported by Dooley et al. (2009). Such commercially operating plants and commercial CO_2 scrubbing solvents are furnished in Tables 1 and 2 for the benefit of our improved understanding.

1.2.1 CO_2 CAPTURE FROM FLUE GAS OF TPPS BY AMINE

The special report of IPCC on Carbon Dioxide Capture and Storage (IPCC 2005) has classically reviewed various methods for CO_2 capture in the light of GHG emission reduction that includes CO_2 capture from flue gas of thermal power plants also. Besides other processes, application of MEA- or amine-based process for CO_2 absorption was discussed therein. It was, however, mentioned earlier in this article that amine-based absorption of CO_2 was not developed for removing CO_2 from the flue gas stream of TPP. In fact, such an amine-based CO_2 absorption process can be adopted for flue gas of TPP, if the quality of the flue gas is similar to the feed gas being treated for the conventional amine-based CO_2 absorption process. However, such a quality matching seldom occurs and as a result, the flue gas of TPP requires special treatment prior to its introduction into the amine-based CO_2 absorption process. Therefore, the aspects of conventional amine-based CO_2 absorption and its restriction on application toward CO_2 capture for TPPs need special attention which is described in this section for improving our understanding. The flue gas of a coal-fired TPP may contain several contaminants, like sulfur dioxide, SO_2 = 300–3,000 ppmv, oxides of nitrogen, NO_x = 100–1,000 ppmv, and particulate matter = 1,000–10,000 mg/m^3. In contrast, natural gas-fired power plants generate contaminants at considerably lower levels and the concentrations here are SO_2 < 1 ppmv, NO_x = 100–500 ppmv, and particulate matter =

~10 mg/m^3 (Chakravarti et al. 2001). The temperatures of flue gases generated from a fossil-fueled power plants are usually above 100°C, which means that they need to be cooled down to the temperature levels required for the absorption process (IPCC 2005). This can be done in a cooler with direct water contact, which also acts as a flue gas washer with additional removal of fine particulates. In addition to the above, flue gas from coal combustion will contain other acid gas components such as NO_x and SO_2. Flue gases from natural gas combustion will mainly contain NO_x as the contaminant as pointed out earlier. These acidic gases, similar to CO_2, will have a chemical interaction with the amine as solvent. This is not desirable as the irreversible nature of this interaction leads to the formation of heat stable salts, and hence a loss in absorption capacity of the solvent and the risk of formation of solids in the solution. It also results in an extra consumption of chemicals to regenerate the amine and the production of a waste stream such as sodium sulfate (Na_2SO_4) or sodium nitrate ($NaNO_3$). Therefore, the pre-removal of NO_x and SO_2 to very low values before CO_2 capture becomes essential. For NO_x , it is the NO_2 which leads to the formation of heat stable salts. In addition, careful attention must also be paid to fly ash and soot present in the flue gas as they might plug the absorber if their concentration levels are too high. The operation of some demonstration facilities of CO_2 capture plant to coal-fired TPP caused several problems associated mainly with these contaminants. It was further suggested (IEA 2007) that the flue gas of the TPP for CO_2 capture should be pretreated to avoid amine degradation so as to achieve the following compositions of the contaminants prior to introducing into the amine-based CO_2 absorption systems:

1. SO_2 concentration: 10–30 mg/Nm^3 [3.82–11.46 ppmv]
2. NO_2 concentration: 40 mg/Nm^3 [21.26 ppmv]
3. particulate matter: <5 mg/Nm^3.

Therefore, the MEA-based CO_2 capture process requires very low SO_2 in the flue gas at its inlet. Since most commercially available SO_2 scrubbing systems, commonly called as the Flue Gas Desulphurization (FGD) systems are not efficient enough to attain such a lower value of SO_2 concentration in the flue gas prior to being introduced into to the CO_2 capture system.

The situation is further complicated on a system having FGD installed previously. Because such a system would, therefore, require an auxiliary SO_2 scrubbing system to make the flue gas stream ready for the CO_2 capture system and in turn increase the capital cost of the CO_2 capture system (McLarnon and Duncan 2009).

Oxygen present in the flue gas would cause another problem of rapid degradation of some of the alkanolamines used for amine absorption. The degradation byproducts lead to corrosion problems and cause significant deterioration in the overall separation performance. For instance, the mixture of MEA and MDEA cannot be made to be oxygen resistant (Nsakala et al. 2001). Therefore, while this process potentially offers an improved system from the standpoint of solvent regeneration as well as energy requirement, it is imperative to separate the excess oxygen from the flue gas stream. This purification has been tested in demonstration plant facilities by converting the O_2 present in the flue gas by burning it with natural gas over a De-Oxy catalyst upstream of the solvent contactor into CO_2. The de-oxygenated flue gas thus leaving the De-Oxy system is introduced to the MEA/MDEA absorption system where CO_2 is removed (~90%+). The constituents thus present in the flue gas leaving the MEA/MDEA absorption system are N_2, H_2O vapor, CO_2, and relatively small amounts of NO_x, SO_2, and CH_4 that are discharged to the atmosphere through stacks above the absorber.

A key feature of post-combustion CO_2 capture process based on absorption is the high energy requirement and the resulting efficiency penalty on power cycles. This is primarily due to the heat necessary to regenerate the solvent (steam stripping) and to a lesser extent the electricity required for liquid pumping, the flue gas fan, and finally compression of the CO_2 product. Therefore, the CO_2 removal/capture plant should be designed in such a way so that the generation of CO_2 from the energy required for operating the CO_2 capture plant should be less than the amount of CO_2 it is programmed to remove with the purpose of classically meeting the basic philosophy of GHG emission reduction.

In the absence of a complete design data for a typical CO_2 capture process using MEA, the range of data generated from a large number of simulated runs reported by Rao (2002) is presented in this article. He has simulated CO_2 capture process using ProTreat, a software package. This package is generally used for simulating processes for the removal of CO_2, H_2S, and mercaptans from a variety of high and low pressure gas streams by absorption into thermally regenerable aqueous solutions containing single or blended amines. The package deals with the separation as a mass transfer rate process through the use of column model. A large number of process simulation runs were conducted to cover a reasonable range of values for the parameters that described the CO_2 capture process. The CO_2 capture and separation system consists of a flue gas compressor, cooler, absorber, heat exchangers, regenerator, sorbent circulation pumps, etc. Values of model parameters estimated are presented in Table 3.

TABLE 3: Typical range of simulated values of parameters for CO_2 capture process by MEA

No	Parameter	Type	Range
1	CO_2 content in the flue gas (mol%)	Input	3.5–13.5
2	Flue gas flow rate, G (kmol/h)	Input	9,000–24,000
3	Inlet flue gas temperature (°C)	Input	40–65
4	MEA concentration (wt%)	Input	15–40
5	Solvent flow rate, L (kmol/h)	Input	16,000–70,000
6	L/G ratio, dimensionless	Input	0.73–5.56
7	Reboiler heat duty, Q (GJ/h)	Input	95–664
8	Q/L (MJ/kmol)	Input	2.4–22.5
9	CO_2 removal efficiency (%)	Output	41.2–99.9
10	CO_2 product flow rate (kmol/h)	Output	333–2,840
11	Lean sorbent CO_2 loading (mol CO_2/mol MEA)	Output	0.05–0.34
12	Rich sorbent CO_2 loading (mol CO_2/mol MEA)	Output	0.27–0.55
13	Absorber diameter (ft)	Output	26–42
14	Regenerator diameter (ft)	Output	12–42
15	Exhaust flue gas temperature (°C)	Output	40.4–71.6
Parameters held constant were			
Absorber height		40 ft	
Absorber packing		Rasching rings, metallic, 1-inch packing size	
Inlet flue gas pressure		3 psi	
Solvent pumping pressure		30 psi	
Number of trays in regenerator		24 (tray spacing = 2 ft, weir height = 3 inches)	
Compressor efficiency		60–100%	

1.2.2 CURRENT STATUS OF RESEARCH ON THE SELECTIVITY OF AMINES FOR CO_2 CAPTURE FROM EXHAUST STREAMS

Critical appraisal of the available literature on the classical amine-based CO_2 absorption systems coupled with the suggested requirements of the level of contaminants for the CO_2 capture plant as described earlier indicates that the CO_2 capture plant would require specially designed amine absorption system, in which the classical amine-based CO_2 absorption systems developed, based on acid gas treatment cannot be replicated to the CO_2 capture plants augmented to the TPPs due to various constraints as described before.

Recent literature (Aronua et al. 2009) also reveals that alkanolamines reported for CO_2 absorption are still deficient for CO_2 absorption due to inherent problems associated with their use in CO_2 capture process. Legion of factors affect the efficacy of a solvent to be selected for CO_2 absorption. The factors for selecting a solvent are solubility, vapor pressure, molecular weight, foaming tendency, degradation properties, and corrosivity. Other associated factors include reaction kinetics, heat of reaction, energy of regeneration, and the capacity of cyclic use of amine. Finally, the environmental and cost factors are also to be taken into consideration. Therefore, the solvent selection is more dependent on techno-enviro-economic feasibility rather than a simple techno-economic feasibility in such applications.

Aronua et al. (2009) further investigated performances of various solvents for CO_2 capture, which were compared to MEA that was chosen as the base case for all comparisons. Solvents selected for their investigation were AMP (1.0, 2.5, and 5.0 M), mixture of 0.42 M N,N′-di-(2 hydroxyethyl) piperazine (DIHEP) and 0.58 M N-2-hydroxyethylpiperzine (HEP), 2.5 M AMP, mixture of 2.5 AMP and 0.5 M piperizine (PZ), 1.0 M tetraethylenepentamine (TEPA), and 2.5 M potassium salt of sarcosine (KSAR) (prepared by neutralizing equimolar amounts of sarcosine). A rapid screening apparatus was used for performing a relative comparison of the CO_2 absorption potentials of the aforementioned selected solvent systems. Experimental results showed that besides absorption data, desorption data were also squarely important to assess the performance of solvents for CO_2 absorption, since different solvents exhibit different desorption behaviors. The absorption–desorption studies reported by them are summarized below:

1. The capacity of MEA in mol CO_2/mol amine was found to decrease with increase in concentration while its CO_2 removal per cycle increases with concentration.
2. The “0.42 M DIHEP + 0.58 M HEP” mixture has shown the lowest CO_2 absorption potential of all the systems investigated despite its best desorption ability.
3. 2.5 M KSAR showed a similar behavior to 2.5 M MEA; however, it was found to have slightly lower performance.
4. The performance of CO_2 absorption in AMP was enhanced with PZ, and the combination also showed high desorption ability.
5. TEPA showed outstanding CO_2 absorption potential by removing a large amount of CO_2 per cycle among the various solvents investigated. For example, 1.0 M TEPA removed three times more CO_2 per cycle than what was removed by 1.0 M MEA per cycle. Such higher absorption capacity was attributed due to presence of five amine (two primary and three secondary) sites in it. However, working with TEPA at higher concentration more than 1.0–2.0 M may be challenging owing to its higher viscosity.

Hakka and Ouimet (2006) developed absorbent based on tertiary amines that also includes a promoter to yield sufficient absorption rates to be used for low pressure flue gas streams. The use of oxidation inhibitors enables this process to be operative in oxidizing environments as also where limited concentrations of oxidized sulfur exist. They have claimed that this process can also simultaneously remove SO_2. The process comprises selecting absorbent from the following tertiary amines either alone or in combination as a mixture:

- methyldiethanolamine (MDEA),
- N,N′-di-(2-hydroxyethyl) piperazine (DIHPA),
- N,N′-di-(3-hydroxypropyl) piperazine,
- N,N,N′,N′-tetrakis (2-hydroxyethyl)-1,6-hexanediamine,
- N,N,N′,N′-tetrakis (2-hydroxypropyl)-1,6-hexanediamine,
- tertiary alkylamine sulfonic acids,
- triethanolamine (TEA).

The tertiary alkylamine sulfonic acid is selected from the group as given below either alone or in combination as a mixture:

- 4-(2-hydroxyethyl)-1-piperazineethanesulfonic acid,
- 4-(2-hydroxyethyl)-1-piperazinepropanesulfonic acid,
- 4-(2-hydroxyethyl)-1-piperazinebutanesulfonic acid,
- 4-(2-hydroxyethyl) piperazine-1-(2-hydroxypropanesulfonic acid),
- 1,4-piperazinedi (ethanesulfonic acid).

The chief advantage of this invention is the stability of certain tertiary amines used in the process that may also be used to remove SO_2 from the flue gas. Thus, the flue gas entering into the CO_2 capture system may contain SO_2 that would not degrade the absorbent as would occur for other generic amine-based absorbents described earlier. For example, pretreatment step is not required, if the flue gas would contain SO_2 under the present circumstances for reducing its concentration to avert excessive absorbent degradation. Simultaneously, the presence of SO_2 may be exploited to restrict oxidative degradation of the absorbent. This process is developed in such a way so that sufficient SO_2 may be either slipped from an upstream SO_2 removal process or added to the feed gas to the process to maintain sufficient sulfite in the CO_2 absorbent to scavenge and react with molecular oxygen effectively which is absorbed from the feed gas. As a result, the molecular oxygen would be unavailable for oxidizing the amine-based solvent. Accordingly, the feed gas undergoing CO_2 capture may be allowed to contain SO_2 ranging between 0 and 1000 ppmv. The process developed for recovering SO_2 and CO_2 from a flue gas stream comprises the following steps of operation:

1. SO_2 scrubbing loop: Treating the flue gas stream in an SO_2 scrubbing loop with a first absorbent stream to obtain a SO_2-rich stream and a SO_2-lean stream. Subsequent treatment of the SO_2-rich stream to obtain a first regenerated absorbent stream which is used back for SO_2 scrubbing;
2. CO_2 scrubbing loop: Treating the SO_2-lean stream in a CO_2 scrubbing loop with a second absorbent stream to obtain a CO_2-rich stream. Subsequent treatment of the CO_2-rich stream to obtain a

second regenerated absorbent stream which is used back for CO_2 scrubbing; and

3. treating at least a portion of one or both of the first and second regenerated absorbent streams to remove heat stable salts.

The absorbent used in each of the scrubbing loops comprises at least one tertiary amine and at least one secondary amine as an activator. In fact, the process developed by Hakka and Ouimet (2006) as described has been patented (assignee: Cansolv Technologies Inc.). Based on the process described above, Cansolv Technologies Inc. (Cansolv 2009) has proposed to integrate SO_2–CO_2 capture system. In this method 80% of heat used for SO_2 stripping is recycled for CO_2 stripping and thereby reducing unit CO_2 capture costs. This method helps in reducing the SO_2 concentration to a very lower level; however, it is more complicated and cost intensive than MEA or simple amine-based processes.

1.3 CO_2 CAPTURE BY NH_3

As described earlier that the major thrust in CO_2 capture or CO_2 absorption from the CO_2 emitting sources, like flue gases from the combustion facilities, mainly concentrated on the cost of capture that can be reduced by finding a low-cost absorbent with minimum energy requirements, and equipment size as also corrosion. Scrubbing with organic amines for CO_2 capture has been elaborated with its associated limitations. These constraints have directed the research on NH_3 scrubbing for CO_2 emitted from the TPPs. As such, NH_3 scrubbing exists in the available literature for acid gas removal like removal of SO_2 from the flue gas (Kohl and Nielsen 1997). Its scrubbing potential for flue gas cleaning is extraordinary in a sense that it acts on multi-pollutant system. The major by-products produced are ammonium sulfate ($(NH_4)_2SO_4$), ammonium nitrate (NH_4NO_3), and ammonium bicarbonate (NH_4HCO_3). Ammonium sulfate and NH_4NO_3 are well-known fertilizers. The application of NH_4HCO_3 is though uncommon as a fertilizer; it is, however, being explored as a fertilizer in China. In the event of its uncertain market conditions, it can be thermally decomposed to recycle NH_3 and to use CO_2 in pure form. This process in general encourages the combustion of coal containing high-S content. These are the most important aspects of NH_3 scrubbing of CO_2 from the flue gas streams of the TPPs.

1.3.1 REACTION OF NH_3 WITH CO_2

Critical appraisal of the existing literature (Valenti et al. 2009) revealed that the species present in the NH_3–CO_2–H_2O ternary system at working conditions of NH_3 scrubbing of CO_2 are: CO_2, H_2O, and NH_3, in both the vapor and the liquid phases, as well as the aqueous ions, like hydronium ion, H_3O^+; hydroxyl ion, OH^-; ammonium ion, NH_4^+; bicarbonate ion, HCO_3^-; carbonate ion, CO_3^{2-}; and carbamate ion (NH_2COO^-). Also included in this group are, but not limited to, NH_4HCO_3, ammonium carbonate ($(NH_4)_2CO_3$), and ammonium carbamate (NH_2COONH_4). All these compounds are either in aqueous solution or in crystal form. CO_2 can be removed by NH_3 scrubbing through chemical absorption at various temperatures and operating conditions (Brooks and Audrieth 1946; Brooks 1953; Hatch and Pigford 1962; Shale et al. 1971; Koutinas et al. 1983; Valenti et al. 2009; He et al. 2010):

$$2NH_{3(g/aq)} + CO_{2(g)} + H_2O_{(l)} \leftrightarrow CO_{3(aq)}^{\;2-} + 2NH_{4(aq)}^{\;+} \leftrightarrow (NH_4)_2CO_{3(aq)}$$
$$[NH_3/CO_2 \text{ molar ratio} = 0.5] \quad (5)$$

$$(NH_4)_2CO_{3(aq)} + CO_{2(g)} + H_2O_{(l)} \leftrightarrow 2NH_{4(aq)}^{\;+} + 2HCO_{3(aq)}^{\;1-} \quad (6)$$

$$NH_{4(aq)}^{\;+} + HCO_{3(aq)}^{\;1-} \leftrightarrow NH_4HCO_{3(aq)} \leftrightarrow NH_4HCO_{3(s)} \quad (7)$$

$$CO_{2(g)} + H_2O_{(l)} \leftrightarrow HCO_{3(aq)}^{\;1-} + H_{(aq)}^{\;+} \quad (8)$$

$$HCO_{3(aq)}^{\;1-} + NH_{3(aq)} \leftrightarrow NH_2COO_{(aq)}^{\;-} + H_2O_{(l)}$$
$$[NH_3/CO_2 \text{ molar ratio} = 1.0] \quad (9)$$

$$NH_2COO_{(aq)}^{-} + CO_{2(g)} + 2H_2O_{(l)} \leftrightarrow 2HCO_{3(aq)}^{1-} + NH_{4(aq)}^{+} \quad (10)$$

$$NH_2COO_{(aq)}^{-} + NH_{4(aq)}^{+} \leftrightarrow NH_2COONH_{4(aq)} \quad (11)$$

All the above reactions occur at atmospheric pressure. Reactions are, however, very temperature sensitive. The scrubbing/absorption/capture of CO_2 occurring inside the absorber is described by Eqs. 5 and 6; while the precipitation of the NH_4HCO_3 salt is represented by Eq. 7 that occurs at low temperature. In contrast, the NH_2COO^- ion formation through Eqs. 8 and 9 may lead to an undesirable CO_2 capture, and in that, the Eq. 10 ascribing the regeneration necessitates a greater enthalpy of reaction. Alternatively, NH_2COO^- ion may combine with NH_4^+ ion per Eq. 11.

It can therefore, be seen from the above reactions that different ammonium salts can be obtained due to different reaction conditions (temperature, pressure, and concentration of reactants). Comprehensively, such different salts can be illustrated (He et al. 2010) by the following Eq. 12:

$$aNH_3 + bCO_2 + cH_2O \rightarrow NH_4HCO_3;\ (NH_4)_2CO_3 \cdot H_2O;\ 2NH_4HCO_3 \cdot (NH_4)_2CO_3;\ NH_4COONH_2 \quad (12)$$

All these products are white solids and may be a single salt or mixed salts.

1.3.2 CLASSIFICATION OF NH3 SCRUBBING OF CO_2 BASED ON TEMPERATURE

The NH_3 scrubbing of CO_2 is available in two different modes, based on the temperature of absorption (Darde et al. 2009). In the first mode, NH_3 absorbs the CO_2 at low temperature ranging between 0 and 10°C and is, therefore, called chilled ammonia process (CAP). This low temperature process has two advantages (i) reduction of the slip of NH_3 in the absorber and (ii) reduction of the flue gas volume. This process permits precipitation of different ammonium salts in the absorber. The second process absorbs CO_2 at ambient temperature in the range of 25–40°C. These two processes are briefly described later.

1.3.2.1 CHILLED AMMONIA PROCESS FOR CO_2 CAPTURE

Gal (2006) patented the use of CAP to capture CO_2 in 2006. The process that was patented is described briefly here in this section. The operating temperature specified should be in the range of 0–20°C and preferably in the range of 0–10°C, while the pressure should be close to atmospheric. Operation of the process at this low temperature prevents the NH_3 from evaporating. Thus, the flue gas containing CO_2 is to be cooled at first. This is usually done in direct contact coolers at the entrance of the process. The stream leaving the cooler contains low moisture content as well as near zero concentration of particulate matter and acidic gases. In fact, the low process temperature reduces the vapor pressure of these compounds leading to their condensation into water. The flue gas, thereafter, enters into the CO_2 capture and regeneration subsystem. similar to the amine-based capture processes, this subsystem consists of absorption and desorption columns. The cold flue gas rich in CO_2 enters the bottom of the absorber while the CO_2-lean solvent stream enters counter-currently from the top of it. The typical concentration of NH_3 in the solvent is up to 28% (w/w). The patent also specifies that the CO_2 loading capacity of the CO_2-lean stream should be ranging between 0.25 and 0.67, and preferably between 0.33 and 0.67.

It was claimed in the patent that more than 90% of the CO_2 could be captured from the flue gas under aforesaid conditions. The cleaned gas, after washing out with cold water and an acidic solution to eliminate the residual NH_3, mainly contains nitrogen (N_2), oxygen (O_2), and low concentration of CO_2. The CO_2-rich stream comprising a solid phase and a liquid phase leaves the bottom of the absorber as slurry. This stream has the range of CO_2 loading of 0.5–1.0 and preferably of 0.67–2.0. The patent further elicits that a part of the CO_2-rich stream should be recycled to the absorber to raise the CO_2 loading of the CO_2-rich stream by producing more solids. Then, the CO_2-rich stream is pressurized and pumped to a desorber, the temperature of which is in the range of 50–200°C and preferably of 100–150°C, while the range of pressure is specified at 2–136 atm. The operating conditions cause CO_2 to evaporate from the solution, and it leaves the top of the desorber as a pure stream at higher pressure. On the other hand, the water vapor and the NH_3 contained in this stream can be recovered by cold washing preferably using weak acid. Though the energy demand highly depends on the composition of the CO_2-rich stream entering the desorber, the patent claimed that the energy needed in the CAP is much lower than for MEA or other amine-based processes.

In 2006, Alstom initiated the development of CAP for its commercialization through a 5-year program (Valenti et al. 2009). This program comprised three phases, for instance, (i) small and large scale testing at SRI in the San Francisco Bay Area, CA, (ii) field pilot testing of capture on coal-fired exhausts at We Energies' plant in Pleasant Prairie, WI, and at E.ON's plant in Karlshamn, Sweden, and (iii) commercial demonstration of capture as well as of storages at the American Electric Power (AEP)'s Mountainer coal plant. The field plant commissioning of CO_2 capture on coal-fired facility at We Energies' was designed (Kozak et al. 2009) to remove CO_2 from the flue gas extracted from the duct that runs between the wet FGD (WFGD) of Unit # 2 and the stack. The treated flue gas is, thereafter, discharged from the stack after it is recombined through the Unit # 2 flue gas duct. A 1.7-MW system is being operated (Blomen et al. 2009) for captured CO_2 from a portion of flue gas generated at the 1224 MW coal-fired boiler. The design capacity of the plant (Kozak et al. 2009) is 1600 kg/h (almost 15 kT per annum). The full scale will require a 400 times scale up. The system is expected to achieve ca. 90% CO_2 removal along with reduction of residual emissions of other gases and particulate matter. In general, the CAP is designed under few operating stages/parameters as described (Valenti et al. 2009; Kozak et al. 2009) in Table 4.

TABLE 4: Typical operating features of CAP

Ambient conditions
Pressure: 1.01 bar
Temperature: 15°C
Conditions of the flue gas after the wet FGD
Pressure: 1.025 bar

TABLE 4: *Cont.*

Ambient conditions
Temperature: 60°C
Gas composition
CO_2: 9.8–12.3 mol%
SO_2: 3.4–9.4 ppmv
NO_x : 28–54 ppmv
N_2: 64.55 mol%
O_2: 4.24 mol%
Ar: 0.43 mol%
H_2O: saturated
PM: filterable/condensable
H_2SO_4 acid mist and HCl: condensable
Absorber operating conditions
Pressure: 1 atm
Temperature: 10°C
Moisture content: <1 mol%
SO_2 concentration: not greater than 5 ppmv
HCl/PM: undetectable
Initial concentration of NH_3 solution [containing NH_4^+, CO_3^{2-}, HCO_3^- ions]: 28 wt%
CO_2 loading in lean solution: 0.33–0.67 mol CO_2/mol NH_3
CO_2 loading in rich solution: 0.67–1.00 mol CO_2/mol NH_3
CO_2 Capture efficiency: 90%
Regenerator operating conditions
Pressure: 20–40 bar [typical value 40 bar]
Temperature: > 100°C [typical value 120°C]

1.3.2.2 CO_2 CAPTURE BY NH3 AT AMBIENT TEMPERATURE

In this section, the recent investigations for absorption of CO_2 using aqueous NH_3 carried out at room temperature (25–40°C) are briefly reviewed. Bai and Yeh (1997) investigated on the CO_2 removal by aqueous NH_3 solution experimentally in a semi-continuous flow reactor under various inlet CO_2 concentrations (8–16% v/v) at a temperature of 25 ± 1°C. The overall CO_2 removal efficiency was reported to be 98% under proper operating conditions. The absorption capacity of NH_3 was around 0.9 kg CO_2/kg of NH_3 and was reported to be higher than MEA-based absorption (0.36 kg CO_2/kg MEA). Ammonium bicarbonate solution and its crystalline solids were the major reaction products that were established through chemical analysis. Yeh and Bai (1999) reported experimental findings for CO_2 scrubbing using two reagents, namely, aqueous NH_3 and MEA. A semi-continuous flow reactor was used for the purpose of experimentation. The temperature of experimentation was varied between 10 and 40°C. The performances of the system in removing CO_2 using these two regents were characterized as a function of various pertinent variables of the system, and were compared. The maximum CO_2 removal efficiency in NH_3 scrubbing was shown to be 99% as against 94% in MEA scrubbing under similar physico-chemical as well as hydrodynamic conditions. The detailed performance of this study is described latter. Corti and Lombardi (2004) reported on the simulated absorption of CO_2 emission from semi-closed gas turbine/combined cycle (SCGT/CC) by aqueous NH_3 by means of Aspen PlusTM. The performance of the model system was characterized as a function of various operating variables. In this investigation, both absorption of CO_2 and desorption were described. A sensitivity analysis with respect to the primary parameters of the system (absorber and desorber pressure and temperature, NH_3 concentration in the absorbing solution, etc.) was carried out. With reference to the SCGT/CC case study, the CO_2 removal, considering a removal efficiency of 89%, dramatically decreased the overall cycle efficiency from 53 to 41%. The main contribution to this decrease was due to the power consumption for flue gas compression up to the absorption unit pressure. The CO_2 specific emissions, under an overall removal of CO_2 of 89%, pass from 390 kg/MWh for the simple SCGT/CC to 57 kg/MWh for SCGT/CC with NH_3-based absorption process. In comparison with the MEA-based absorption process, the efficiency reduction that must be paid to decrease CO_2 emissions was greater for the system that used NH_3, due primarily to the power required for flue gas compression, since the NH_3-based process must work at a higher pressure level in the absorber. It was concluded from this consideration that the NH_3-based process was probably more suitable for CO_2 removal applications where the gas to be treated was already in a pressurized conditions. Diao et al. (2004) described an experimental investigation on CO_2 removal by NH_3 scrubbing in an open continuous flow reactor (height: 0.45 m; diameter: 0.25 m) with five layers of sieve plates under various reaction temperatures (25–55°C), concentration of CO_2 (10–14% v/v) and concentration of NH_3 solution (0.07–0.14 mol/l). It was reported that the removal efficiencies were quite stable in the range from 95 to 99% under proper conditions. The reaction temperature played a key role in the CO_2 removal. The CO_2 removal efficiency reached its highest at 33°C. The effect of inlet CO_2 concentration on its removal was minimal when the removal efficiencies were more than 90% for inlet CO_2 concentrations in the range of 10–14% (v/v). This indicated that the NH_3 solution had a high potential scrubbing capacity with a fast absorption rate. It was also shown that the NH_4HCO_3 was the main product of the CO_2–NH_3 reaction. Experimentation further showed that the CO_2–NH_3 reaction conformed to Arrhenius' law in the range of temperatures investigated. The activation energy of chemical reaction was 26.73 kJ/mol and the exponential factor, A, was 2.4 $\times 10^5$ dm^3/mol s. Wang et al. (2007) investigated on the scrubbing of CO_2 using aqueous NH_3 in a counter-current reactor (height: 0.45 m; diameter: 0.25 m) with five layers of sieve plates. The concentration of CO_2 was ranging between 10 and 14%, while NH_3 concentration in the scrubbing liquor was varied from 0.066 to 0.140 mol/l. The operating temperature was ranging between 25 and 55°C. The overall CO_2 removal efficiency was experimentally shown to be 95%. It was proved that NH_4HCO_3 was the main product of the CO_2–NH_3 reaction. A life cycle analysis was finally carried out from the experimental findings that are discussed latter. Kim et al. (2009) investigated CO_2 removal from the simulated flue gas using the low concentration NH_3 liquor ranging from 2 to 7 wt% in a packed bed absorber (diameter: 0.05 m; maximum packing height: 0.8 m; random packing materials: Raschig ring with specific surface area of 712 m^2/m^3 and Pro-Pak with specific surface area of 1,890 m^2/m^3). The simulated flue gas consisted of 25% CO_2 and balance N_2. Other parameters studied were gas flow rate (20–30 lpm) and liquid flow rate (0.25–0.5 lpm). The temperatures of gas and liquid were kept at 40°C and the pressure was maintained at 1 atm. The observed CO_2 removal efficiencies were more than 90% with proper gas to liquid flow rate ratio. The concentration of NH_3 should be higher than 5% (w/w) for achieving the similar level of removal efficiency in MEA process. The optimum regeneration temperatures and transfer capacities for 2, 5, and 7% (w/w) NH_3 solutions were 96, 86, 83°C and 0.02, 0.032, 0.036 kg CO_2/kg NH_3, respectively. The rise in temperature due to the heat of absorption could be reduced, and the CO_2 removal efficiency could be increased to maximum 8% with the cooling of NH_3 solution by installing side stream and cooler. The reboiler temperature for CO_2 stripping was decreased from 96°C at 2% (w/w) solution to 83°C at 7% (w/w) solution which was an important advantage over conventional MEA-based CO_2 absorption process. The exit gas concentrations of NH_3 in the absorber and regenerator were as high as 3% indicating the requirement of water wash of NH_3 for practical application. The NH_3 solution was highly recommended for the CO_2 absorption, e.g., in iron and steel-making plants, where the low and medium waste heat recovery was available due to its low regeneration temperature. Derks and Versteeg (2009) studied the kinetics of the carbamate formation reaction from the absorption of CO_2 into aqueous NH_3 in a stirred cell reactor (100 rpm) at temperatures between 5 and 25°C and NH_3 concentrations ranging from 0.1 to 7 kmol/m^3. Experiments were carried out such that the so-called pseudo first-order mass transfer regime was obeyed—and hence the kinetics of the reaction between CO_2 and NH_3 could be derived. The overall kinetic rate data obtained were interpreted using the well-known zwitterion mechanism. It was observed from the rate data that the rate of the reaction between CO_2 and NH_3 was in the same order of magnitude as the conventional alkanolamines like, MEA and DEA and was substantially faster than the reaction between CO_2 and MDEA. Liu et al. (2009) investigated on the absorption of CO_2 in aqueous NH_3 in a semi-batch reactor. A small wetted wall column (WWC) with a contact area of about 41.45 cm^2 was also built for studying the absorption rate, diffusion, and solubility of CO_2 in aqueous NH_3. CO_2 removal efficiencies at different concentrations of aqueous NH_3 were studied under similar operating conditions to compare the characteristics of these two systems in the CO_2 absorption process. The reaction rate and overall gas transfer coefficient were studied by them. The concentrations of NH_3 were ranging from 1 to 15%, whereas the concentrations of CO_2 were ranging from 5 to 20%. The reaction temperatures were 20 and 40°C. The initial CO_2 removal efficiency could reach 90% for 5% or higher concentration of NH_3. This showed that low concentration of NH_3 could also have a higher rate of removal. Experimentation revealed that a lot of NH_3 vapor will volatilize from the solution if the concentration of the aqueous NH_3 was more than 15%. Thus, the concentration of aqueous NH_3 was recommended to be selected from 5 to 10%. Aqueous NH_3 showed a high flux in WWC, which was three times higher than that of MDEA + PZ, under the same operating conditions. It was also found that the overall gas side mass transfer coefficient was appropriate to be used in the mass transfer process of the CO_2 absorption by aqueous NH_3. Qi et al. (2010) investigated CO_2 capture in a scrubber using fine spray of aqueous NH3 solution. The spray scrubber (height: 1.3 m; diameter: 0.12 m) comprised two atomizers each of which was capable of generating droplets of Sauter Mean Diameter 30–40 μm. The CO_2 removal efficiencies were measured at different concentrations of aqueous NH3 solution (0.8–10%), liquid flow rates (120–200 ml/min), gas flow rates (7.6–24.7 lpm), initial temperatures (28–38°C), and CO_2 inlet concentrations (7–15%). Results showed that the CO_2 removal efficiency was increased with increasing aqueous NH_3 concentration and liquid flow rate, while it decreased with increased gas flow rate and initial CO_2 concentration. Besides, an increase in temperature resulted in a higher absorption performance and CO_2 removal efficiency. The maximum removal of CO_2 achieved was 96% within the framework of the experimentation.

1.4 AMMONIA AND AMINES AS ADDITIVES TO SOLVENTS FOR CO_2 SCRUBBING

In order to improve the overall performance of absorption as well as to improve the operating features of CO_2 scrubbing, NH_3 was used as an additive in amine scrubbing of CO_2, and likewise, amines were used as additives in NH_3 scrubbing of CO_2. These issues are elaborately discussed in this section.

1.4.1 AMMONIA AS ADDITIVE IN AMINE SCRUBBING OF CO_2

As discussed earlier, MEA as a solvent for CO_2 absorption led to several problems for which sterically hindered amines were evolved. For instance,

CO_2 removal efficiency of aqueous solution of 30% AMP was lower than that of aqueous solution of 30% MEA but the CO_2 loading was approximately two-fold higher because the stoichiometry (0.5 mol CO_2/mol amine) of MEA was limited (Choi et al. 2009a). Blended alkanolamines were thus investigated to explore absorbents with unique absorption characteristics that have elaborated earlier in this article. Blended amines have shown to enhance the rate of absorption of CO_2 and the absorption capacity in spite of maintaining the stripping characteristics of sterically hindered amines (Li and Chang 1995). Unlike the previous studies as mentioned, the impact of adding NH_3 as an additive into aqueous amine solutions as available in the literature is described below.

Lee et al. (2008) investigated on the CO_2 removal efficiency, amount absorbed, and CO_2 loading into a blend of aqueous AMP and NH_3 solutions using an absorber (height: 0.9 m; diameter: 0.0508 m) and a regenerator (height: 0.9 m; diameter: 0.0508 m) with ¼″ Rasching rings as packing materials. The conditions maintained in the experiment were as follows: absorber temperature: 40°C; regenerator temperature: 80–110°C; gas flow rate: 5–15 lpm; CO_2 concentration at the inlet: 12%; liquid recirculation rate: 110 ml/min; and AMP + NH_3 blend: 30% AMP + n% NH_3 (where n = 0, 1, 3, 5, 7, 10, and 20). Experiments were conducted to determine the effects of blending AMP with an increasing amount of NH_3 under various operating conditions. The absorption of CO_2 was increased as the mixing ratio of NH_3 increased. However, the concentration of NH_3 more than 5% was not adequate because of the problem associated with the scale formation. With regard to the concentration of NH_3 investigated in the blends, 5% NH_3 was shown to be the optimum concentration because the CO_2 removal efficiency was almost 100%, and absorption capacity was 2.17 kg CO_2/kg solvent. Also the problem of scale formation under this condition was insignificant. The CO_2 loadings of rich- and lean amines were analyzed by using 1, 3, and 5% NH_3 at regenerator temperatures of 80–110°C. As the additive concentration of NH_3 was increased, both rich- and lean loadings were increased. However, with the increase in the regenerator temperature, the rich loading was almost constant while lean loading was decreased. The characteristics of absorption/regeneration reaction were assessed by analyzing values of CO_2-rich and -lean loadings. The optimum experimental conditions when NH_3 is added to AMP, therefore, could be presented. Furthermore, addition of 5% NH_3 to 30% AMP was shown to improve the CO_2 removal efficiency owing to the improvement of the absorption capacity. For instance, increase in the gas flow rate from 5 to 15 lpm reduced the CO_2 removal efficiency from 97.2 to 58.4% with 30% AMP alone, and 100 to 85.2% with 30% AMP + 5% NH_3. Choi et al. (2009a) investigated on the absorption and regeneration of CO_2 and SO_2, in an absorber and a regenerator of dimensions similar to that reported by Lee et al. (2008). The conditions of the absorption experiments were as follows: absorber and regenerator temperatures: 40°C; gas flow rate: 7.5 lpm; simulated gas I: 15% CO_2 + air balanced; simulated gas II: 15% CO_2 + 2,000 ppm SO_2 + air balanced; and liquid circulation rate: 90 ml/min. The conditions of the absorption and regeneration experiments were as follows: absorber temperature: 40°C; regenerator temperature: 110°C; gas flow rate: 5–15 lpm; simulated gas I: 15% CO_2 + air balanced; simulated gas II: 15% CO_2 + 2,000 ppm SO_2 + air balanced; and liquid circulation rate: 70–130 ml/min. The AMP and NH_3 blend used in these experiments was 30% AMP + n% NH_3 (where n = 0, 1, 3, and 5). The absorption rate of CO_2 and the CO_2 loading ratio were increased with the addition of NH_3 into the aqueous AMP solution, however, maintaining the removal efficiencies of CO_2 and SO_2 over 90 and 98%, respectively. Considering the CO_2 removal efficiency, CO_2 loading, energy requirements, and corrosiveness in the absorption and regeneration processes, the experimental conditions were optimized at a gas flow rate of 7.5 lpm and a liquid circulation rate of 90 ml/min. It was shown that the simultaneous absorption of CO_2 and SO_2 was achievable without any impact on CO_2 removal process by the SO_2 removal because the removal efficiencies of CO_2 and SO_2 were maintained at a high level and the CO_2 loading was increased in the blend of 30% AMP + 3% NH_3 solution. In another investigation, Choi et al. (2009b) measured the CO_2 absorption rates of AMP and a blend of AMP and NH_3 at different partial pressures (5, 10, and 15 kPa), temperatures (20, 30, 40, and 50°C) and aqueous concentrations (NH_3 additive concentrations were 1, 3, and 5% for the 30% AMP) using a stirred cell reactor (height: 0.16 m, diameter: 0.095 m, batch liquid volume: 500 ml, and stirring speed: 50 rpm). Results showed that the CO_2 absorption rate was increased by 144% with increasing NH_3 concentration and partial pressure. The CO_2 absorption rate on the addition of NH_3 into 30 wt% AMP solution at partial pressure of 15 kPa increased by 16.3–67.6% compared to that of the aqueous 30 wt% AMP solution without NH_3. In addition, the rate of absorption of CO_2 into aqueous blend of AMP/NH_3 was directly proportional to the NH_3 concentration that was attributed to the high absorption capacity of $NH3_3$ for CO_2. From the absorption studies, kinetics were derived. Results of the reaction rate constant for a blend of AMP and NH_3 with CO_2 as a function of temperature was 1.6- to 2.4-fold higher than that of AMP without NH_3 demonstrating that the NH_3 addition enhanced the AMP reactivity.

1.4.2 AMINES AS ADDITIVES IN NH_3 SCRUBBING OF CO_2

The advantages of using NH_3 over MEA (alkanolamine) for CO_2 removal have already been discussed earlier. However, the application of NH_3 in such a situation could have problems of NH_3 vaporization. Attempt of increasing NH_3 concentration to increase the CO_2 removal efficiency may lead to loss of NH_3 vapor at higher initial NH_3 concentration. As a result, CO_2 removal efficiency is reduced due to lowering of NH_3 concentration in the absorbent. In addition, NH_4HCO_3 crystals would also be formed by the reaction between NH_3 vapor and CO_2 gas that could have a potential to plug the flow path in the equipment of CO_2 capture system by forming scales. In order to reduce the loss of NH_3 vapor, i.e., the NH_3 slip, the use of CAP described earlier could be a plausible option. As an alternative to the CAP, NH_3 as a solvent with functional additives, e.g., alkanolamines (like primary, secondary, tertiary, and sterically hindered amines) have been studied extensively (Seo and Hong 2000; Ali et al. 2002; Ali 2005; Paul et al. 2007). This section deals with the modification of NH_3 as absorbent with amine additives for CO_2 capture. You et al. (2008) investigated the performance of NH_3 amine additive–CO_2 absorption system especially keeping in mind the loss of NH_3 and maintaining or enhancing the CO_2 removal efficiency of NH_3. They have selected sterically hindered amines like AMP, 2-amino-2-methyl-1,3-propandiol (AMPD), 2-amino-2-ethyl-1,3-propandiol (AEPD), and tri(hydroxymethyl) aminomethane (THAM) considering the lower heat required for regeneration of the modified absorbents. Experiments were conducted in a batch bubbling reactor (diameter: 0.045 m) having 200 ml of aqueous absorbent solutions using

15% CO_2 and the rest N_2. The concentration of NH_3 was 10% in the case of NH_3 absorbent while blended absorbents were all composed of 10% NH_3 + 1% respective amine. The loss of NH_3 by vaporization was reported to be decreased irrespective of the existence of the CO_2 in the sequence as NH_3 > (NH_3 + AMPD) > (NH_3 + AEPD) > (NH_3 + AMP) > (NH_3 + THAM). Results further showed that the total CO_2 removal efficiencies of aqueous blended NH_3 were slightly increased in the order as NH_3 < (NH_3 + THAM) < (NH_3 + AEPD) < (NH_3 + AMP) < (NH_3 + AMPD). This enhancement of CO_2 removal efficiencies was attributed to the intermolecular interactions between the additives and CO_2. This investigation clearly demonstrates that amines as additives to NH_3 could be useful for CO_2 capture from the perspective of NH_3 slippage as well as CO_2 removal efficiency.

1.5 COMPARISON OF NH_3- AND AMINE-BASED CO_2 CAPTURE SYSTEMS

Existing literature indicates that the CO_2 capture from the exhaust of coal fired TPPs by NH_3 offers many advantages over the commercially available amine-based processes. The relative demerits of the amine-based process as well as the merits of the aqueous NH_3-based process were well reported in the existing literature (Resnik et al. 2004; Ciferno et al. 2005; McLarnon and Duncan 2009; Zhou et al. 2009), and these are summarized here in this section.

The disadvantages of the conventional MEA-based absorption process for removing CO_2 are as follows:

1. low CO_2 loading capacity (mol CO_2 absorbed/mol absorbent);
2. oxidative degradation of the amine by SO_2, NO_x, hydrogen chloride (HCl), hydrogen fluoride (HF), and O_2 in the flue gas requiring a high liquid to gas flow rate ratio;
3. high corrosion rate requiring exotic materials of construction for both absorber and stripper columns; and
4. high energy requirement for regenerating the absorbent.

In contrast, aqueous NH_3 offers benefits such as

1. having higher loading capacity (mol CO_2 absorbed/mol of absorbent);
2. being free from corrosion problem;
3. being stable in the environment of the flue gas;
4. requiring lower liquid to gas flow ratio;
5. having multi-pollutant capture capability—especially SO_2 and NO_x removal could be integrated with the process of CO_2 removal, thus eliminating the necessity of pretreatment of the flue gas in respect of these pollutants as is required for amine-based processes;
6. consuming much less energy for regenerating the solvent, if necessary; otherwise, the process can be routed to yield valuable products like NH_4HCO_3, $(NH_4)_2SO_4$, and NH_4NO_3 that can be put into soil as fertilizer;
7. being more economic than MEA as well host of other amines; and
8. ease of transportation of NH_4HCO_3 (major product) produced as a white crystalline solid; similar to other solid products, it can easily be transported without any extra investment as is often required for "pipeline transportation of compressed CO_2."

Thus, this solid product can be

1. shipped to site for deep ocean or geologic storage;
2. sold for EOR and Enhanced Coal Bed Methane (ECBM) recovery applications; and
3. sold to agriculture for conventional fertilizing.

Compared to other CO_2 shipping options, transporting NH_4HCO_3 would be cost effective. Therefore, NH_4HCO_3 can suitably function as a "CO_2 Carrier" for other sequestration options. On the other hand, NH_4HCO_3 thus synthesized can be applied on land as fertilizer (Zhou et al. 2009). It is necessary to discuss further the performances of these processes for improved understanding, which is done in the next section.

1.5.1 COMPARISON OF PERFORMANCES OF NH_3- AND AMINE-BASED PROCESSES

Experimental investigations were carried out by Yeh and Bai (1999) using a laboratory bubble column to evaluate the detailed performances of NH3- and MEA-based CO_2 absorption processes. The performances of these two processes were also compared in terms of CO_2 removal efficiency and absorption capacity. The detailed performance-based comparison of these two processes is presented in Table 5 (see p. 46). It can be seen from the table that the CO_2 removal efficiency for the NH_3 solvent could be as high as 99%, and the CO_2 absorption capacity was over 1.0 kg CO_2/kg NH_3, while the maximum CO_2 removal efficiency and absorption capacity for the MEA solvent were 94% and 0.40 kg CO_2/kg MEA, respectively. The rise in temperature in the NH3 scrubbing was less than that in the MEA scrubbing indicating requirement of lesser heat for regeneration in the NH_3 process than that for the MEA process. Reportedly, the absorption capacity of NH_3 was 2.4–3.2 times higher than MEA, and the cost of the industrial grade aqueous NH3 was one-sixth than that of the MEA solvent indicating NH_3 to be the most economical solvent for CO_2 scrubbing. They have recommended the installation of a mist eliminator at the upper cross section of the scrubber with water wash arrangement to reduce the slippage of $NH3_3$into the atmosphere. A post-scrubber reheater was also recommended to be installed for reheating the flue gas to produce buoyancy as NH_3 scrubbing was carried out at room temperature. Further, quenching of flue gas prior to scrubbing could be achieved by a cooler that might be integrated with the reheater to reduce the cost of energy. Yeh and Bai (1999) were motivated to take up the investigation by the fact that a properly designed NH_3 scrubbing process can additionally remove other acidic gases like SO_2 and NO_x present in the flue gas to produce $(NH_4)_2SO_4$ and $(NH_4)NO_3$. The qualities of these products as fertilizers are proven to be better than NH_4HCO_3; however, the quantum generation would be considerably lower in the case of scrubbing of flue gas generated from the TPPs.

The laboratory results reported by McLarnon and Duncan (2009) were analyzed for studying the impact on output of a coal-fired TPP. The performances of NH_3- and MEA-based processes on an existing coal-fired TPP achieving 90% CO_2 removal were estimated, and the results are shown in Table 6. The gross and net power outputs for the base plant with WFGD, an MEA-based process with enhanced WFGD, and the NH_3-based process for CO_2/SO_2 removal were estimated. Their analyses showed that the gross output of the power plant was reduced from 463 to 449 MW with the installation of the NH_3-based process, whereas the gross output was reduced to 388 MW with an MEA-based process. Thus, the net output of the plant would be reduced by 16% and 30% when using an NH_3- and an MEA-based processes, respectively.

TABLE 6: Comparison of performances of CO_2 capture processes

	Base plant without CO_2 capture	MEA with enhanced WFGD	NH3-based process with CO_2/SO_2 capturea
Gross output (MW)	463	388	449
Balance of plant (MW)	30	30	30
CO_2 capture and compression (MW)	–	55	57
Net output (MW)	434	303	362
% Loss from base plant	–	30	16

[a]*Developed by Powerspan*

The energy usage patterns as compared by McLarnon and Duncan (2009) in an NH_3-based as well as an MEA-based process for CO_2 capture and regeneration are shown in Table 7. The consumption of process energy was analyzed based on sensible heat, reaction energy, and energy for steam stripping. The sensible heat was the energy required to raise the temperature of the scrubbing liquid to the stripper temperature that would be lost. Reaction energy was the heat of reaction for absorption of CO_2. Finally, the energy for steam stripping was the energy associated with the steam needed during the stripping operation. The estimated total energy requirement for the NH_3-based process was 1,147 kJ/kg-CO_2 that was only about 27% of the energy requirement of MEA-based process of 4,215 kJ/kg-CO_2. This estimate clearly demonstrates as to what extent the NH_3-based process is energetically favorable.

TABLE 7: Comparison of energy usages between MEA- and NH3-based processes for CO_2 Capture

	MEA-based process (kJ/kg-CO_2)	NH3-based processa (kJ/kg-CO_2)
Sensible heat	865	298
Reaction energy	1,920	644
Stripping steam	1,430	205
Total energy	4,215	1,147

[a]*Developed by Powerspan*

TABLE 8: Performances of TPPs under various scrubbing conditions

Case	1	2	3	4	5	6
Absorbent	None	MEA	Aq. NH_3	Aq. NH_3	Aq. NH_3 (USC)[a]	Aq. NH_3 (USC)[a]
Pollutant(s) removed	None	CO_2	CO_2	CO_2, SO_x, NO_x, Hg	CO_2	CO_2, SO_x, NO_x, Hg
Total gross power (MW)	425	492	478	482	473	476
Auxiliary load (MW)						
Base plant	22.1	28.3	27.3	27.5	25.1	25.3
CO_2 capture	–	21.4	14.5	10.3	13.6	10.2
CO_2 compression	–	35.3	30.0	30.2	28.1	28.3
NO_x and SO_x	3.1	4.4	3.8	11.0[b]	3.5	10.3[b]
Transport and storage	–	2.9	2.5	2.5	2.3	2.3
Total	25	92	78	82	73	76
Net power (MW)	400	400	400	400	400	400
Coal consumption rate (TPD)	3,480	4,895	4,172	4,200	3,904	3,935
CO_2 captured (TPD)	–	10,240	8,727	8,789	8,168	8,233
Net heat rate (kJ/kWh)	8,918	12,550	10,697	10,773	10,010	10,091
Fertilizer production (TPD)	–	–	–	443	–	415
Overall efficiency (%)	40	29	34	34	36	36
Energy penalty (%)	–	29	17	17	16[a]	16[a]

Energy penalty: percent decrease in power plant efficiency due to CO_2 capture
[a]*Ultra-supercritical steam cycle; USC base case with no capture was 43% efficient*
[b]*Auxiliary load for the multi-pollutant removal was ~11 MW*

Ciferno et al. (2005) reported on the overall performances of the following six cases:

Case 1
Power Plant operation without air pollution control options;
Case 2
Power Plant operation with MEA-based process for CO_2 Capture;
Case 3
Power Plant operation with aqueous NH_3-based process for CO_2 Capture;
Case 4
Power Plant operation with aqueous NH_3-based process for removing CO_2, SO_x, NO_x, and mercury (Hg);
Case 5
Power Plant (Ultra Super Critical) operation with aqueous NH_3-based process for CO_2 Capture; and
Case 6
Power Plant (Ultra Super Critical) operation with aqueous NH_3-based process for removing CO_2, SO_x, NO_x, Hg.

The detailed performance-related data are furnished in Table 8. It can be seen from the table that Case 2 (MEA–CO_2 capture option) was very energy intensive and was requiring 57 MW for both capture and compression with an additional coal consumption of 1,415 TPD that decreased efficiency approximately by 30%. The benefits of higher CO_2 absorption capacity as well as lower heat of reaction was reported for aqueous NH_3-based process compared to MEA-based process that resulted in a 15% decrease in parasitic load from 92 to 78 MW, and 15% decrease in net power plant heat rate for Case 3. The same proportional amount of energy savings was also reported in the ultra-supercritical cases (i.e., Case 5 and Case 6) with the use of aqueous NH_3. They also highlighted the potential for selling the $(NH_4)_2SO_4$, NH_4NO_3 by-products as an additional source of revenue which would help recoup some of the cost of capture. Ammonia-based CO_2 capture process has, therefore, been found to be more techno-enviro-economically feasible than the MEA-based process. Furthermore, considering the complexity and higher cost inherently associated with other amine-based processes developed for CO_2 capture, NH_3-based process still seems superior in comparison to MEA-based process elicited earlier.

1.6 LIFE CYCLE CO_2 EMISSIONS IN MEA SCRUBBING OF CO_2

The life cycle CO_2 emissions of three pulverized coal-fired TPP with and without post-combustion MEA-based CO_2 capture, transport, and storage were compared (Koornneef et al. 2008) in respect of the power plants that were operating in the Netherlands. The three cases that were assessed were

Case 1

The reference case representing the average sub-critical pulverized coal-fired TPP;

Case 2

An ultra-supercritical pulverized coal-fired TPP with best available technology proposed to be installed in the near future in the Netherlands;

Case 3

A coal-fired TPP, equal to case $_2$, equipped with a post-combustion MEA-based CO_2 capture facility assuming it would be available in the near future, with the CO_2 Capture and Storage (CCS) chain further comprising compression, transport, and underground storage of the CO_2.

The performance parameters for these three cases are presented in Table 9. The study was carried out to assess whether the implementation of post-combustion capture with transport and storage results in benefits toward GHG emission reduction for the coal-fired TPP in the Netherlands. The life cycle inventory data used were specific for the Netherlands. However, worldwide average data were used in the event of non-availability of adequate data for Netherlands. In order to assess the life cycle CO_2 emission, GHG balance was studied (Koornneef et al. 2008) for the three cases that are presented in Table 10. It can be seen from the table that the total life cycle CO_2 emission for Case 3 estimated was 243.2 gCO_2-

equivalents as against 1092.9 and 838.1 gCO_2-equivalents for Cases 1 and 2, respectively. Thus, a significant reduction in CO_2-equivalents for the CCS case (Case 3) took place compared to Cases 1 and $_2$. It was concluded by Koornneef et al. (2008) that CO_2 capture and its compression with the associated efficiency penalty resulted in the increase in the total emission of CO_2 equivalents contributed by up- and down-stream processes.

TABLE 9: Performance parameters for the three coal-fired TPPs investigated for life cycle CO_2 emissions

Parameter	Case 1	Case 2	Case 3
Net generating efficiency (based on LHV)			
Without capture (%)	35	46	46
With capture (%)	–	–	35
Thermal capacity, MWth	1,303	1,303	1,303
Net generating capacity, MWe	460	600	455
Full load hours (h/year)	7,800	7,800	7,800
ESP + FGD efficiency for particulate matter (%)	99.95	99.98	
FGD efficiency for SO_2 (%)	90	98	
FGD limestone and quicklime use (kg/kg SO_2 removed)	1.2/0.3	1/0	
FGD gypsum product/limestone use (kg/kg)	–	1.85	
SCR efficiency for NO_x (%)	60	85	
SCR ammonia use (kg/kg NO_x removed)	0.3	0.35	
SCR ammonia slip (% of ammonia use)	1	1	
HCl reduction efficiency (%)	90	98	
HF reduction efficiency (%)	70	98	
Hg reduction efficiency (%)	56	90	
Emission factors without flue gas cleaning			
NO_x (kg/MJ)	2.76×10^{-4}	1.35×10^{-4}	
SO_2 (kg/MJ)	5.71×10^{-4}	6.40×10^{-4}	
CO_2 (kg/MJ)	0.0947	0.0947	
HF (kg/MJ)	3.77×10^{-6}	6.59×10^{-6}	
HCl (kg/MJ)	1.06×10^{-5}	3.30×10^{-5}	
Hg (kg/MJ)	4.18×10^{-9}	5.47×10^{-9}	
Particulate matter (kg/MJ)	4.29×10^{-3}	8.29×10^{-3}	

TABLE 10: Greenhouse gas balance for the three coal-fired TPPs investigated

Emission sources	Case 1		Case $_2$		Case 3	
	gCO_2 equiv.	%	gCO_2 equiv.	%	gCO_2 equiv.	%
First-order emissions						
Electricity generation	976	89	749	89	107	44
Second-order emissions						
MEA production chain	–	–	–	–	6	3
Reclaimer bottoms disposal	–	–	–	–	7	3
Coal supply chain total	98	9	75	9	99	41
Coal mining	33	3	25	3	33	14
Coal transport	41	4	31	4	41	17
Remaining coal chain	25	2	19	2	25	10
Remaining processes	6	1	4	0.5	10	4
Subtotal	105	10	80	9	123	50
Third-order emissions						
Power plant	1.3	0.11	1.0	0.11	1.3	0.52
CO_2 capture installation	–	–	–	–	1.4×10^{-2}	0.006
CO_2 compressor	–	–	–	–	5.4×10^{-3}	0.002
CO_2 pipeline	–	–	–	–	3.6×10^{-1}	0.15
CO_2 injection facility	–	–	–	–	3.6×10^{-1}	0.15
Infrastructure coal supply chain	10.4	0.95	8	0.95	10.5	4.31
Remaining processes	0.2	0.02	0.1	0.02	0.7	0.29
Subtotal	11.9	1.08	9.1	1.08	13.2	5.42
Total life cycle CO_2 emission	1092.9	100%	838.1	100%	243.2	100%

1.7 LIFE CYCLE CO_2 EMISSIONS IN NH_3 SCRUBBING OF CO_2

The life cycle CO_2 emissions were described (Wang et al. $_2$007) based on two systems viz. (i) System-I: CO_2 absorption by NH_3 scrubbing and (ii) System-II: industry production of NH_4HCO_3. System-I comprises NH_3 production by steam reforming of methane (SRM) in coal-fired TPP with installed capacity of 300 MW and CO_2 absorption by NH_3 scrubbing with the formation of NH_4HCO_3; while System-II includes syngas production by coal gasification, NH_4HCO_3 production by carbonation and a coal-fired

TPP of 300 MW. Coal was used to produce NH_4HCO_3 through gasification and carbonation of natural gas. Coal gasification could produce enough CO_2 that was used to produce NH_4HCO_3. Since SRM cannot afford enough CO_2 for the production of NH_4HCO_3, CO_2 must be produced in some other ways for SRM. For simplification, coal gasification was used in System-II.

The consumption of water, steam, and catalysts in both systems was ignored in the analysis due to sparse information. Their production would also emit CO_2 from the standpoint of life cycle analysis. Furthermore, the CO_2 emissions due to manufacturing of equipment and transportation were also ignored. The CO_2 emissions from these ignored factors were, however, generally assumed (Wang et al. 2007) well within 5% of the total emissions of the system, and this assumption would not affect the results. The yearly yield of NH_4HCO_3 and the yearly electricity generated were the functional units for the purpose of their study. Detailed lifecycle analysis data are shown in Table 11 (see p. 48). It can be seen from the table that NH_3 scrubbing considerably reduces the CO_2 emission potential from a TPP. Furthermore, the resultant CO_2 emission from the NH_4HCO_3 plant using CO_2 from a TPP would significantly be lower than the CO_2 emission from an equal capacity NH_4HCO_3plant having a TPP of a same size.

Ammonia scrubbing of CO_2 requires a large amount of NH_3. In case where it is desired to use NH_3 back into the scrubbing operation, then it is necessary to decompose the reaction product (NH_4HCO_3) to regenerate NH_3 and CO_2 through a properly designed regenerator. This will reduce NH_3 consumption in the scrubbing system. Carbon dioxide gas thus regenerated is in pure form, and can be used commercially meeting purity specifications, viz., medical, general commercial (including geological sequestration, ECBM recovery), food processing, beverage, dry ice, and refrigeration. The CO_2 purity specifications for various classes of use are shown in Table 12 (Sass 2002) (see p. 49). Besides, NH_4HCO_3 can be used as a source of fertilizer if its regeneration is not necessitated by the availability of NH_3. In fact, NH_4HCO_3 is a cheap and widely used fertilizer in China (Yeh and Bai 1999). However, NH_4HCO_3 cannot compete with other ammonium compounds as a fertilizer due to its lower value of nitrogen content (18% N) as compared with urea (45% N). Ammonium bicarbonate is, therefore, not an ideal source of fertilizer from the economical standpoint.

However, the use of NH_4HCO_3 as a source of synthetic N-fertilizer could be feasible based on its recycling ability. Recent researches have shown that the geological solidification of NH_4HCO_3 in comparison to NH_4HCO_3 is considerably higher in so far as GHG emission is concerned. Considerable progresses have also been made in the area of long-term effect of NH_4HCO_3 so as to restrict its decomposition into NH_3 and CO_2 making it as a promising fertilizer. However, application of NH_4HCO_3 as a fertilizer may release CO_2 back into the atmosphere and also it may lead to emission of N_2O a GHG, stemming from the microbiological process in the soil. Considering the fact that the Global Warming Potential of N_2O is 310 as against 1.0 for CO_2, a comparison must be made between CO_2 captured by NH_3 scrubbing and the emission of CO_2 (combining the release of CO_2 and emission of N_2O) in terms of CO_2-equivalent. These issues are elaborated further.

Lee and Li (2003) investigated NH_3 scrubbing of flue gas containing CO_2, NO_x, and SO_2 producing mainly NH_4HCO_3 fertilizer at Oak Ridge National Laboratory. The fate of NH_4HCO_3 whence put into soil for fertilization was investigated with great care from environmental standpoints. On putting deep into the soil, the N-fertilization effect of NH_4HCO_3 on crops was similar to that of other N-fertilizer like, $(NH_4)_2SO_4$ and urea $(NH_2)_2CO$ (Xi et al. 1985). It is dissociated into NH_4^+ and HCO_3^- ions when put into aqueous soil matrix. Soil particles containing alkaline minerals are carrying negative surface charges that are attracting positively charged ions. Thus, they are showing much higher retaining affinity for positively charged species such as NH_4^+ ions than for negatively charged ions like HCO_3^- ions, thereby allowing HCO_3^- ions to percolate with rainfall runoff and/or irrigation down into groundwater. This enables carbonates to react with the alkaline earth metals like Ca^{2+} ions available in the groundwater and be deposited as carbonated materials, such as $CaCO_3$, in the sub-soil of earth layers. It was further noted that when NH_4HCO_3 was put into the soil as a fertilizer, its NO_3^- ion could easily release with water from soils that would result in the loss of the fertilizer and would contaminate the groundwater with NO_3^- content. The release of NO_3^- ions could also lead to emission of N_2O back into the atmosphere contributing to GHG emission. On the other hand, the results could be very different when NH_4HCO_3 and urea were used as fertilizers. Unlike NO_3^- ion, in this case carbonate (e.g., $CO3^{2-}$ and HCO_3^-) ions were released and resulted in carbonated groundwater as described earlier. Solid carbonated materials like $CaCO_3$ are perfectly stable form of sequestered CO_2. Also carbonates are harmless species and would not cause any health problems as would occur in case of NO_3^-. Moreover, groundwater movement could carry carbonates as deep as from 500 to over 1,000 m down to the earth subsurface where they could be deposited by the carbonation reaction with minerals containing alkaline earth metals. Importantly, the residence time of groundwater in many geological areas could be on the order of hundred, even thousands of years (Plummer et al. 1983). Once the fertilizer-released carbonates thus entered into groundwater, they would remain fixed for hundreds of years. As long as the percolation of carbonates would occur effectively from soil with rainfalls and/or irrigation down into groundwater, the sequestration of CO_2 could occur potentially. Considering 55.7% CO_2 and 17.7% N contents by weight in NH_4HCO_3, it was further estimated by Lee and Li (2003) that the annual production of 100 Mt of NH_4HCO_3 as N-fertilizer utilizing the CO_2 from smokestacks of fossil fuel (e.g., coal)-fired TPPs by NH_3 scrubbing, would require 315 Mt CO_2/year. This amount of CO_2 equals to 18.6% of the CO_2 emissions from the fossil fuel-fired TPPs in the United States or 4.8% of the CO_2 emissions from the world's coal fired TPPs. The concept developed has integrated the pollutant removing fertilizer production reactions from the flue gas of coal fired TPPs and other energy related operations resulting in a clean energy system keeping in harmony with the earth's ecosystem. The technology was claimed to contribute importantly to the global CO_2 sequestration and clean air production. Finally, it was concluded that although the practical potential might be somewhat smaller than estimated, CO_2 solidification in terms of NH_4HCO_3 production process would be a promising approach that could be worth exploring for efficient carbon management.

In a recent study conducted by Cheng et al. (2007), the fate of carbon using the 14C tracer technique in ecosystems was detailed experimentally by applying NH_4HCO_3, an economic source of nitrogen fertilizer that was synthesized from NH_3 scrubbing of CO_2. The unique features of NH_4HCO_3 as N-fertilizer have already been discussed. The decomposition of NH_4HCO_3 is not uncommon similar to other ammonium salts. This in effect could recharge CO_2 into the atmosphere on application of

NH_4HCO_3 into the soil as fertilizer. As a result, the very purpose of producing NH_4HCO_3 would go baffled. Thus, in order to improve the stability of NH_4HCO_3, dicyandiamide could be added during NH_4HCO_3 production. The modified NH_4HCO_3 thus produced is called the long-term effect NH_4HCO_3 (LE-NH_4HCO_3). The synthesized product was used as a "carrier" for transporting atmospheric CO_2 to the crops and the soil. Reportedly, an indoor greenhouse containing wheat as the study plant was built. The investigated ecosystem was composed of wheat (the study-plant), soils with three different pH values (alkaline, neutral, and acidic), and three types of underground water having varied concentrations of Ca^{2+} and Mg^{2+} ions. It was reported (Cheng et al. 2007) from a separate post-experiment demonstration that carbon in soil could go in four directions after NH_4HCO_3 or LE-NH_4HCO_3 was put into the soil, for instance, air, plant, soil, and water or groundwater. The majority of the unused carbon source, up to 76% from LE-NH_4HCO_3 and 75% from NH_4HCO_3 , percolated into the soil as environmentally benign calcium carbonate ($CaCO_3$). It was shown that up to 88% of the carbon from NH_4HCO_3 existed in insoluble salts such as, $CaCO_3$ in the alkaline soil. However, for acidic soil, the percentage of $CaCO_3$ in the soil was low and a reduced level was of 10–25% was reported. Moreover, there was an abundance of Ca^{2+} and Mg^{2+} ions in the soil studied. The concentrations of Ca^{2+} and Mg^{2+} ions were 857 and 1,238 ppm, respectively. The availabilities of Ca^{2+} and Mg^{2+} ions in the soil were the key sources for storing carbon in the soil. Carbon trapped into the soil was increased with the increase in alkalinity, i.e., alkaline soil showed higher capacity to store carbon. It was shown that a considerable amount, up to 10% of the carbon source from NH_4HCO_3 or LE-NH_4HCO_3, was absorbed by the wheat with increased biomass production after biological assimilation and metabolism in wheat. It was further demonstrated that 0.64–4.74% carbon from NH_4HCO_3 or LE-NH_4HCO_3 percolated down to be stored in the groundwater. The losses of carbon into the air would be calculated by difference. However, a maximum loss of carbon into the air was reported to the tune of 50% from NH_4HCO_3 or LE-NH_4HCO_3 for acidic soil. Clearly, these findings demonstrated that NH_4HCO_3 or LE-NH_4HCO_3 themselves did not show any difference in carbon storage capacity for the same soil, even though there was an advantage for LE-NH_4HCO_3 owing to its greater stability. It was finally concluded from this research that NH_4HCO_3 synthesized from NH_3 scrubbing of CO_2 had promoted the development of plant roots that helped plant growth. The plant photosynthesis was, therefore, enhanced to absorb more CO_2 from the atmosphere.

These investigations revealed that NH_4HCO_3 synthesized, from the NH_3 scrubbing of CO_2 from the flue gas of fossil fuel (e.g., coal)-fired TPP, could be gainfully utilized as a fertilizer. This geological solidification of CO_2 could contribute positively to the global CO_2 sequestration and clean air production.

The prime concern of the foregoing discussion was the emission of CO_2 after NH_4HCO_3 was put into the soil as a fertilizer. However, the synthetic N-fertilizer, on land application, has a tendency to emit N_2O through nitrification and denitrification processes occurring into the soil utilizing the NH_4^+ ions released from the (i) ammonium-based synthetic N-fertilizers [like NH_4HCO_3, NH_4HCO_3, $(NH_4)_2SO_4$ etc.] and (ii) reduction of urea under conditions prevailing in the soil. Such emission of N_2O is something different from the emissions that originate from NO_3^- ions in the case of NH_4HCO_3 as elucidated earlier. The emissions of N_2O from synthetic N-fertilizers depend mainly on water or moisture content on field or soil, rate of N-application, soil properties, and humidity. Under this circumstance, the emission factor of N_2O is generally larger in uplands than in rice-based ecosystems (Zucong 2009). According to the revised IPCC Guidelines for National Greenhouse Gas Inventories (IPCC 1997), the default value of the emission factors are 1% for uplands and 0.3% for rice fields. The overall emission factor of synthetic N-fertilizer estimated by Lu et al. (2007) was ranging from 0.25 to 2.25% with an average value of 0.92%, using a statistical model for uplands, and default emission factor of IPCC Guidelines for rice fields of 0.3%. This estimate was in closest agreement with the average value of 0.93% estimated by Yan et al. (2009) for East Asia, South-East Asia, and South Asia. However, the average emission factor of synthetic N-fertilizer in China calculated from the estimates of Lu et al. (2007) was 1.15% and was higher than the default value of 1.0% suggested in the IPCC Guidelines (IPCC 1997). Thus, the estimate of Lu et al. (2007) seems to be a conservative one and can be used for the present analysis. In this study, it has already been described that 100 Mt NH_4HCO_3 per annum can be produced by NH_3 scrubbing of 315 Mt CO_2/year emitted from the flue gas of fossil fuel (e.g., coal)-fired TPPs. The calculation of

N_2O emission after 100 Mt NH_4HCO_3 put into the soil as a synthetic N-fertilizer considering the upper value of the emission factor of 2.25% following the methodology of Lu et al. (2007) (worst scenario) results in 0.4 Mt N_2O. This equals to a value of 124 Mt CO_2-equivalent. Further, considering a maximum loss of 50% of carbon of NH_4HCO_3 (Cheng et al. 2007) into the air after NH_4HCO_3 was put into the soil (worst scenario), 100 Mt NH_4HCO_3 in the present analysis would release 27.5 Mt CO_2. Therefore, a total emission of about 152 Mt CO_2-equivalent could occur after use of 100 Mt NH_4HCO_3 as synthetic N-fertilizer that is about 50% of the total CO_2 captured (315 Mt) for producing the fertilizer, NH_4HCO_3. Clearly, this estimate demonstrates that the synthetic N-fertilizer, NH_4HCO_3, produced by NH_3 scrubbing of CO_2 from fossil fuel (e.g., coal)-fired TPP could have a significant beneficial environmental impact so far as GHG emission is concerned.

1.8 CONCLUSIONS

Carbon dioxide is a GHG, and its emission from the burning of fossil fuels in addition to other industrial sources is affecting the climate of the earth adversely. Increased burning of fossil fuels will continue to enhance the atmospheric CO_2 levels. Estimates indicated that power production contributes to the tune of 70% of the total CO_2 released into the atmosphere from fossil fuel combustion worldwide. Legion of researchers have thus far developed absorbent to remove CO_2 from combustion facilities that are currently recognized globally as most effective. The cost of capturing CO_2 can be reduced by finding a low-cost solvent that can minimize energy requirements, equipment size, and corrosion. Monoethanolamine is being used in removing CO_2 from exhaust streams and is a subject inculcated for a period of about last 80 years. Host of such other amines are being investigated and put into practice. However, commercialization of such operating plants for capturing CO_2 from power plants in the world are few and far between. On the other hand, NH_3 scrubbing is the other chemical solvent for capturing CO_2 from an exhaust stream, like flue gas emitted from the power plant. Detailed processes of NH_3 absorption of CO_2 were described. The NH_3 absorption of CO_2 has proven experimentally to be more effective than amine-based absorption that is so far the most acceptable method. This method was shown to more effective than the amine-based process due to its several advantages: (i) having higher loading capacity (mol CO_2 absorbed/mol of absorbent); (ii) being free from corrosion problem; (iii) being stable in the environment of the flue gas; (iv) requiring lower liquid to gas flow ratio; (v) having multi-pollutant capture capability—especially SO_2 and NO_x removal could be integrated with this process for CO_2 removal, thus eliminating the pretreatment of the flue gas in respect of these pollutants as is required for amine-based processes; (vi) consuming much less energy for regenerating the solvent, if necessary; else, the process can be routed to yield valuable products like NH_4HCO_3, $(NH_4)_2SO_4$, or NH_4HCO_3 that can be put into soil as fertilizer; (vii) being more economic than MEA as well host of other amines, and (viii) being associated with ease of transportation of the NH_4HCO_3produced as a white crystalline solid; similar to other solid products, it can be easily transported without any extra investment as is often required for "pipeline transportation of compressed CO_2." It can be (a) shipped to site for deep ocean or geologic storage, (b) sold for EOR and ECBM recovery applications, and (c) sold to agriculture for conventional fertilizing.

Plant operating data further suggested that the net output of the plant would be reduced by 16% when using an NH_3-based process and by 30% when an MEA-based process. The estimated total energy requirement for the NH_3-based process was 1,147 kJ/kg-CO_2 that was only about 27% of the energy requirement of MEA-based process of 4,$_2$15 kJ/kg-CO_2. The benefits of higher CO_2 absorption capacity as well as lower heat of reaction was reported for aqueous NH_3-based process compared to MEA-based process that resulted in a 15% decrease in parasitic load from 92 to 78 MW and 15% decrease in net power plant heat rate.

The literature further revealed that NH_3 as an additive to amine-based CO_2 capture process and amine as an additive to NH_3-based CO_2 capture process had the potential to improve the rate of CO_2 absorption as well as the CO_2 absorption capacity. Life-cycle CO_2 emission analyses were elaborated for the two CO_2 capture systems for registering first hand information. Such analyses revealed beneficial impacts of CO_2 capture process. Finally, it was estimated that a total emission of about 152 Mt CO_2-equivalent could occur after use of 100 Mt NH_4HCO_3 as synthetic N-fertilizer that was about 50% of the total CO_2 captured (315 Mt) for producing the fertilizer, NH_4HCO_3. Clearly, this estimate demonstrated that the synthetic N-fertilizer, NH_4HCO_3, produced by NH_3 scrubbing of CO_2 from fossil fuel (e.g., coal)-fired TPP could have a significant beneficial environmental impact so far as GHG emission was concerned.

APPENDIX 1: ADDITIONAL TABLES

TABLE 5: Detailed performance based comparison of NH3- and MEA- based CO2 capture processes

Laboratory absorber specification:			
Laboratory Bubble Column (height: 0.21m; diameter: 0.06 m)			
Quiescent liquid volume studied: 200 ml NH_3 or MEA			
Initial concentration of reagents used were: NH_3 solution: 35% (w/w); MEA solution: 99% (w/w)			
Range of solvent concentration: 7 – 35 % (w/w) [made by dilution]			
Range of CO_2 concentration: 8 – 16 % (v/v) [air diluted]			
Range of Temperature: 10 – 40 OC			
Range of gas (CO2 + clean air) flow rate: 2 – 10 lpm			
Effect of operating time:			
Reaction rate of NH_3 scrubbing was faster than that of MEA scrubbing and NH_3 scrubbing of CO_2 provides a wider operating range than by using MEA solvent.			
Effect of solvent concentration[a]:			
The removal efficiency of CO_2 was proportional to the solvent concentration.			
The removal efficiency in NH_3 scrubbing was 6 – 7 % higher than in MEA scrubbing.			
Optimum solvent concentration reported was 28 % since beyond this concentration there was no improvement in the increasing rate of CO2 removal efficiency.			
[solvent]	Absorption capacity, A, kg CO_2/kg solvent		Conclusions
	NH_3 scrubbing	MEA scrubbing	
7%	1.20	0.38	"A" was influenced by [NH_3]
35%	0.85	0.36	A(NH_3) = [2.4 – 3.2] times A(MEA)

[[solvent]: solvent concentration; A(NH3) and A(MEA): A for NH3 and MEA scrubbing]

Temperature variation:			
Maximum reaction temperature for CO_2-MEA system: 50 OC [35% MEA]			
Maximum reaction temperature for CO_2-NH_3 system: 40 OC [35% NH_3]			
Thus CO_2-MEA requires higher energy than CO_2-NH_3 during regeneration.			
Effect of temperature:			
Temperature, T, OC	CO_2 Removal efficiency, η, %		Conclusions
	NH_3 scrubbing	MEA scrubbing	
10	92	88	$\eta \propto T$
40	99	94	
	Absorption capacity, A, kg CO_2/kg solvent		
10	1.1	0.35	$A \propto [1/T]$
40	0.82	0.40	
Effect of inlet CO_2 concentration:			
CO_2 at inlet, Cin, % v/v	CO_2 Removal efficiency, η, %		Conclusions
	NH_3 scrubbing	MEA scrubbing	
8	94	88	$\eta \propto [C_{in}]$
16	97	92	
	Absorption capacity, A, kg CO_2/kg solvent		
8	0.9	0.38	A was independent of C_{in}
16			
Effect of gas flow rate:			
Gas flow rate, Q_g, lpm	CO_2 Removal efficiency, η, %		Conclusions
	NH_3 scrubbing	MEA scrubbing	
2	97	92	$\eta \propto [1/Qg]$
10	72	62	
	Absorption capacity, A, kg CO_2/kg solvent		
2	0.9	0.38	$A \propto [1/Qg]$
10	0.76	0.26	

[a] *[[solvent]: solvent concentration; A(NH_3) and A(MEA): A for NH_3 and MEA scrubbing]; [NH_3]: NH_3 concentration*

TABLE 11: Life cycle CO_2 emission for NH_3 scrubbing of CO_2

Thermal Power Plant (TPP) capacity	**300 MW**
Yearly rated capacity	7000 hr
Total electricity generation	$300 \times 7000 \times 10^3$ kWh = 2.1×10^9 kWh
CO_2 emission factor (EF) in China for TPP	0.8 kg/kWh
Annual CO_2 emission from TPP	$2.1 \times 10^9 \times 0.8$ kg = 1.68×10^6 t
System – I Analysis	
Ammonia scrubbing plant CO_2 removal efficiency	95%
Total annual ABC produced in NH_3 scrubbing	$1.68 \times 10^6 \times 0.95 \times [79/44] = 2.87 \times 10^6$ t
Absorption factor of NH_3	2.5 t CO_2/t NH_3
NH_3 required for CO_2 absorption	$[(1.68 \times 10^6 \times 0.9)/2.5]$ t = 0.64×10^6 t
Steam Methane Reformer (SMR):	
NH_3 Production	
CO_2 EF	1.22 t CO_2/t NH_3
NH_3 required	0.64×10^6 t
CO_2 emission	$0.64 \times 10^6 \times 1.22$ t = 0.78×10^6 t
CO_2 emission for electricity from NH_3 scrubbing or ABC Plant	
NH_3 scrubbing requires	130 kWh/t ABC
CO_2 emission from ABC plant [CO_2 EF = 0.8 kg/kWh]	0.8×130 kg/t ABC = 0.104 t/t ABC
Total CO_2 from ABC Plant [ABC produced = 2.87×10^6 t]	$[2.87 \times 10^6 \times 0.104]$ t = 0.3×10^6 t
CO_2 Emission: System – I	
CO_2 unabsorbed in NH_3 scrubbing plant [5% CO_2 emitted]	$0.05 \times 1.68 \times 10^6$ t = 0.084×10^6 t
CO_2 emission from NH_3 Production Plant	0.78×10^6 t
CO_2 emission for electricity from ABC Production Plant	0.30×10^6 t
Total CO_2 emission from System - I	1.164×10^6 t
System – II Analysis	
CO_2 EF for coal gasification	3.59 t/t NH_3
1 t ABC production requires	0.25 t NH_3 + 0.65 t CO_2
0.25 t NH_3 emits /t ABC	(3.59/4) t CO_2 = 0.898 t CO_2
Excess CO_2 emission / t ABC	(0.898 – 0.65) = 0.248 t
Total annual ABC produced from coal gasification	2.87×10^6 t [same as in System – I]
Total annual CO_2 produced due ABC production	$0.248 \times 2.87 \times 10^6$ t = 0.71×10^6 t
CO_2 Emission: System – II	
Total annual CO_2 produced due ABC production	0.71×10^6 t
CO_2 from 300 MW TPP	1.68×10^6 t
CO_2 emission for electricity from ABC Production Plant	0.30×10^6 t
Total CO_2 emission from System - II	2.69×10^6 t

TABLE 12: CO_2 purity specification for different uses [Values are in ppmv unless otherwise specified. Blank indicates no maximum limiting characteristic.]

Limiting characteristics	Medical	General commercial	Food processing	Beverages	Dry ice, refrigeration
Carbon dioxide (min % v/v)	99	99	99.5	99.9	
Acetaldehyde		0.5	0.5	0.2	
Ammonia	25			2.5	
Acidity				To pass JECFA Test	
Benzene				0.02	
Carbon monoxide	10 (vapor)		10 (vapor)	10	
Carbonyl sulfide			0.5		
Hydrogen cyanide				None detected	
Methanol				10	
Nitric oxide	2.5 (vapor)		5 (total NO + NO_2)	2.5	
Nitrogen dioxide	2.5			2.5	
Oxygen		50	50	30	
Phosphene				0.3	
Sulfur dioxide	5		5		
Total sulfur		0.5	0.5	0.1	
Total hydrocarbon content (as methane)		50	50	50 max including 20 max of non-methane hydrocarbons	
Hydrogen sulfide	1 (vapor)		0.5 (vapor)		
Color					White opaque
Nonvolatile residues (wt/wt)		10	10	10	500
Oil/grease				5	
Odor/taste	Free from foreign odor or taste				
Water	200	32	20	20	
Dew point (OC)	-36	-51	-55	-55	

REFERENCES

1. Ali SH (2005) Kinetics of the reaction of carbon dioxide with blends of amines in aqueous media using the stopped-flow technique. Int J Chem Kinet 37:391–405

2. Ali SH, Merchant SQ, Fahim MA (2002) Reaction kinetics of some secondary alkanolamines with carbon dioxide in aqueous solutions by stopped flow technique. Sep Purif Technol 27:121–136
3. Aronua UE, Svendsena HF, Hoffb KA, Juliussenb O (2009) Solvent selection for carbon dioxide absorption. Energy Procedia 1:1051–1057
4. Astarita G (1967) Mass transfer with chemical reaction. Elsevier, Amsterdam
5. Bai H, Wei JH (1996) The CO_2 mitigation options for the electric sector: a case study of Taiwan. Energy Policy 24:221–228
6. Bai H, Yeh AC (1997) Removal of CO_2 greenhouse gas by ammonia scrubbing. Ind Eng Chem Res 36(6):2490–2493
7. Banks FE (2000) The Kyoto negotiations on climate change—an economic perspective. Energy Sources 22:481–496
8. Blok K, Worrell E, Caelenaere R, Turkenburg W (1993) The cost effectiveness of CO_2 emission reduction achieved by energy conservation. Energy Policy 25:656–667
9. Blomen E, Hendriks C, Neele F (2009) Capture technologies: improvements and promising developments. Energy Procedia 1:1505–1512
10. Bolin B (1998) The Kyoto negotiations on climate change: a science perspective. Science 279:330–331
11. Bottoms RR (1930) U.S. process for separating acid gases. U.S. Patent 1,783,901
12. Brooks R (1953) Manufacture of ammonium bicarbonate. British Patent 742,386
13. Brooks LA, Audrieth LF (1946) Ammonium carbamate. Inorg Synth 2:85–86
14. Cansolv (2009) Breakthrough CO_2 capture technology. Cansolv Technologies, Inc. http://www.cansolv.com/. Retrieved October 9, 2009
15. Chakma A (1995) Separation of CO_2 and SO_2 from flue gas streams by liquid membranes. Energy Convers Manag 36:405–410
16. Chakravarti S, Gupta A, Hunek B (2001) Advanced technology for the capture of carbon dioxide from flue gases. In: Proceedings of the 1st national conference on carbon sequestration, Washington, DC, May 15–17
17. Cheng Z, Ma Y, Li X, Zhang WPZ (2007) Investigation of carbon distribution with 14C as tracer for carbon dioxide (CO_2) sequestration through NH4HCO3 production. Energy Fuels 21(6):3334–3340
18. Chludzinski GR, Stogryn EL, Weichert S (1986) Commercial experience with Flexsorb SE absorbent. Presented at the 1986 Spring AIChE national meeting, New Orleans, LA
19. Choi W-J, Min B-M, Shon B-H, Seo J-B, Oh K-J (2009a) Characteristics of absorption/regeneration of CO2–SO2 binary systems into aqueous AMP + ammonia solutions. J Ind Eng Chem 15:635–640
20. Choi W-J, Min B-M, Seo J-B, Park S-W, Oh K-J (2009b) Effect of ammonia on the absorption kinetics of carbon dioxide into aqueous 2-amino-2-methyl-1-propanol solutions. Ind Eng Chem Res 48:4022–4029
21. Ciferno JP, DiPietro P, Tarka T (2005) An economic scoping study for CO_2 capture using aqueous ammonia. DOE/NETL final report, revised, February
22. Corti A, Lombardi L (2004) Reduction of carbon dioxide emissions from a SCGT/CC by ammonia solution absorption—preliminary results. Int J Thermodyn 7(4):173–181

23. Danckwerts PV (1970) Gas–liquid reactions. McGraw-Hill, New York
24. Danckwerts PV, Sharma MM (1966) The absorption of carbon dioxide into solutions of alkalies and amines. Chem Eng 44:244–280
25. Darde V, Thomsen K, van Well WJM, Stenby EH (2009) Chilled ammonia process for CO2 capture. Energy Procedia 1:1035–1042
26. Dave N, Do T, Puxty G, Rowland R, Feron PHM, Attalla MI (2009) CO2 capture by aqueous amines and aqueous ammonia—a comparison. Energy Procedia 1:949–954
27. Derks PWJ, Versteeg GF (2009) Kinetics of absorption of carbon dioxide in aqueous ammonia solutions. Energy Procedia 1:1139–1146
28. Diao YF, Zheng XY, He BS, Chen CH, Xu XC (2004) Experimental study on capturing CO2 greenhouse gas by ammonia scrubbing. Energy Convers Manag 45:2283–2296
29. Dooley JJ, Davidson CL, Dahowski RT (2009) An assessment of the commercial availability of carbon dioxide capture and storage technologies as of June 2009. Prepared for the U.S. Department of Energy under Contract DE-AC05-76RL01830. Pacific Northwest National Laboratory, Richland
30. Gal E (2006) Ultra cleaning combustion gas including the removal of CO2. World Intellectual Property, Patent WO 2006022885
31. Goldstein AM (1983) Commercialization of a new gas treating agent. Presented at petro-energy '83 conference, Houston, TX, September 14
32. Hakka LE, Ouimet MA (2006) Method for recovery of CO2 from gas streams. US Patent and Trademark Office Granted Patent. Cansolv Technologies Inc., Assignee. US Patent 7056482 B2. June 6
33. Harald W, Randall S, Lwazikazi T (2002) Comparing developing countries under potential carbon allocation schemes. Clim Policy 2:303–318
34. Hatch TF, Pigford RL (1962) Simultaneous absorption of carbon dioxide and ammonia in water. Ind Eng Chem Fundam 1:209–214
35. He Q, Chen M, Meng L, Liu K, Pan W-P (2010) Study on carbon dioxide removal from flue gas by absorption of aqueous ammonia. http://www.netl.doe.gov/publications/proceedings/04/carbon-seq/158.pdf. Accessed March 18, 2010
36. Huang JP (1993) Energy substitution to reduce carbon dioxide emission in China. Energy 18:281–287
37. IEA (2007) IEA greenhouse gas R&D programme (IEA GHG). CO2 Capture Ready Plants, 2007/4, May
38. IPCC (1990) Policymaker's summary of the scientific assessment of climate change. Report to IPCC from working group. Meteorological Office, Branknell
39. IPCC (1997) Revised 1996 IPCC guidelines for national greenhouse gas inventories. Meteorological Office, Branknell
40. IPCC (2005) Special report on carbon dioxide capture and storage. Prepared by working group III of the intergovernmental panel on climate change, Cambridge University Press, New York, USA
41. Kaplan LJ (1982) Cost-saving process recovers CO2 from power-plant flue gas. Chem Eng 89(24):30–31
42. Kim JY, Han K, Chun HD (2009) CO2 absorption with low concentration ammonia liquor. Energy Procedia 1:757–762

43. Kimura N, Omata K, Kiga T, Takano S, Shikisma S (1995) Characteristics of pulverized coal combustion in O2–CO2 mixtures for CO2 recovery. Energy Convers Manag 36:805–808
44. Kohl AL, Nielsen RB (1997) Gas purification, 5th edn. Gulf Publishing Company Book Division, Texas, chap 2, pp 40–186
45. Koornneef J, von Keulen T, Faajj A, Turkenburg W (2008) Life cycle assessment of a pulverized coal power plant with post-combustion capture, transport and storage of CO2. Int J Greenh Gas Control 2:448–467
46. Koutinas AA, Yianoulis P, Lycourghiods A (1983) Industrial scale modelling of the thermochemical energy storage system based on CO2 + 2NH3 ↔ NH2COONH4 equilibrium. Energy Convers Manag 23:55–61
47. Kozak F, Petig A, Morris E, Rhudy R, Thimsen D (2009) Chilled ammonia process for CO2 capture. Energy Procedia 1:1419–1426
48. Lee JW, Li R (2003) Integration of fossil energy systems with CO2 sequestration through NH4HCO3 production. Energy Convers Manag 44:1535–1546
49. Lee D-H, Cho W-J, Moon S-J, Ha S-H, Kim I-G, Oh K-J (2008) Characteristics of absorption and regeneration of carbon dioxide in aqueous 2-amino-2-methyl-1-propanol/ammonia solutions. Korean J Chem Eng 25(2):279–284
50. Li M-H, Chang B-C (1995) Solubility of mixtures of carbon dioxide and hydrogen sulfide in water + monoethanolamine + 2-amino-2-methyl-1-propanol. J Chem Eng Data 40(1):328–331
51. Liu J, Wang S, Zhao B, Tong H, Chen C (2009) Absorption of carbon dioxide in aqueous ammonia. Energy Procedia 1:933–940
52. Lu YY, Huang Y, Zhang W, Zheng XH (2007) Estimation of chemical fertilizer N-induced direct N2O emission from China agricultural fields in 1991–2000 based on GIS technology. Yingyong Shengtai Xuebao 18:1539–1545 (in Chinese)
53. Lynn S, Straatemeier JR, Kramers H (1955) Absorption studies in the light of penetration theory—I. Long wetted-wall columns. Chem Eng Sci 4:49–57
54. Martin MH, Meyer S (1999) Greenhouse gas carbon dioxide mitigation. Lewis Publishers, Boca Raton
55. McLarnon CR, Duncan JL (2009) Testing of ammonia based CO2 capture with multi-pollutant control technology. Energy Procedia 1:1027–1034
56. Mshewa MM, Rochelle GT (1994) Carbon dioxide absorption/desorption kinetics in blended amines. In: Proceedings of 44th annual Laurance Reid gas conditioning conference, University of Oklahoma, Norman, OK, February 27–March 2, p 251
57. Nishikawa N, Hiroano A, Ikuta Y (1995) Photosynthetic efficiency improvement by microalgae cultivation in tubular type reactor. Energy Convers Manag 36:681–684
58. Nsakala N, Marion J, Bozzuto C, Liljedahl G, Palkes M, Vogel D, Gupta JC, Guha M, Johnson H, Plasynski S (2001) Engineering feasibility of CO2 capture on an existing US coal fired power plant. In: Proceedings of the 1st national conference on carbon sequestration, Washington, DC, May 15–17
59. Paul S, Ghoshal AK, Mandal B (2007) Removal of CO2 by single and blended aqueous alkanolamine solvents in hollow-fiber membrane contactor: modeling and simulation. Ind Eng Chem Res 46:2576–2588
60. Pauley CR, Simiskey PL, Haigh S (1984) N-Ren recovers CO2 from flue gas economically. Oil Gas J 82(20):87–92

61. Plummer LN, Parkhurst DL, Thorstenson DC (1983) Development of reaction models for ground-water systems. Geochim Cosmochim Acta 47:665–686
62. Qi NZ, Cheng GY, Yi LW (2010) Experimental studies on removal of carbon dioxide by aqueous ammonia fine spray. Sci China Technol Sci 53(1):117–122
63. Rao AB (2002) Details of a technical, economic and environmental assessment of amine-based CO2 capture technology for power plant greenhouse gas control; Appendix to Annual technical progress report [DE-FC26-00NT40935]. Prepared for U.S. Department of Energy, National Energy Technology Laboratory, Morgantown, West Virginia, USA
64. Resnik KP, Yeh JT, Pennline HW (2004) Aqua ammonia process for the simultaneous removal of CO2, SO2 and NO x . Int J Environ Technol Manag 4(1/2):89–104
65. Sartori G, Savage DW (1983) Sterically hindered amines for carbon dioxide removal from gases. Ind Eng Chem Fundam 22:239–249
66. Sass BM (2002) Purity specifications for commodity uses of carbon dioxide in the United States. Battelle. October 25
67. Seo DJ, Hong WH (2000) Effect of piperazine on the kinetics of carbon dioxide with aqueous solutions of 2-amino-2-methyl-1-propanol. Ind Eng Chem Res 39:2062–2067
68. Shale CC, Simpson DG, Lewis PS (1971) Removal of sulfur and nitrogen oxides from stack gases by ammonia. Chem Eng Prog Symp Ser 67:52–57
69. Sun JW (2002) The Kyoto negotiations on climate change—an arithmetic perspective. Energy Policy 30:83–85
70. Sun JW (2003) The natural and social properties of CO2 emission intensity. Energy Policy 31:203–209
71. UNFCCC (1997) Kyoto protocol to the United Nations Framework Convention on Climate Change (UNFCCC). UNFCCC/CP/1997/L.7/Add.1, Bonn
72. Valenti G, Bonalumi D, Macchi E (2009) Energy and exergy analyses for the carbon capture with the chilled ammonia process (CAP). Energy Procedia 1:1059–1066
73. Wang S, Liu F, Chen C, Xu X (2007) Life cycle emissions of greenhouse gas for ammonia scrubbing technology. Korean J Chem Eng 24(3):495–498
74. Wolsky AM, Daniels EJ, Jody BJ (1994) CO2 capture from the flue gas of conventional fossil-fuel-fired power plants. Environ Prog 13:214–219
75. Xi Z, Shi X, Liu M, Cao Y, Wu X, Ru G (1985) Agrochemical properties of ammonium bicarbonate. Turang Xuebao 22(3):223–232
76. Yan XY, Akiyama H, Yagi K, Akimoto H (2009) Global estimations of the inventory and mitigation potential of methane emissions from rice cultivation conducted using the 2006 intergovernmental panel on climate change guidelines. Glob Biogeochem Cycle 23:GB2002. doi:10.1029/2008GB003299
77. Yeh AC, Bai H (1999) Comparison of ammonia and monoethanolamine solvents to reduce CO2 greenhouse gas emissions. Sci Total Environ 228:121–133
78. You JK, Park H, Yang SH, Hong WH, Shin W, Kang JK, Yi KB, Kim J (2008) Influence of additives including amine and hydroxyl groups on aqueous ammonia absorbent for CO2 capture. J Phys Chem B 112(14):4323–4328
79. Zhou J, Shang J, Der V, Li Z, Zhang J, Li X, Zhang Z (2009) A feasibility study on a two stage benefits CO2 sequestration technology for fossil fuel power generation. www.netl.doe.gov/publications/proceedings/01/carbon_seq/p2.pdf. Accessed June 14, 2009

80. Zucong C (2009) Contribution of N-fertilization to N2O emissions from crop lands of China and mitigation options. In: 7th IFA annual conference, Shanghai, China, May 25–27
81. Over 8.3 million scientific documents at your fingertips

There are several tables that do not appear in this version of the article. To view these tables, please visit the original article as cited on the first page of this chapter.

CHAPTER 2

CO_2 CAPTURE IN A SPRAY COLUMN USING A CRITICAL FLOW ATOMIZER

AMITAVA BANDYOPADHYAY AND MANINDRA NATH BISWAS

2.1 INTRODUCTION

The emission of CO_2 into the atmosphere is assumed greatest adverse impact on the observed green house effect causing approximately 55% of the global warming [1]. The growing awareness on the risks associated with the climate change has drawn attention of the researchers for curbing the emission of CO_2 from various fixed point stationary sources since 1989 [2]. Given the 30% of the total global fossil fuel used for power generation that emits considerable amount of CO_2, the thrust on the researches of CO_2 absorption is being imparted to the flue gas containing CO_2 emitted from the fossil fuelled thermal power plants (TPPs). An estimate suggests [3] that burning fossil fuels emits around 6 billion tons of carbon globally in which about 1.8 billion tons is contributed from TPPs alone. As a result, various end-of-pipe treatment methods have evolved to capture and recover CO_2 from the flue gas streams. These methods are gas–liquid absorption, adsorption, cryogenic separation, membrane separation, biological fixation and oxyfuel combustion with CO_2 recycling [4], [5] and [6]. Each of these methods has its own merits and demerits. However, the amine based absorption method, being commercially adopted since early part of the last century especially for separation of H_2S and CO_2 in the hydrocarbon industry, has been studied extensively for reducing CO_2 emission from fossil fuel fired TPPs [7].

In fact, such an amine based CO_2 absorption process can not be adopted following a pick-and-choose principle for capturing CO_2 from the flue gas of TPP unless its quality is similar to the feed gas being treated in a conventional amine based CO_2 absorption process. But such a quality matching seldom occurs and as a result, the flue gas of TPP requires special treatment prior to its introduction into the amine based CO_2 absorption process. The flue gas of a coal-fired TPP may contain several contaminants, like 300–3000 ppmv of SO_2, 100–1000 ppmv of NO_X and 1000–10,000 mg/m3 of particulate matter. In contrast, the level of concentration of these pollutants for natural gas firing power plants are considerably lower, for instance, $SO_2 < 1$ ppmv, 100–500 ppmv of NO_X and nearly 10 mg/m^3 of particulate matter [8]. Alike CO_2, these gases (SO_2 and NO_X) being acidic in nature, will react with the amine to form heat stable salts irreversibly and hence a loss in absorption capacity of the solvent. Amines could however, be regenerated with the consumption of additional chemicals leading to lowering of overall performance of the system. Therefore, the pre-treatment of the flue gas to achieve very low values of NO_X and SO_2 before CO_2 capture becomes essential. Besides these acidic gases, the presence of particulate matter (e.g., fly ash or soot) in the flue gas at a very high concentration might plug the absorber and the stripper in tandem. International Energy Agency [9] therefore, suggested that the flue gas of the TPP for CO_2 capture should be pretreated prior to introducing into the amine based CO_2 absorption systems to achieve 4–11 ppmv of SO_2, ~21 ppmv of NO_2 and a level of particulate matter well below 5 mg/Nm^3.

Oxygen on the other hand, present in the flue gas would cause another problem of rapid degradation of some of the alkanolamines used as solvent for absorption. In order to avoid this problem it was suggested [10] to convert the O_2 of flue gas by burning it with the natural gas over a De-Oxy catalyst upstream of the solvent contactor into CO_2. Furthermore, evaporative loss of the amines to the gas stream during operation would cause environmental pollution owing to their toxicity and corrosivity [11] and [12]. To avoid the corrosivity, dilution of amines with water would be necessary that would result in the increase in the heat duty of the stripper column for desorption of CO_2. In addition, the water dilution would diminish the absorption capacities of amines for CO_2 [11], [12], [13] and [14]. The temperature of flue gas of a fossil fuelled TPP, on the other hand, plays an important role in the amine based absorption process, for instance, the temperature is usually ~120 °C, and would thus require a cooler to cool down the temperature required (~40 °C) for such an absorption process [15]. The storage and pipeline transportation of CO_2 recovered from the amine based process would further require exorbitantly costly cryogenic systems. These are the reasons that the amine based absorption process available commercially for CO_2 absorption are not techno-enviro-economically suitable for capturing CO_2 from the flue gas of fossil fuelled TPPs. In order to avoid the problems encountered in the amine based systems described above, absorption of CO_2 in hydroxides/carbonates of alkali metals could be a better option primarily because reactions between them could be carried out under ambient conditions and this method does not require a stripper column. In such a method, the solid reaction products collected from the absorption column could easily be stored and transported to the site of application as in the case of EOR or geologic sequestration. Fundamentally, such a scheme avoids many of the problems associated with the amine based systems.

The absorption of CO_2 in hydroxide and carbonates of alkali metals was developed since middle of the last century while determining the interfacial area of contact and true liquid side mass transfer coefficient in gas–liquid absorptive mass transfer operations [16]. In these deterministic studies researchers have classically demonstrated the uses of reactions of CO_2 with various metal hydroxides and carbonates like sodium hydroxide (NaOH) and sodium carbonate/sodium bicarbonate (Na_2CO_3/$NaHCO_3$) mixture or potassium carbonate/potassium bicarbonate K_2CO_3/$KHCO_3$ mixture [16], [17] and [18]. These methods are also being extensively used at present for characterizing the mass transfer design parameters of gas–liquid absorbers [19]. Interestingly, such studies seldom reported on the performance of the absorber in terms of CO_2 capture. In this light, researches have beginning to develop recently to characterize the CO_2 absorption in dilute NaOH solution [20]. Thus dilute NaOH solution has been selected for the present study of CO_2 absorption/capture.

Besides selection of the absorbent, choosing the type of absorption column assumes considerable importance in the case of CO_2 capture from the flue gas of TPPs. The amine based system generally deploys a packed absorber and a stripper in tandem. The operation of a packed column for gases containing particulate matters and/or for solids being produced as a reaction product would not be techno-enviro-economically viable on a long term basis, since the bed would become clogged by the solid deposits. The spray column can however, be operated in such a situation easily avoiding this difficulty. The other advantages of a spray column are its high turn down ratio, low pressure drop and low installation as well as maintenance cost. Considering a very high volumetric flow rate (~10^6 Nm^3/hr) of the flue gas of a 250 MW Indian coal fired TPP with ~8 to 10% CO_2 concentration at ~120 °C to be treated [21], large and empty spray chamber can be selected for CO_2 capture using dilute NaOH solution. As a result, the temperature of the flue gas will cool down, but the exact drop in temperature depends of course, upon the detailed design of the column. In the light of these observations, spray column has been selected in the present study for CO_2 capture using dilute NaOH solution.

Survey of literature further indicates that the investigations on the spray column mainly were carried out for the purpose of determining the interfacial area and other mass transfer parameters and as such the system performance in terms of CO_2 capture does not seem to be available. Based on these findings, the existing literature on CO_2 absorption in water/alkaline solution using spray column for determining the interfacial area has been reviewed and presented here for our better understanding. Mehta and Sharma [22] reported on the experimental determination of interfacial area and mass transfer coefficients of spray column of various dimensions using shower nozzles (droplet diameters: 1500–2500 μm) and solid cone nozzles (droplet diameters: 250–430 μm). The values of interfacial area reported using the solid cone nozzle were in the range of 30–50 m^2/m^3. Ruckenstein et al. [23] theoretically investigated the absorption of trace gas followed by chemical reaction in single moving drop. The validation of their models was however, not studied experimentally under similar conditions. Absorption of a gas from a mixture of acid gases by a water droplet was theoretically investigated by Adewuyi and Carmichael [24]. Pinilla et al. [25] investigated mass transfer coefficients for the gas and liquid phase, interfacial areas and axial dispersion parameters in a spray column using a solid cone nozzle. Liquid phase mass transfer coefficients and interfacial area were evaluated for the absorption of CO_2 into a carbonate-bicarbonate solution with arsenite as catalyst and using Danckwerts' plot. They have observed interfacial area in the range of 0.95–1.30 m^2/m^3. Altwicker and Lindhjem [26] experimentally investigated the absorption of CO_2 on a falling water drop of diameters 1200 and 600 μm separately. Taniguchi and Asano [27] studied the rates of absorption of CO_2 using water in a co-current spray column having droplet SMDs ranging between 137 and 185 μm. A theoretical model based on penetration theory was developed by them ignoring the internal circulation inside droplets in which the model showed good agreement with the experimental data. Taniguchi et al. [28] measured the rates of absorption of CO_2 from CO_2–air mixtures with a drop of water and aqueous NaOH solution falling freely. The initial concentration of aqueous NaOH solution investigated was ranging between 0.05 and 0.4 mol/l and the feed gas concentration of CO_2 was varied from 5% to 100%. Model equation was developed for predicting the dimensionless rates of absorption that was validated with experimental values. Experimental results were in good agreement with the theoretical values. Taniguchi et al. [29] developed an expression for predicting the rate of absorption of CO_2 in NaOH solution assuming a solid sphere penetration model with second order chemical reaction. The predicted values were verified with experimentally determined values on the same system that showed a good agreement. The values of interfacial area reported taking all these studies together was in the range of 0.95–50 m^2/m^3. The ranges of other pertinent operating variables influencing the values of the interfacial area are discussed later in details in Section 6. Available literature therefore, indicates that the conventional spray columns offer lower values of interfacial area. However, the values of interfacial area could be improved by selecting an energy efficient atomizer for the purpose of operating a spray column.

Critical appraisal of the existing literature further indicates that the interfacial area of a spray column deploying a two-phase critical flow atomizer capable of producing very fine sprays (droplet size in the range of ~30 to 150 μm) with high degree of uniformity and moving at a very high velocity (20–30 m/s), is yet to be reported in the literature. The performance of such a spray column in terms of CO_2 removal efficiency also does not seem to be available in the existing literature. However, the performances of such a spray column had been characterized for the abatement of particulate and/or SO_2 emission from simulated waste gas streams [30], [31] and [32]. These studies showed very high degree of removal (~100%) for particulates and/or SO_2 in the spray column studied using a two-phase critical flow atomizer that can generate very fine drops as mentioned above. An attempt has therefore, been made in this article to investigate into the CO_2 removal efficiency from an air laden CO_2 mixture using dilute NaOH solution experimentally in a spray column deploying a two-phase critical flow atomizer. The atomizer is capable of generating solid cone sprays having droplet SMDs ranging from 30 to 150 μm moving at very high velocities (20–30 m/s) with uniformity [33].

Since the conventional spray columns reported in the literature did not report on the CO_2 removal efficiency while determining the interfacial area using the absorption of CO_2 in dilute NaOH solution, comparison of our presented study with the existing systems would become difficult. However, such comparison could be possible if the present study, determines the interfacial area in addition to determining the CO_2 removal efficiency. An attempt has also been made to compare the performance of the presented system in terms of the interfacial area with those available in the literature. As discussed earlier, the atomizer plays an important role in operating a spray column and thus the development of the two-phase critical flow atomizer along with its spray hydrodynamics are seemingly important. These are elucidated in details elsewhere in the literature [30], [31] and [32]. However, the salient features are briefly described here for improved understanding. In the Section 2 the fundamentals of such an atomizer is briefly discussed.

2.2 OPERATING PRINCIPLE OF THE TWO-PHASE CRITICAL FLOW ATOMIZER

Central to any spray operation, is the atomization or the disintegration of the liquid stream into droplets or sprays. For spray scrubbing, a good atomizer should produce a fairly uniform spray with drop diameters small enough to generate large interfacial area of contact at the same time large enough to prevent excessive entrainment. Uniformity of spray, small drop size with high velocity and low energy of atomization are the desired criteria for the atomizer. Existing commercial atomizers are requiring very high energies for desired spray hydrodynamics. Therefore, an energy efficient and cost-effective critical flow atomizer [33] satisfying the above desired criteria is used in the present study. The critical flow atomizer utilizes the fact that the critical velocity of two-phase mixture is much lower than the critical velocities of either of the single phases. The critical or mass limiting flow occurs in an atomizer when an increase in pressure drop across the atomizer does not increase the mass flow rate through the atomizer. This flow condition is also called the choked flow. When a mass limiting flow is achieved in a two-phase mixture, any increase in the pressure drop further results only in restructuring of the gas-in-liquid flow geometry i.e., readjustment of the drop sizes in liquid-in-gas dispersion. Thus when a well mixed two-phase dispersion, with mixture velocities very near to the critical flow conditions is expanded, a pressure jump takes place very near the tip of the atomizer. Since this pressure drop cannot be utilized in increasing the flow rate, the energy released by the difference in drop in pressure is quantitatively utilized in atomizing the liquid into a fine spray. The advantages of this design consist in lower frictional acceleration losses compared to the external mixing atomizer. In the present atomizer liquid and gas are brought together in a dispersing chamber at relatively low velocity in comparison to the critical velocity of individual phases because the critical velocity of a two-phase mixture is lower than that of the individual critical velocities. The gas–liquid mixing volume and mixing area can be adjusted so that either sprays with different droplet Sauter Mean Diameters (SMDs) at a fixed mean droplet velocity, or a fixed droplet SMD with different mean velocities can be generated.

2.3 THEORETICAL CONSIDERATION OF DETERMINATION OF INTERFACIAL AREA

As described in Section 1 earlier, the comparison of the performance of the presented system in terms of CO_2 capture is not possible due to sparse data. However, the absorption of CO_2 in dilute NaOH solution had been extensively studied in order to determine the interfacial area of spray columns. Therefore, the interfacial area can be determined in the presented spray column so as to compare its performance with the values reported in the literature. In this section, the mathematical considerations for determining the interfacial area in a spray column are described.

The determination of interfacial area of any gas–liquid contactor has been a subject of extensive research over the years. In fact, three methods, for instance, high-speed photography, light attenuation and chemical method have widely been used for the determination of interfacial area as revealed from the existing literature [34], [35], [36], [37] and [38]. The measurement of interfacial area by physical methods results in obtaining only local values of the specific area and are thus indicative of the total interfacial area in the dispersion regardless of whether all or merely part of the total area is participating in the mass transfer process. Conversely, the chemical method determines an integral (volume average) value of the interfacial area which represents the sum of only those surface area elements across which mass transfer actually takes place during the measurement period [39]. The determination of interfacial area by the chemical method necessitates low gas concentrations of the absorbent to avoid plug flow in the gas phase. Chemical method is thus characterized by gas absorption accompanied by chemical reaction [37] and [40]. Moreover, chemical method is often used where the intense mixing between the phases generates high turbulence as in the present system. Sharma and Danckwerts [40] further reported that interfacial area determined by the chemical method would yield more reliable values than that obtainable from physical methods. In the light of these observations, the chemical method is chosen for the present study.

In the chemical method, the rate of absorption of a gas into a liquid with which it undergoes a reaction is measured. The conditions of the reaction are carefully chosen so that the rate of gas absorption per unit area of interface is constant, independent of contact time or surface age and may be calculated from the physico-chemical constant of the reaction. If the total absorption rate is measured experimentally then the interfacial area may be calculated from the known kinetics of the reaction. Of the varying chemical reactions with known kinetics, the liquid phase reaction between dissolved CO_2 and NaOH is a convenient system and as such it has been selected for the present study. The reaction steps are:

$$CO_{2(dissolved)} + OH^{-} \rightarrow HCO^{-3} \quad (1)$$

$$HCO^{-3} + OH^{-} \rightarrow CO_2^{-3} + H_2O \quad (2)$$

Since the second reaction is ionic, it occurs much faster than the first one and as a result, the first reaction becomes rate-controlling. Thus the overall reaction may be considered to take place as

$$CO_{2(dissolved)} + 2OH^{-} \rightarrow CO_2^{-3} + H_2O \quad (3)$$

Now if the concentration of OH^- remains practically the same as in the bulk of the liquid then the reaction becomes pseudo-first order [41]. Under these circumstances, the interfacial area, a, has been calculated in the present study from the measured chemical rate of absorption of CO_2 in NaOH solution from the following relation:

$$R_a \cdot V_T = a \cdot V_T \cdot C^{*}_{CO2} \sqrt{D_{CO_2} \cdot k_2 \cdot C^{O}_{OH^-}} \quad (4)$$

subject to the following condition [40]

$$3 < \left[\frac{D_{CO_2} \cdot k_2 \cdot C^O_{OH^-}}{k_L^2}\right]^{1/2} \ll 1 + \frac{C^O_{OH^-}}{2 \cdot C^*_{CO_2}} \tag{5}$$

The rate expression given in Eq. (4) for the absorption of CO_2 in NaOH solution is pseudo-first order with respect to CO_2 and the effect of fluid dynamic condition is negligible. The interfacial area can be calculated from Eq. (4). A step-wise method of calculation is presented through a flow chart (Fig. 1) for better understanding. The physico-chemical parameters required for calculating interfacial area are DCO_2, k_2, $C_{CO2}{}^*$. D_{CO2} in alkaline solution (e.g., in NaOH solution) has been calculated by one of the most convenient methods suggested by Nijsing et al. [4_2] as follows

$$D_{CO_2} = D_{CO_2,water}\left(\frac{\mu_w}{\mu_{soln.}}\right)^{0.85} \tag{6}$$

in which the diffusivity of CO_2 in water has been calculated from

$$\frac{D_{CO_2,water}}{T} = \text{constant} \tag{7}$$

The second order rate constant, k_2, has been determined by the following equation [43]

$$\log_{10} k_2 = 13.635 - \frac{2895}{T} + 0.132I \tag{8}$$

where I, the ionic strength of the solution, has been calculated from

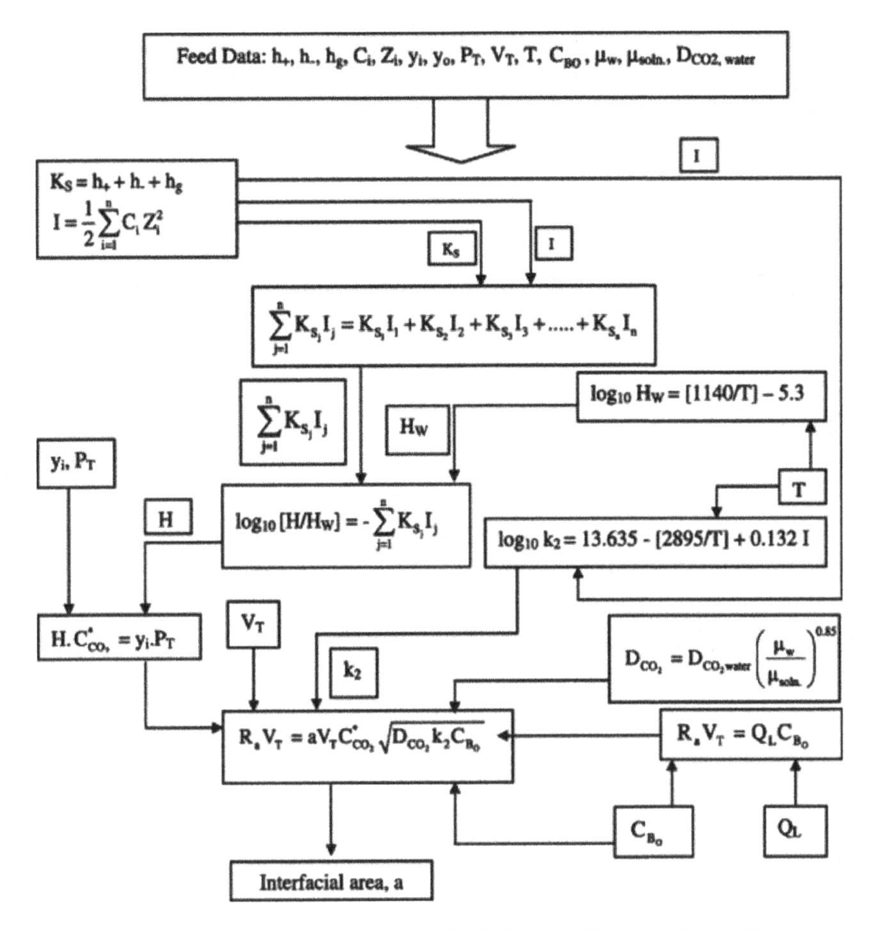

FIGURE 1: Flow chart showing the method of calculation of interfacial area of contact by chemical method.

$$I = \frac{1}{2}\sum_{i=1}^{n} C_i Z_i^2 \tag{9}$$

The concentration of CO_2 at the interface $C_{CO2}{}^*$, has been calculated from the method suggested by Danckwerts and Sharma [16] taking into consideration the ionic strength of the solution as well as ignoring the gas phase resistance

$$H \cdot C^*_{CO_2} = y_i \cdot P_T \tag{10}$$

where H, the solubility of CO_2 in non-reacting electrolyte solutions has been calculated from

$$\log_{10}\lfloor H/H_W \rfloor = -\sum_{j=1}^{n} K_{S_j} I_j \tag{11}$$

The values of H_w (the solubility of CO_2 in water) has been calculated from the following relationship

$$\log_{10} H_W = \left\lfloor \frac{1140}{T} \right\rfloor - 5.3 \tag{12}$$

The expression on the right hand side of Eq. (11) has been calculated from

$$\sum_{j=1}^{n} K_{S_j} I_j = K_{S_1} I_1 + K_{S_2} I_2 + K_{S_3} I_3 + \cdots + K_{S_n} I_n \tag{13}$$

in which K_S, the salting-out coefficient for CO_2 in NaOH solution, is calculated from

$$K_S = h_+ + h_- + h_g \tag{14}$$

2.4 EXPERIMENTAL METHODS

The experimental setup is schematically shown in Fig. 2. The experimental column is a vertical cylindrical Perspex column, 0.1905 m in diameter and 2.0 m long. At the top of the column the energy efficient two-phase critical flow atomizer was provided for generating alkaline sprays. The simulated gas was generated by mixing air and CO_2 in an air-jet ejector (E) assembly for intense mixing of the components. Compressed air at the desired motive pressure and flow rate was forced through the air nozzle and simultaneously CO_2 from a cylinder (CL), was routed through a regulator (R), into the ejector at point P. The air and CO_2 mixed intensely in the mixing throat of the ejector and the mixture was allowed to feed into the spray tower at point F. The CO_2 feed rate was varied by changing its flow rate with the help of a rotameter (R_1). The values of interfacial areas were determined from the absorption of CO_2 into NaOH solution [18] besides studying the CO_2 removal efficiency.

Under steady operating conditions of the spray column, both gas and liquid samples were collected and analyzed. The inlet and outlet liquid

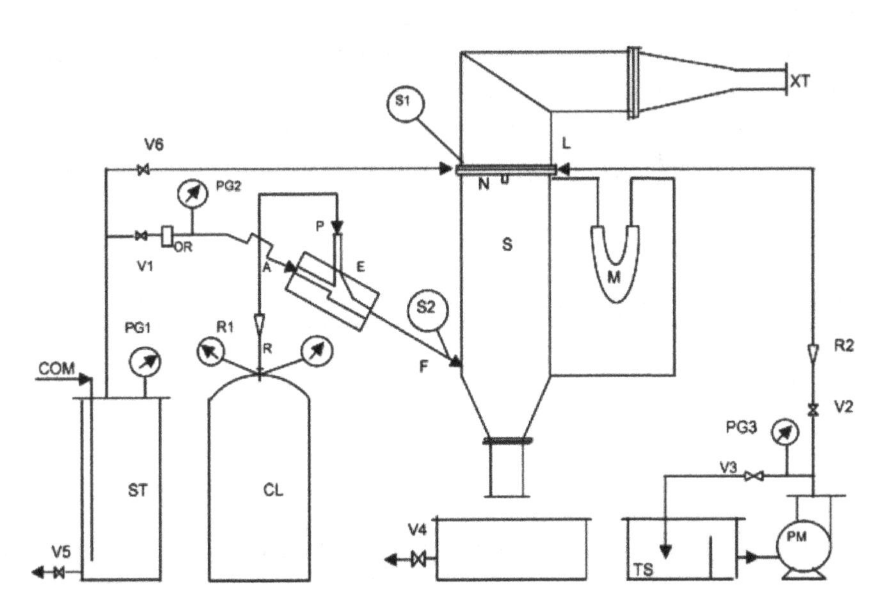

FIGURE 2: Schematic of the experimental setup. A, Primary air line; CL, Gas Cylinder; COM, Air from compressor; E, Air jet ejector; F, Air diluted gas entry; L, Liquid entry; M, Manometer; N, Two-phase critical flow atomizer; OR, Orifice meter; P, Gas entry into the ejector; PG1–PG3, Pressure gauges; PM, Pump; R, Gas regulator; R1–R_2, Rotameters; S, Spray column; S1–S_2, Sampling points; ST, Surge tank; TS, Absorbing liquid tank; V1–V6, Gate valves; XT, Exhaust.

samples to the spray column were collected and were analyzed by standard methods of titration [44]. On the other hand, gas samples were drawn at a flow rate of about 1.67×10^{-5} m^3/s (1.0 lit/min) at sampling points S_1 and S_2 as shown in the figure and were analyzed by titration methods as detailed elsewhere in the literature [45]. It was described in Section 3 earlier that satisfying the conditions of pseudo-first order chemical reaction would require knowledge of k_L. As a result, a few experimental runs were taken for its determination in the counter-current mode using the diffusion controlled slow reaction, for instance, the absorption of CO_2 into a (CO_3^{2-}/HCO_3^{-1}) buffer solution as detailed in the literature [17]. After the system had attained steady state, the flow rate of the gas and liquid, total pressure of the column, system temperature were noted and the gas and liquid samples were collected at the sampling points for subsequent analysis. The total pressure of the column was determined using a pressure gauge.

In the actual experiment, alkaline solutions were pumped into the column through the atomizer routed through valve (V_2) and rotameter (R_2). Low pressure [$1.19 \times 10^5 - 1.68 \times 10^5$ N/m^2 (abs)] air was used to convert the liquid into fine sprays at high velocities. A simulated air–CO_2 mixture was then introduced into the spray column. The experiments were carried out without liquid recirculation under various operating conditions (Table 1).

TABLE 1: Experimental conditions.

Temperature (T)	296 ± 1
CO_2 flow rate (l/h)	150, 200, 250, 300
Initial concentration of CO_2 (%)	0.676 – 2.50
Concentration of NaOH (kmol/m³)	0.2721 – 0.3123
Gas flow rate, Q_G (m³/s)	3.33×10^{-3}, 3.75×10^{-3}, 4.20×10^{-3}, 5.00×10^{-3}, 6.20×10^{-3}
Superficial gas velocity, U_g (m/s)	0.1169, 0.1316, 0.1474, 0.1755, 0.2176
Liquid flow rate, Q_L (m³/s)	1.11×10^{-5}, 1.83×10^{-5}, 2.50×10^{-5}, 3.11×10^{-5}
Liquid flow rate (kg/h m²)	1403, 2313, 3159, 3930
Liquid to gas flow rate ratio (m³/1000 ACMM)	1.77 – 9.34
Droplet diameter, D_d (μm)	72.2, 102, 119, 139.8
Second order rate constant, k_2 (m³/kmol s)	7502.13 – 7598.29

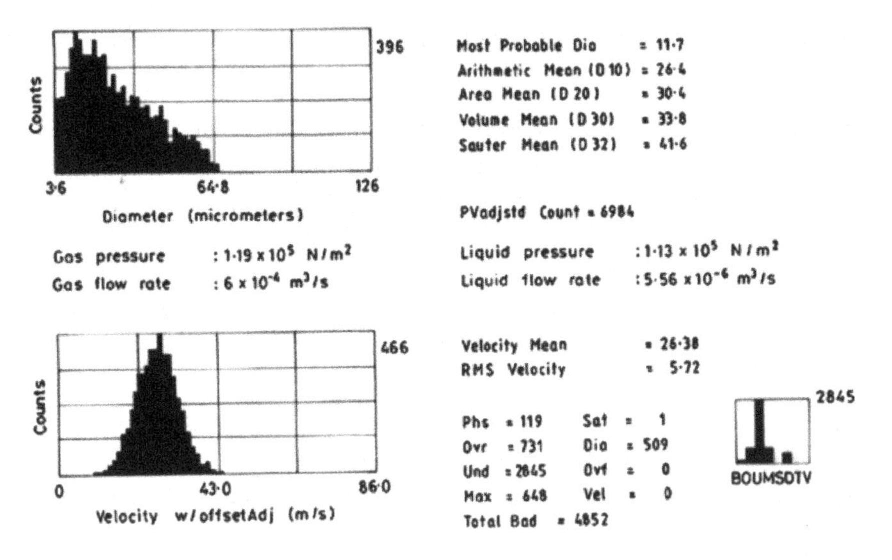

FIGURE 3: Typical drop size distribution generated by the two-phase critical flow atomizer [velocities are all in m/s; droplet diameters are all in μm].

The atomizing air introduced into the spray column was about 1% of the total gas flow rate. Therefore, necessary corrections were made for the dilute atomized air (~1%) that was introduced into the gas phase. The effects of various flow and operating variables on the CO_2 absorption were studied at a fixed atomizing pressure of 1.19×10^5 N/m^2 (abs). The droplet size was measured with a Phase Doppler Analyzer (PDA). The PDA was programmed to evaluate the droplet Sauter Mean Diameter (SMD) in situ while measuring the size distribution. The mean velocity and the root mean square (r.m.s.) velocity of the droplets was measured by Laser Doppler Velocimetry (LDV) method. Mean droplet velocities were measured 1 m downstream from the nozzle exit. A typical drop size distribution along with its velocity distribution is shown in Fig. 3.

2.5 RESULTS AND DISCUSSION

The spray hydro dynamics are described initially and then the trends of variation of percentage removal as well as the interfacial area as functions of different operating variables are discussed in details in the respective sections.

2.5.1 SPRAY HYDRODYNAMICS

The effect of atomizing air pressure on droplet SMD at different liquid flow rates is shown in Fig. 4. It can be seen from the figure that the droplet SMD was reduced uniformly with the increase in the atomizing air pres-

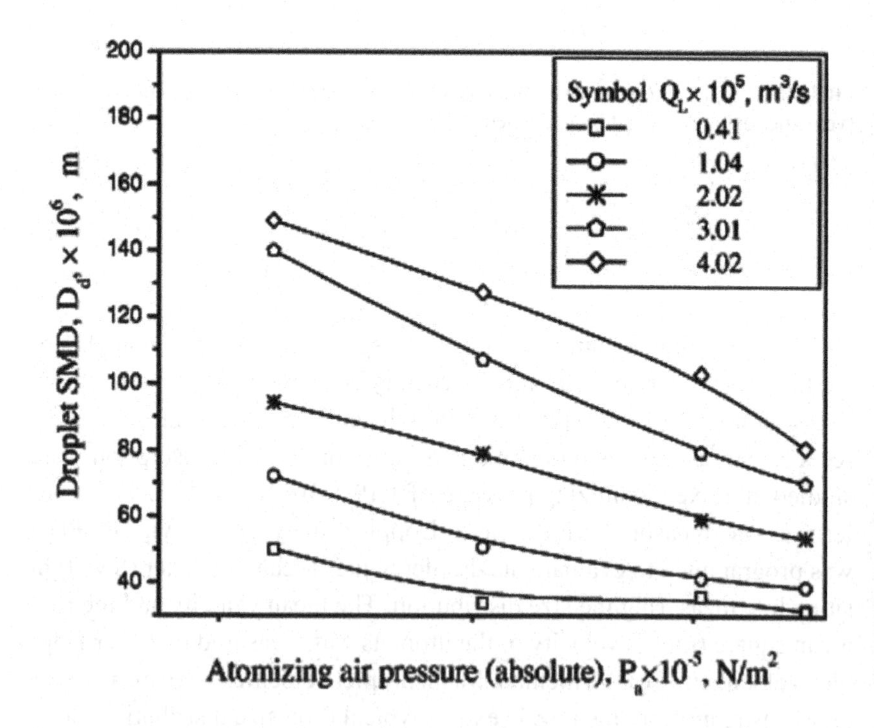

FIGURE 4: Effect of atomizing air pressure on the droplet SMD at constant liquid flow rates.

sure at constant liquid flow rates. It might be attributed due to the fact that the atomizing air at higher pressure enabled the mass of the definite amount of liquid to accelerate at a higher rate whence the liquid was being disintegrated into smaller droplets. The droplet SMD was however, increased with the increase in the liquid flow rate at a constant atomizing air pressure. It might be perhaps due to the increase in the definite mass of liquid at a constant atomizing air pressure that retarded the disintegration process and larger droplets were produced.

The effect of atomizing air pressure on the droplet velocity is shown in Fig. 5 at various liquid flow rates. It can be seen from the figure that the droplet velocity was increased sharply with the increase in the atomizing air pressure for all the liquid flow rates studied and finally reached almost a constant value of 1.68×10^5 N/m^2 (abs) beyond which no significant change in droplet velocity was observed. It might be attributed due to at-

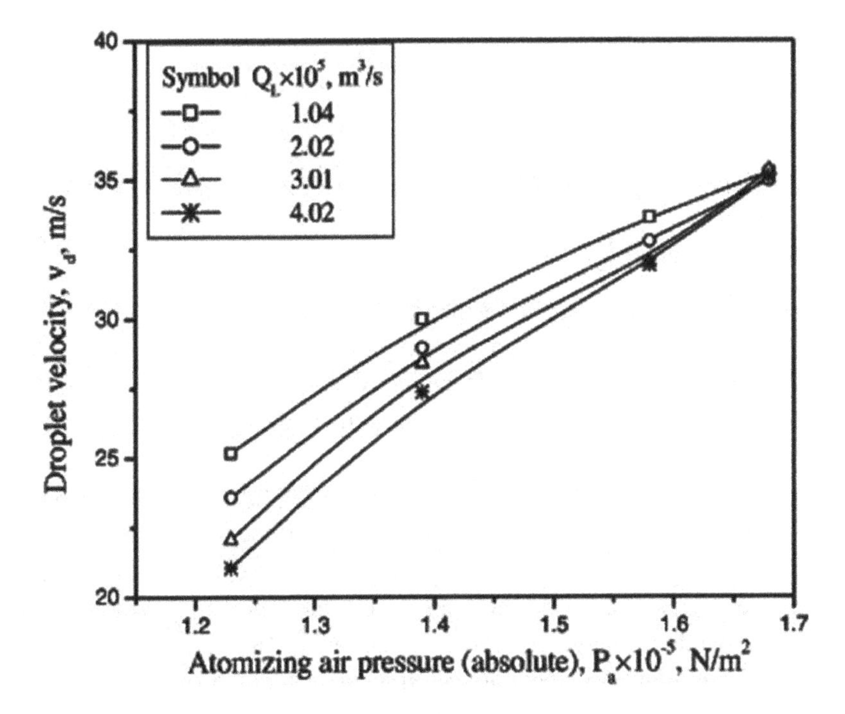

FIGURE 5: Effect of atomizing air pressure on the droplet velocity at constant liquid flow rate.

taining the critical velocity of the two-phase mixture at about 35 m/s. It can also be seen from the figure that the increase in the liquid flow rate, reduced the droplet velocity for a fixed atomizing air pressure. It might be perhaps due to the fact that the process of disintegration of liquid into drops was retarded at higher liquid flow rates for a constant atomizing air pressure that caused droplets to become larger as discussed earlier and thereby reduced the droplet velocity.

2.5.2 CO_2 CAPTURE

The experiments on CO_2 capture were conducted at various process operating conditions and the percentage removal of CO_2 was calculated in each run by the following generalized expression,

$$\eta = \left[\frac{(y_i - y_o)}{y_i}\right] \times 100 \tag{15}$$

2.5.2.1 EFFECT OF GAS FLOW RATE ON THE PERCENTAGE REMOVAL

The effect of gas flow rate on the percentage removal is shown in Fig. 6 at various CO_2 feed rates and at a fixed liquid flow rate of 1.11×10^{-5} m^3/s. It can be seen from the figure that the increase in gas flow rate reduced the percentage removal. The reduction in the percentage removal at higher gas flow rate might be due to the reduction in the gas–liquid contact time, i.e., the reduction in the rate of CO_2 absorption. Further, CO_2 accumulated on the droplet surface did not get enough time to diffuse into the bulk liquid phase and this phenomenon hindered the CO_2 absorption at higher gas flow rate, which resulted in the reduction in the percentage removal. On the other hand, at higher gas flow rate, the rate of flow of CO_2 molecules into the scrubber was also higher. But the rate of liquid discharge from the spray column might not be sufficiently fast in our system to accommodate fresh CO_2 molecules to get absorbed quickly while the gas flow rate was increased. There might have been an over-crowding of CO_2 molecules in

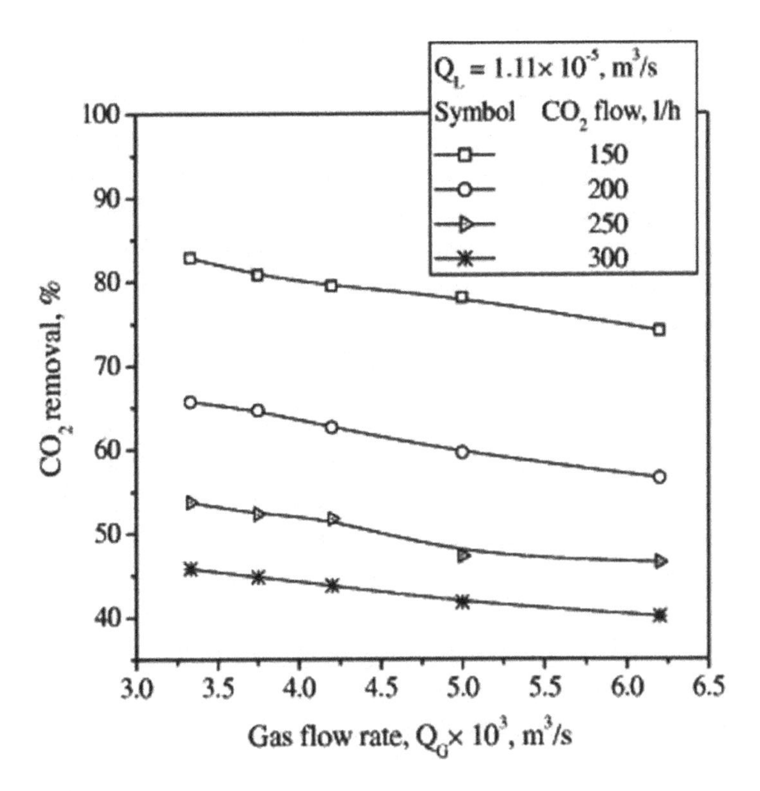

FIGURE 6: Effect of gas flow rate on the percentage removal of CO_2 at various CO_2 feed rate and at a constant liquid flow rate.

the bulk liquid phase in such a situation, thereby causing a hindrance to the overall diffusion of CO_2 into the bulk liquid phase that resulted in reduction in the percentage removal at higher gas flow rate. It can also be seen from the figure that the effect of the gas flow rate beyond 4.5×10^{-3} m^3/s on the percentage removal was not significant compared to its initial influence.

2.5.2.2 EFFECT OF LIQUID FLOW RATE ON THE PERCENTAGE REMOVAL

The liquid flow rate directly influences the number of droplets introduced into the spray tower for any gas–liquid absorption process. Thus the effect of liquid flow rate has been studied. The effect of liquid flow rate on the

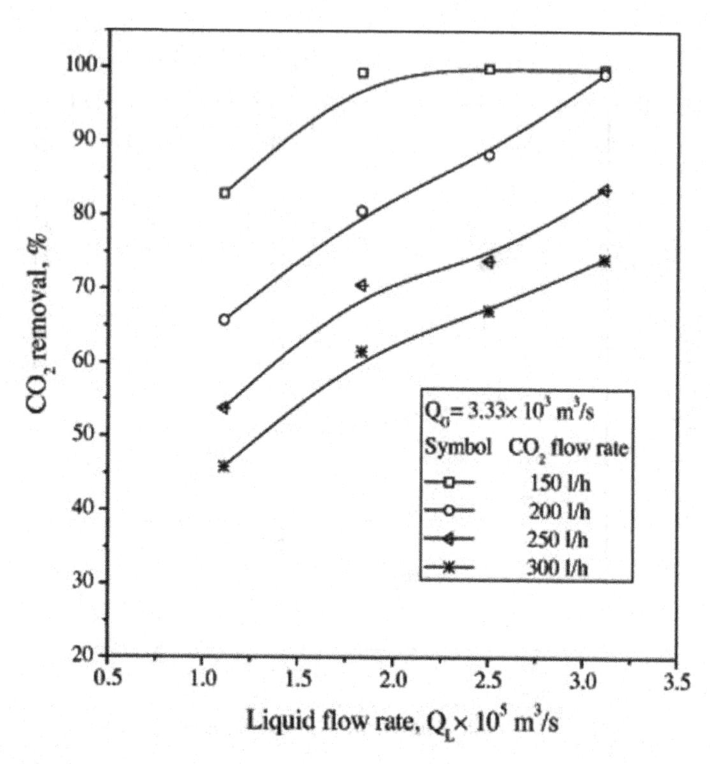

FIGURE 7: Effect of liquid flow rate on the percentage removal of CO_2 at various CO_2 feed rates.

percentage removal is shown in Fig. 7 at various CO_2 feed rates and at a fixed gas flow rate of 3.33×10^{-3} m^3/s. It can be seen from the figure that the percentage removal was increased with the liquid flow rate. As the liquid flow rate was increased, the drop diameter was also increased (since atomizing air pressure was kept constant in these experiments), and the total drop surface area to sweep the gas stream was also increased (Table 2). Although increasing the liquid flow rate reduced the number of droplet per second, the total droplet surface area per second was however, increased as can be seen from Table 2.

TABLE 2: Experimental data on liquid flow rate, droplet SMD and calculated values of number of droplet per second, droplet surface area, total droplet surface area generated per second.

Liquid flow rate (m^3/s)	Atomizing air pressure (N/m^2)	Droplet SMD $D_d \times 10^6$ (m)	Number of droplets, N_d (s^{-1})	Droplet surface area, S (m^2)	Total droplet surface area generated $S_T = N_d \times S$ (m^2/s)
1.11E-05		7.22E−05	5.63E+07	1.64E−08	0.9224
1.83E-05	1.19E + 05	1.02E−04	3.26E+07	3.29E−08	1.0733
2.50E-05		1.19E−04	2.84E+07	4.44E−08	1.2616
3.11E-05		1.40E−04	2.17E+07	6.14E−08	1.3348

This phenomena coupled with the enhanced droplet oscillation at higher liquid flow rate have caused the droplet-gas collision more effective and hence the percentage removal was increased with the increase in the liquid flow rate. However, the trend of variation in case of CO_2 feed rate of 150 l/h was somewhat different than the other cases. It might be perhaps due to achieving the maximum CO_2 absorption. The maximum percentage removal observed was about 99.96% for $Q_L = 1.83 \times 10^{-5}$ m³/s, $Q_G = 3.33 \times 10^{-3}$ m³/s and CO_2 feed rate = 100 l/h within the framework of the present investigation.

2.5.2.3 EFFECT OF LIQUID TO GAS FLOW RATE RATIO ON THE PERCENTAGE REMOVAL

From the economic point of view, the liquid-to-gas flow rate ratio in terms of m³/1000 ACM (actual cubic meter) has been found to be one of the most important criteria for reporting the scrubbing performance. In the light of this observation, the effect of Q_L/Q_G ratio on the percentage removal of CO_2 is shown in Fig. 8 for different CO_2 feed rates. It can be seen from the figure that the percentage removal of CO_2 increased very sharply with the increase in Q_L/Q_G ratio but beyond a Q_L/Q_G ratio of 6.0 m³/1000 ACM the increase in percentage removal was not that rapid. The reason for such observation may be explained as follows. With the increase in Q_L/Q_G ratio the hydraulic loading, i.e., Q_L was increased in the scrubber. As the liquid flow rate was increased, the drop diameter and the total droplet surface area were also increased (Table 2). This phenomena coupled with enhanced droplet oscillation at higher liquid flow rate might have caused the droplet-gas collision more effective and hence the percentage removal was increased with the increase in Q_L as well as Q_L/Q_G ratio. It can also be seen from the figure that the increase in Q_G reduced the percentage removal. It might be due to the reduction in gas–liquid contact time i.e., the reduction in the overall mass transfer. Also the CO_2 accumulated at the surface of the droplet phase did not get enough time to diffuse into the bulk liquid phase and this phenomenon hindered the absorption rate of CO_2 at higher CO_2 feed rate, which resulted in the reduction in the percentage removal. On the other hand, at higher CO_2 feed rate with relatively lower liquid flow rate, the rate of flow of CO_2 molecules into the scrubber was

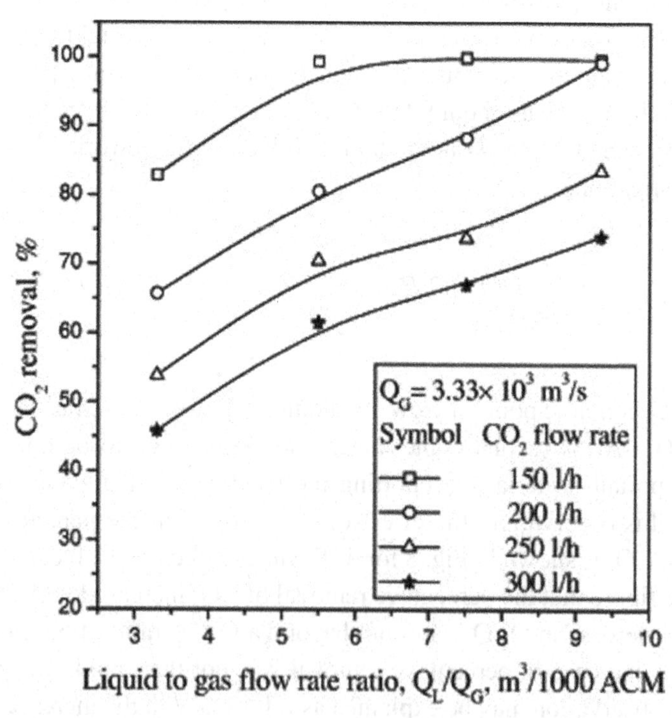

FIGURE 8: Effect of liquid to gas flow rate ratio on the percentage removal of CO_2 at various CO_2 feed rates.

also higher and there might have been an over-crowding of CO_2 molecules in the bulk liquid phase causing a hindrance to the overall diffusion of CO_2 which resulted in reduction in the percentage removal at higher gas flow rate. That the rate of change of percentage removal beyond a Q_L/Q_G ratio of 6.0 m^3/1000 ACM, might be due to the fact that CO_2 concentration in the liquid was approaching its equilibrium value in the gas phase. It could also be due to the result of increased rate of coalescence of the droplets, whereby both the number and the total droplet surface area might decrease. The maximum percentage removal observed was about 99.96% for a Q_L/Q_G ratio of 6.0 m^3/1000 ACM and for CO_2 feed rate of 100 l/h within the framework of the present investigation. Under this operating condition the liquid and the gas flow rates were 1.83×10^{-5} and 3.33×10^{-3} m^3/s, respectively.

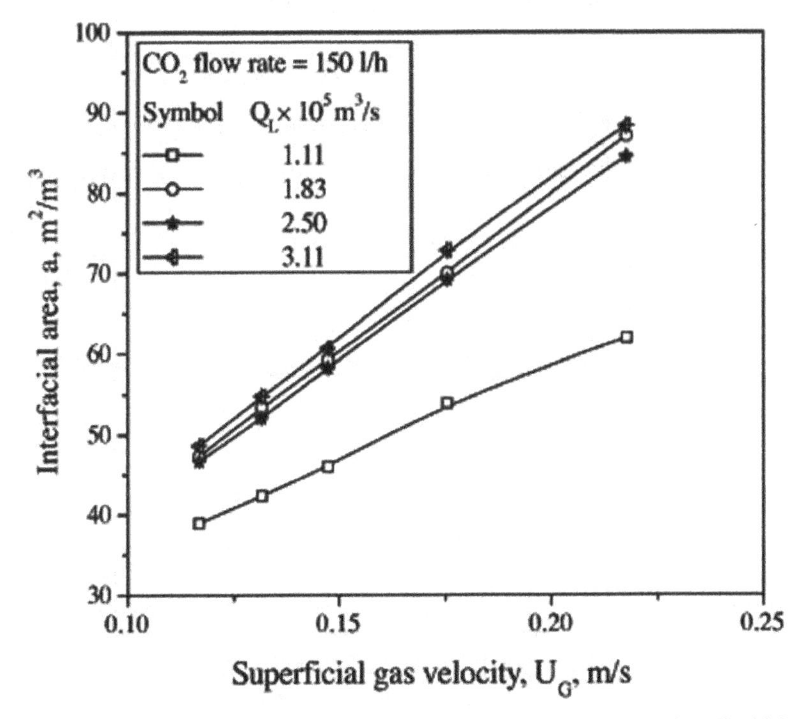

FIGURE 9: Effect of superficial gas velocity on the interfacial area at various liquid flow rates and at constant CO_2 feed rates.

2.5.2.4 EFFECT OF SUPERFICIAL GAS VELOCITY ON THE INTERFACIAL AREA

The effect of superficial gas velocity on the interfacial area at various liquid flow rates and at constant CO_2 feed rate is shown in Fig. 9. It can be seen from the figure that the interfacial area was increased with the superficial gas velocity almost linearly. It may be explained as follows. The interfacial area has been calculated from Eq. (4) as described earlier. It can be seen there from that the values of calculated interfacial area depend on various factors besides the rate of absorption. The percentage removal or the rate of absorption was observed lower at higher gas flow rates while interfacial area was observed higher at higher superficial gas velocities (or gas flow rates). It is attributed due to the fact that the fall in the rate of absorption at higher gas flow rates was less than the fall in the values of the denominator of rearranged Eq. (4) at the same gas flow rate.

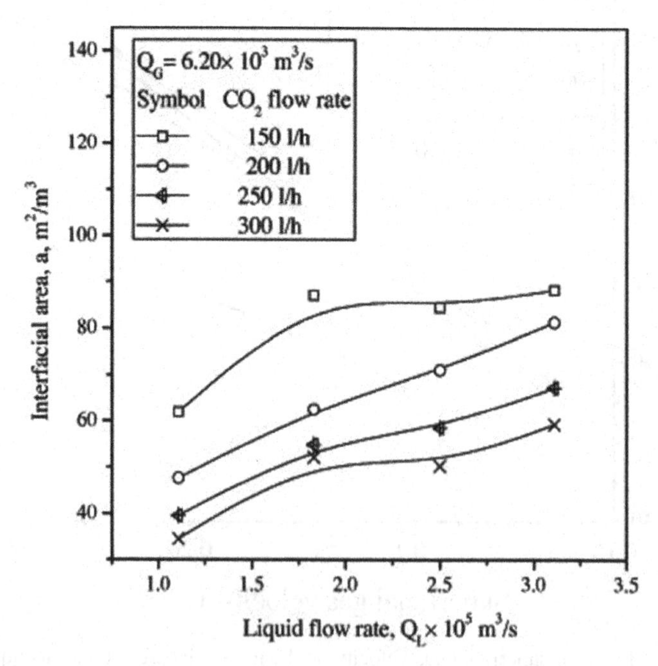

FIGURE 10: Effect of liquid flow rate on the interfacial area at different CO_2 feed rate and at a constant gas flow rates.

2.5.2.5 EFFECT OF LIQUID FLOW RATE ON THE INTERFACIAL AREA

The effect of liquid flow rate on the interfacial area at constant air flow rate at various CO_2 feed rates is shown in Fig. 10. It can be seen from the figure that the interfacial area was increased with the increase in the liquid flow rate. It might be due to the increased interfaces generated owing to increased liquid flow rate. In this case the increased liquid flow rate at constant Pa was associated with increased drop size. As a result, the number of droplets was decreased while total surface generated by the droplets (S_T) was increased (Table 2) slowly. The increase in S_T caused the interfacial area to increase with the liquid flow rate. The values of interfacial area observed in the present study were ranging from 22.62 to 88.35 m^2/m^3 within the framework of the experimentation.

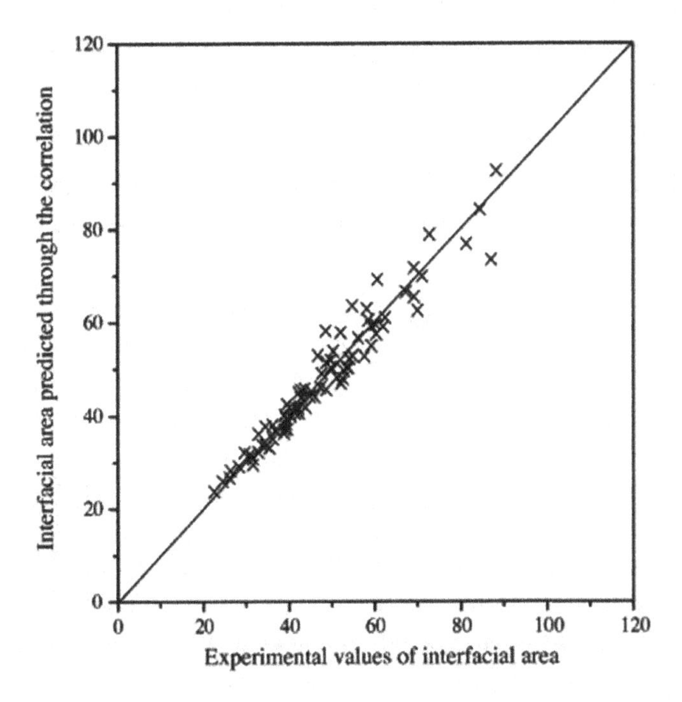

FIGURE 11: Comparison between experimental values of interfacial area and values predicted through the correlation.

2.5.2.6 PREDICTION OF INTERFACIAL AREA THROUGH CORRELATION

The dependence of the interfacial area on the CO_2 feed rate, the superficial gas velocity and the liquid flow rate in the present system was found to be of the following form (correlation coefficient = 0.9767; standard error = 0.0651; 99% confidence range)

$$a = 2 \times 10^2 (Q_{CO2})^{-0.6465} (U_g)^{0.7481} (Q_L)^{0.4362} \quad (16)$$

The values of interfacial area predicted by Eq. (16) have been plotted against the experimental values of interfacial area in Fig. 11. It can be seen from the figure that predicted data fitted excellently well with experimental values. The percentage deviation between predicted and experimental values is well within ±10% excepting a few data points. The correlation developed is thus highly significant.

2.6 COMPARISON OF THE PERFORMANCE OF THE PRESENT SYSTEM WITH SYSTEMS REPORTED IN THE LITERATURE

The present system yielded higher values of interfacial area compared to the existing systems (Table 3). It might be attributed due to the formation of smaller droplets in the presented system at relatively very lower liquid flow rates. Furthermore, the droplets generated by the two-phase critical flow atomizer was falling down at a very high velocity that prevented coalescence as occurred in conventional devices producing relatively larger droplets with smaller velocities. The presented system used 10–15 times lower values of liquid flow rates to obtain interfacial area of 22.62–88.35 $m_2/m3$ than used by Mehta and Sharma [22] to obtain interfacial area of 30–50 m^2/m^3. On the other hand, Pinilla et al. [25] though used higher but comparable liquid flow rates to obtain interfacial area in the range of 0.95–1.30 m^2/m^3. From the foregoing, it is clear that the present system produced higher values of interfacial area than the existing systems and hence the performance of the system was better than the existing systems.

TABLE 3: Comparison of values of interfacial area obtained in the present study with values reported in existing systems.

Parameters	Mehta and Sharma [22]	Pinilla et al. [25]	Present study
	Solid cone nozzle	Solid cone nozzle	Critical flow atomizer
Column dimensions (m)	$D_C = 0.205$	$D_C = 0.45$	$D_C = 0.1905$
	$H_C = 1.23$	$H_C = 0.95$	$H_C = 2.00$
Chemical system for "a"	CO_2–NaOH	CO_2–Na_2CO_3/$NaHCO_3$ + Na-arsenite as catalyst	CO_2–NaOH
U_g (m/s)	0.25–0.51	0.097–0.143	0.1169–0.2176
Q_L (kg/h m^2)	24,000–40,000	3708–5508	1403–3930
D_d (μm)	250–430	3400–3800	72.2–168
a (m^2/m^3)	30–50	0.95–1.30	22.62–88.35

2.7 CONCLUSIONS

The emission of CO_2 into the atmosphere is causing majority of the global warming and thus various end-of-pipe treatment methods have evolved to capture CO_2 from fixed point sources. The present article deals with CO_2 capture from a simulated gas stream using dilute NaOH solution in a spray column using a two-phase critical flow atomizer capable of producing very fine sprays with high degree of uniformity and moving at a very high velocity. The performances of such a spray column for the abatement of particulate and/or SO_2 emission from simulated waste gas streams showed very high degree of removal (~100%) for particulates and/or SO_2.

Preliminary hydrodynamic studies revealed that the droplet SMD was reduced with the increase in atomizing air pressure and decrease in the liquid flow rate while the droplet velocity was increased with the increase in atomizing air pressure and decrease in the liquid flow rate. Results revealed that the increase in gas flow rate reduced the percentage removal while increased liquid flow rate increased it. The maximum percentage removal observed was about 99.96% for a Q_L/Q_G ratio of 6.0 m^3/1000 ACM and for CO_2 feed rate of 100 l/h within the framework of the present investigation. Under this operating condition the liquid and the gas flow rates were 1.83×10^{-5} and 3.33×10^{-3} m^3/s, respectively.

Experimental results further indicated that the interfacial area was increased with the superficial gas velocity and the liquid flow rate. The values of interfacial area observed in the present study were ranging from 22.62–88.35 m^2/m^3 within the framework of the experimentation. A simple correlation was put forward for predicting the interfacial area which was highly functional. The comparison of the interfacial area observed between the present system and the existing systems revealed that the present system produced higher values of interfacial area than the existing systems and hence the performance of the system was better than the existing systems.

REFERENCES

1. IPCC, Policymaker's Summary of the Scientific Assessment of Climate Change, Report to IPCC from working group, Meteorological Office, Branknell, UK, 1990.
2. Y.F. Diao, X.Y. Zheng, B.S. He, C.H. Chen, X.C. Xu Experimental study on capturing CO_2 greenhouse gas by ammonia scrubbing Energy Convers. Manag., 45 (2004), pp. 2283–2296
3. M.H. Martin, S. Meyer Greenhouse Gas Carbon Dioxide Mitigation. Lewis Publishers (1999)
4. A.M. Wolsky, E.J. Daniels, B.J. Jody. CO_2 capture from the flue gas of conventional fossil-fuel-fired power plants. Environ. Prog., 13 (1994), pp. 214–219
5. N. Kimura, K. Omata, T. Kiga, S. Takano, S. Shikisma. Characteristics of pulverized coal combustion in O_2–CO_2 mixtures for CO_2 recovery Energy Convers. Manag., 36 (1995), pp. 805–808
6. N. Nishikawa, A. Hiroano, Y. Ikuta. Photosynthetic efficiency improvement by microalgae cultivation in tubular type reactor. Energy Convers. Manag., 36 (1995), pp. 681–684
7. A. Bandyopadhyay. Amine versus ammonia absorption of CO_2 as a measure of reducing GHG emission: a critical analysis. Technol. Environ. Policy, 13 (2) (2011), pp. 269–294
8. S. Chakravarti, A. Gupta, B. Hunek, Advanced Technology for the Capture of Carbon Dioxide from Flue Gases, Proc. 1st National Conference on Carbon Sequestration. Washington, DC, May 15–17, 2001.

9. IEA: IEA Greenhouse Gas R&D Programme (IEA GHG), CO_2 Capture Ready Plants, 2007/4, May 2007.
10. N. Nsakala, J. Marion, C. Bozzuto, G. Liljedahl, M. Palkes, D. Vogel, J.C. Gupta, M. Guha, H. Johnson, S. Plasynski, Engineering Feasibility of CO_2 Capture on an Existing US Coal Fired Power Plant, Proc. of the 1st National Conference on Carbon Sequestration May 15–17, Washington, DC, 2001.
11. M. Hasib-ur-Rahman, M. Siaj, F. Larachi. Ionic liquids for CO_2 capture–development and progress. Chem. Eng. Process., 49 (2010), pp. 313–322
12. F. Karadas, M. Atilhan, S. Aparicio. Review on the use of ionic liquids (IL) as alternative fluids for CO_2 capture and natural gas sweetening. Energy Fuels, 24 (2010), pp. 5817–5828
13. M.C. Duke, B. Ladewig, S. Smart, V. Rudolph, J.C.D. da Costa. Assessment of postcombustion carbon capture technologies for power generation. Front. Chem. Eng. China, 4 (2010), pp. 184–195
14. H. Yang, Z. Xu, M. Fan, R. Gupta, R.B. Slimane, A.E. Bland, I. Wright. Progress in carbon dioxide separation, capture: a review. J. Environ. Sci. (Beijing, China), 20 (2008), pp. 14–27
15. IPCC, Special Report on Carbon Dioxide Capture and Storage, Prepared by Working Group III of the Intergovernmental Panel on Climate Change, Cambridge University Press, New York, USA, 2005.
16. P.V. Danckwerts, M.M. Sharma. The absorption of carbon dioxide into solutions of alkalies and amines. Chem. Eng. (1966), pp. 244–280
17. G. Astarita. Mass Transfer with Chemical Reaction. Elsevier, Amsterdam (1967)
18. P.V. Danckwerts. Gas-Liquid Reactions. McGraw-Hill, NY (1970)
19. A. Mandal, G. Kundu, D. Mukherjee. Interfacial area and liquid-side volumetric mass transfer coefficient in a downflow bubble column. Can. J. Chem. Eng., 81 (2003), pp. 212–219
20. J.K. Stolaroff, D.W. Keith, G.V. Lowry. Carbon dioxide capture from atmospheric air using sodium hydroxide spray. Environ. Sci. Technol., 42 (2008), pp. 2728–2735
21. Plant operating data. Accessed through private communication to Kolaghat Thermal Power Plant, West Bengal, India, 2010.
22. K.C. Mehta, M.M. Sharma. Mass transfer in spray columns. Brit. Chem. Eng., 15 (Pt. I) (1970), pp. 1440–1444. K.C. Mehta, M.M. Sharma. Mass transfer in spray columns. Brit. Chem. Eng., 15 (Pt. II) (1970), pp. 1556–1558
23. E. Ruckenstein, V.D. Dang, W.N. Gill. Mass transfer with chemical reaction from spherical one or two component bubbles or drops. Chern. Eng. Sci., 26 (1971), pp. 647–668
24. Y.G. Adewuyi, G.R. Carmichael. A theoretical investigation of gaseous absorption by water droplets from SO, –HNO, –NH, –CO, –HCI mixtures. Atmos. Environ., 16 (1982), pp. 719–729
25. E.A. Pinilla, J.M. Diaz, J. Coca. Mass transfer and axial dispersion in a spray tower for gas–liquid contacting. Can. J. Chem. Eng., 62 (1984), pp. 617–622
26. E.R. Altwicker, C.E. Lindhjem. Absorption of gases into drops. AIChE J., 34 (1988), pp. 329–332

27. I. Taniguchi, K. Asano. Experimental study of absorption of carbon dioxide from carbon dioxide–air gas mixtures by water spray. J. Chem. Eng. Jpn., 25 (1992), pp. 614–616
28. I. Taniguchi, J. Kawabata, K. Asano. Absorption of carbon dioxide with a freely falling aqueous sodium hydroxide solution drop. Chem. Eng. Commun., 159 (1) (1997), pp. 119–135
29. I. Taniguchi, Y. Takamura, K. Asano. Experimental study of gas absorption with a spray column. J. Chem. Eng. Jpn., 30 (3) (1997), pp. 427–433
30. A. Bandyopadhyay, M.N. Biswas. Spray scrubbing of particulates with a critical flow atomizer. Chem. Eng. Technol., 30 (12) (2007), pp. 1674–1685
31. A. Bandyopadhyay, M.N. Biswas. Critical flow atomizer in SO_2 spray scrubbing. Chem. Eng. J., 139 (1) (2008), pp. 29–41
32. A. Bandyopadhyay, M.N. Biswas. spray scrubbing of particulate-laden-SO_2 using a critical flow atomizer. J. Environ. Sci. Health, Part A, A43 (10) (2008), pp. 1115–1125
33. M.N. Biswas. Atomization in two-phase critical flow. Proceedings of the $_2$nd International Conference on Liquid Atomization and Spray Systems-II, Madison, 5–1, Madison, 5–1, Madison, 5–1, Wisconsin, USA (1982), pp. 145–151
34. P.H. Calderbank. The interfacial area in gas-liquid contacting with mechanical agitation. Trans. IChemE., 36 (1958), pp. 443–460
35. J.L. York, H.E. Stubbs. Photographic analysis of sprays. Trans. ASME., 74 (195_2), pp. 1157–1161
36. K.D. Cooper, G.F. Hewitt, B. Pinchin. Photographic method to measure interfacial area in liquid and gas dispersed system. J. Photobiol. Sci., 12 (1964), pp. 269–273
37. R.A. Mashelkar. Bubble columns. Brit. Chem. Eng., 15 (10) (1970), pp. 1297–1304
38. S.A. Patel, J.G. Daly, D.B. Bukur. Holdup and interfacial area measurements using dynamic gas disengagement. AIChE J., 6 (35) (1989), pp. $931–94_2$
39. I.T.M. Hassan, C.W. Robinson./ Mass-transfer-effective bubble coalescence frequency and specific interfacial area in a mechanically agitated gas–liquid contactor. Chem. Eng. Sci., 35 (6) (1980), pp. 1277–1289
40. M.M. Sharma, P.V. Danckwerts. Chemical methods of measuring interfacial area and mass transfer coefficients in two fluid systems. Brit. Chem. Eng., 15 (4) (1970), pp. 522–528
41. P.L.T. Brian, R.F. Baddour, D.C. Matiatos. An ionic penetration theory for mass transfer with chemical reaction. AIChE J., 10 (5) (1964), pp. 727–733
42. R.A.T.O. Nijsing, R.H. Hendriksz, H. Kramers. Absorption of CO_2 in jets and falling films of electrolyte solutions, with and without chemical reaction. Chem. Eng. Sci., 10 (1959), pp. 88–104
43. K.E. Porter, M.B. King, K.C. Varshney. Interfacial areas and liquid-film mass-transfer coefficients in a 3 ft diameter bubble-cap plate derived from absorption rates of carbon dioxide into water and caustic soda solutions. Trans. IChemE., 44 (1966), pp. T274–T283
44. A.I. Vogel. Quantitative Inorganic Analysis. (first ed.)Longman Green, London, UK (1955) pp. 3_27
45. M.M. Sharma, R.A. Mashelkar, V.D. Mehta. Mass transfer in plate columns. Brit. Chem. Eng., 14 (1) (1969), pp. 70–76

CHAPTER 3

CHARACTERISTICS OF CO_2 HYDRATE FORMATION AND DISSOCIATION IN GLASS BEADS AND SILICA GEL

MINGJUN YANG, YONGCHEN SONG, XUKE RUAN, YU LIU, JIAFEI ZHAO, AND QINGPING LI

3.1 INTRODUCTION

The greenhouse effect is leading to a significant climate warming and weather changes [1]. CO_2 is considered to be one of the most important greenhouse gases, and the disposition of CO_2 has become an issue of worldwide concern [2]. International Energy Agency (IEA) proposed that if the target of climate change control was obtained without CO_2 capture and storage (CCS), the total cost will increase 70% more than with CCS by 2050 [3].

Gas hydrate technology is a new subject based on ice-like crystalline compounds where gas molecules are held within cavities formed by water molecules [4], and it is developed and used in some industrial field recently, including refrigeration, gas storage and transportation, gas separation. Hydrate-based CO_2 separation as a promising option for fossil fuel power plant CO_2 capture (the first step of CCS) is attracting people's attention, which is a novel concept that aims to use CO_2 hydrate to trap CO_2 molecules in a lattice of water molecules [5]. The other proposed scheme is to sequester CO_2 in form of gas hydrates in ocean and marine sediment (the last step of CCS) [6]. To make CO_2 hydrate formation quick and economical during the capture process, or stabilized in marine sediments, it is necessary to understand the thermodynamic characters for CO_2 hydrate formation and dissociation, especially the phase equilibrium conditions.

CO_2 hydrate equilibrium conditions have been widely investigated. Wendland et al. [7], Yang et al. [8], Englezos et al. [9], Breland et al. [10], Dholabhai et al. [11,12], Kang et al. [13], Mohammadi et al. [14] investigated the CO_2 hydrate phase equilibrium in water with different additives. The studies concerning CO_2 hydrate equilibrium in porous medium are also familiar. Handa and Stupin [15] observed the shift that hydrates capillary equilibria from that in the bulk water. Uchida, et al. [16] likewise observed an equilibria shift in silica glass with pores of radii 100 Å, 300 Å, and 500 Å, and they estimated that the apparent interfacial free energy between methane hydrates and water in the confined condition was approximately 3.9×10^{-2} $J \cdot m^{-2}$. At the same time, Clennell et al. [17] and Clarke et al. [18] also concluded that observed hydrate equilibrium shifts in the ocean floor were caused by capillary effects in small sediment pores. Zatsepina et al. [19] measured CO_2 hydrate stability in porous media. They concluded that when the vapor phase of CO_2 was absent, the volume of hydrate was limited by the transport of CO_2 from solution. Smith et al. [20] measured equilibrium pressures for CO_2 hydrate in silica gel pores with nominal radii of 7.5, 5.0, and 3.0 nm, and observed they were higher than those for CO_2 hydrate in bulk water. Anderson et al. [21] presented experimental methane, carbon dioxide, and methane-carbon dioxide hydrate equilibrium and ice-melting data for meso-pores silica glass, and determined similar values of interfacial tensions for ice-water, methane clathrate-water, and carbon dioxide clathrate-water. Following their studies, Kumar [22] collected experimental equilibrium conditions data for CO_2 hydrate in porous medium and measured the permeability of the porous medium in the presence of hydrate by flowing through the system. Turner et al. [23] reported that any shift in pores larger than 600 Å in radius cannot be distinguished from errors of the thermocouples in their equilibrium apparatus (with thermocouple error of ±0.5 K). Aladko et al. [24] investigation hydrates equilibrium of ethane, propane, and carbon dioxide dispersed in silica gel meso-pores at pressures up to 1 GPa. The result showed that the experimental dependence of hydrate decomposition temperature on the size of pores can be described on the basis of the Gibbs-Thomson equation only if one takes into account changes in the shape coefficient that is present in the equation.

In this work, considering the limited data available in macro porous medium, we carried out experiments in glass beads which enable the study of the impact of porosity-related properties like capillary effects on the equilibrium conditions. The characteristics of CO_2 hydrate formation and dissociation in porous medium were investigated by experimental observations and numerical modeling. MRI was also used in this study to determine the priority formation position of CO_2 hydrate in different pore sizes.

3.2. EXPERIMENTAL

3.2.1 EXPERIMENTAL APPARATUS AND MATERIALS

The experimental apparatus used in this study is shown in Figure 1 and for further details of the experimental apparatus the reader can be referred to the previous publications of our research team [25,26]. A high-pressure resistant vessel made of 316-stainless steel with a volume of 476 mL is used as the reactor. Thermocouples (produced by Yamari Industries, Osaka, Japan) and two pressure transducers (produced by Nagano Keiki, Tokyo, Japan) are connected to the vessel. The estimated errors of temperature and pressure measurements are ±0.1 K and ±0.1 MPa, respectively. Glass beads (produced by As-One Co., Ltd., Japan) and silica gel (produced by Anhui Liangchen Silicon Material Co. Ltd., Huoshan, Anhui, China) were used to form porous medium. CO_2 (mass fraction 0.999) was provided by Dalian Guangming Special Gas Co. Ltd., China. Table 1 summarizes the specifications of all the components. All the chemicals were unpurified and the de-ionized water was used in all the experiments.

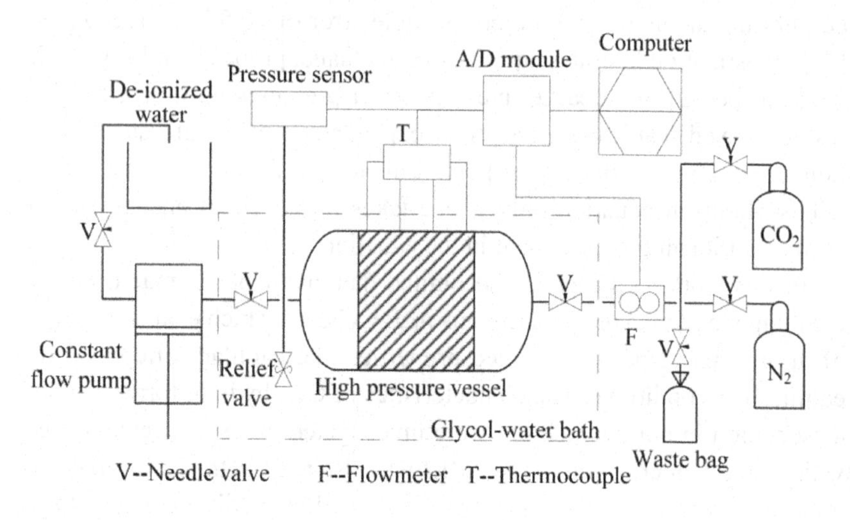

FIGURE 1: Scheme of the gas hydrate experimental apparatus.

TABLE 1: Property and supplier of materials.

Material	Purity/Composition	Particle size	Supplier
CO_2	99.9%	—	Dalian Guangming special gas Co., Ltd., China
BZ-01	Soda glass	0.105–0.125 mm	As-One Co., Ltd., Japan
BZ-02	Soda glass	0.177–0.250 mm	As-One Co., Ltd., Japan
BZ-04	Soda glass	0.350–0.500 mm	As-One Co., Ltd., Japan
Mix glass beads	Soda glass and clay	0.125–1.0 mm	Self-made
Silica gel	Silica gel	0.42–0.84 mm	Anhui Liangchen Silicon Material Co., Ltd., China (mean particle pore size 8.0–10 nm)

3.2.2 EXPERIMENTAL PROCEDURES

The graphical method was used to measure hydrate phase equilibrium conditions by keeping one of the three parameters of pressure (p), volume (V) and temperature (T) constant and changing one of the remaining parameters to form or decompose the hydrate. In this study, experiments were carried out using the graphical method by keeping volume constant. CO_2

hydrate was formed and decomposed by cooling and heating the closed system (volume constant). Neither gas nor water was added to the system during each experimental cycle.

Dry glass beads were tightly packed into the vessel with de-ionized water to simulate the porous medium. Then the vessel was reconnected to the system, and CO_2 was injected to discharge the water partly from the vessel to obtain residual water saturation. After the outlet valve was closed, CO_2 was slowly injected continuously into the vessel to a designated pressure and the pressure was kept constant. The amounts of residual water and injected CO_2 were all recorded. When the temperature was steady and there were no leaks, the temperature was decreased. Once a temperature increase appeared, we confirmed that CO_2 hydrates were formed in the vessel due to the exothermic (hydrate formation) reaction. The formation process finished when there was no pressure change. Then the bath was warmed slowly to dissociate the CO_2 hydrate. The pressure and temperature (p-T) conditions at the end of the hydrate decomposition was considered to be CO_2 hydrate phase equilibrium conditions.

3.3 RESULTS AND DISCUSSION

3.3.1 CO_2 HYDRATE FORMATION AND DISSOCIATION PROCESS

The p-T curve during CO_2 hydrate formation and dissociation was dependent on the kinds of porous medium. Two cases of CO_2 hydrate formation process were examined experimentally (only the vapor CO_2, initial state, was discussed in this study). The p-T curve when hydrate formed in the BZ-01, BZ-02, BZ-04 is shown in Figure 2.

In this case, the pore size of porous medium was approximately uniform, and there is only one hydrate formation stage, point B-E. After the bath temperature was decreased to the target value, which was usually more than 4 K below the estimated equilibrium temperature (A-B), the sample temperature rose suddenly later (B-C), which was caused by CO_2

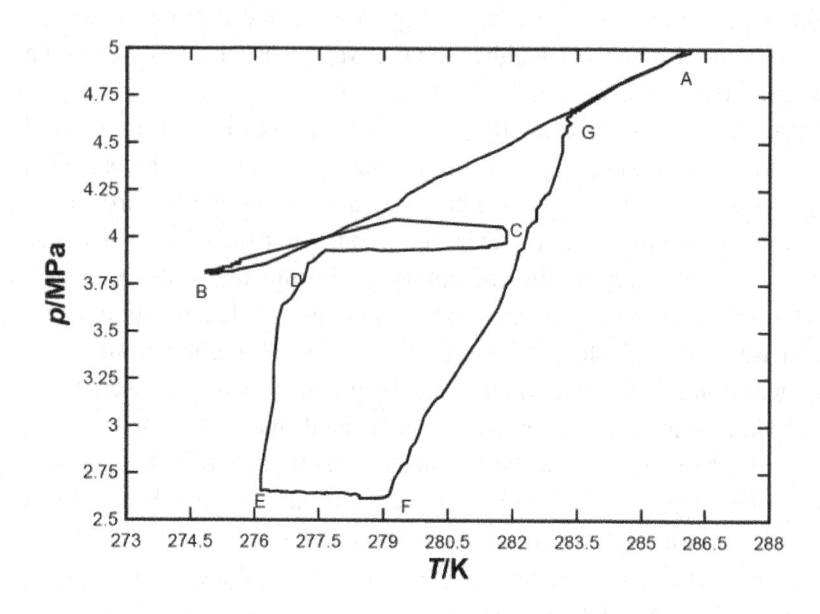

FIGURE 2: Typical p-T curves for CO_2 hydrate formation in BZ-01, BZ-02 and BZ-04 glass beads with uniform pore size distribution.

hydrate formation, and the pressure soon decreased (due to gas consumption as the gas was encaged in the hydrate lattice). Since the CO_2 hydrate formation rate was low and there was a high temperature difference between the porous medium and bath, the temperature decreased slowly (C-D). When the CO_2 hydrate formation process finished, the temperature decreased down to the initial setting value (D-E). Then the vessel was warmed gradually, when the p-T condition reached to F, the hydrate began to decompose, which caused a significant pressure increase (F-G). Point G was considered as the end of hydrate decomposition, which implied the equilibrium condition for this case. After intersecting with A-B, the p-T curve was back to point A along the temperature reduction process.

When CO_2 hydrate formed in mixed glass beads and silica gel, the hydrate formation process can be divided into two phases, as shown in Figures 3 and 4. This was caused by the different kinds of pore sizes present in these pores medium. Figure 3 showed the experiment carried out in mixed glass beads. The first formation stage (A-E) was the same as that discussed

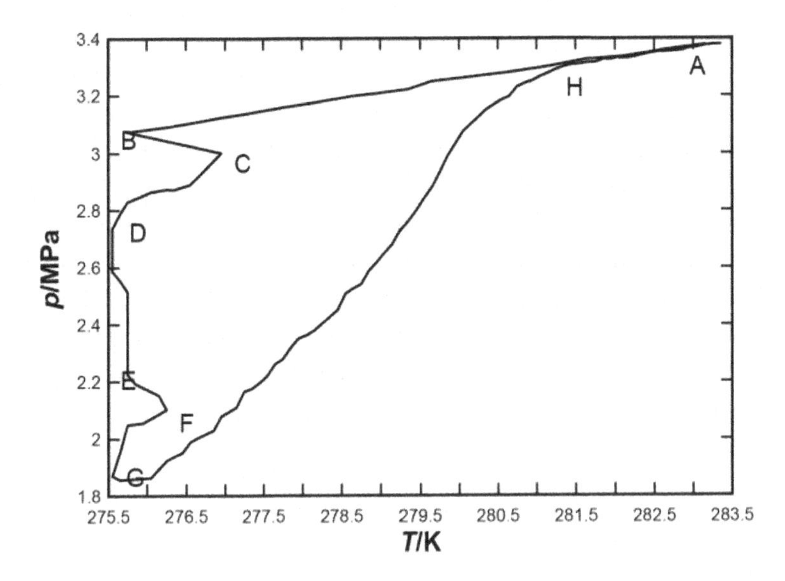

FIGURE 3: Typical p-T curves for CO_2 hydrate formation in mixed glass beads with non-uniform pore size distribution.

for Figure 2. In this stage, CO_2 hydrate may form in bigger or smaller size pores, the conclusion could not be educed. A significant pressure drop and temperature rise caused by the formation of a big amount of hydrates were observed from B to C. Since the thermostat temperature was still low, the temperature will decrease to the set value due to the heat transfer (C-D). After that, the second formation stage occurred at E-F, where temperature also showed a dramatic increase. In this stage, hydrate formed in the other sized pores. In order to determine the first CO_2 hydrate formation site, MRI was used in this work. The experimental procedure and results are discussed in the following paragraph.

When silica gel was used in the experiments, there are two size pores in the sediment, the larger pore is inter-particle porosity and the smaller one is the inherent pore size of silica gel. The hydrate formation process was shown in Figures 4 and 5, and the notes (A-G) in them correspond with each other. Once the pressure had dramatically decreased (B-C-D, in Figure 4), the hydrate began to nucleate and grow (the first formation stage), and the temperature showed a small increase during this time. Then the pressure remained constant and the temperature decreased further (D-E). As the temperature reached 273.6 K, the pressure decreased dramatically again (E-F), which meant the second formation stage of CO_2 hydrate. The hydrate dissociation process was the same as that in Figures 2 and 3. In this case, the first hydrate formation site was also uncertain. Usually, we proposed that the CO_2 hydrate formed firstly in bigger pores, because the two formation stage occurred at 1.95 MPa and 275 K, and 1.75 MPa and 273 K, respectively, during the formation process. The sub-cool temperature of second stage was high than the first one, which was the same as the equilibrium condition of hydrate in different size pores. The smaller the pore size corresponds to the lower hydrate equilibrium temperature at the same pressure.

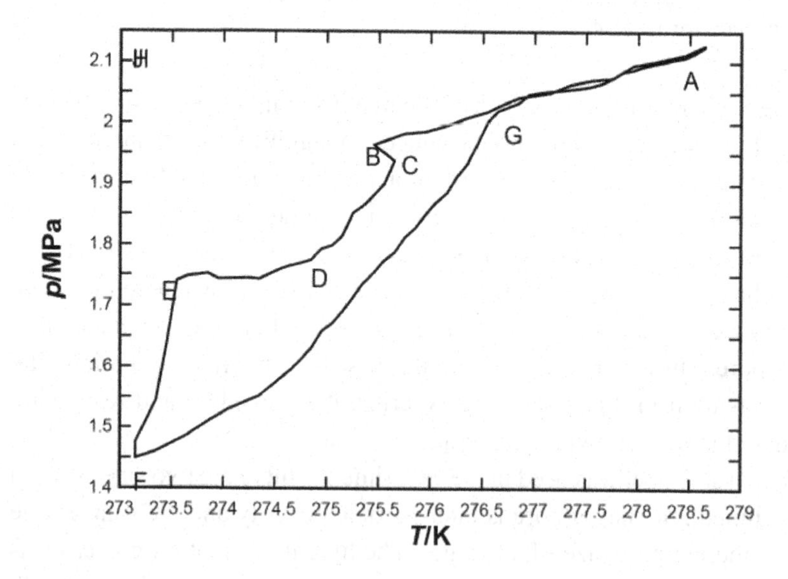

FIGURE 4: Typical p-T curves for CO_2 hydrate formation in silica gel with non-uniform pore size distribution.

3.3.2 DETERMINATION OF CO_2 HYDRATE PRIORITY FORMATION SITE USING MRI

MRI was used to test the hypothesis that CO_2 hydrate formed firstly in the bigger pores. In order to get close to the study conditions, the different pore sizes were built up with BZ-01 and BZ-02 glass beads. During the experiments, BZ-01 glass beads were packed into the vessel tightly with de-ionized water firstly. The height of BZ-01 was about half of the vessel. Then a plastic slice was put into the vessel, which was used to divide the two glass kinds of beads. Then BZ-02 was packed into the vessel fully. The other procedure was similar with that discussed before, the main difference being that the vessel was put into the MRI to obtain images during the formation process. The schematic diagram of the MRI apparatus and

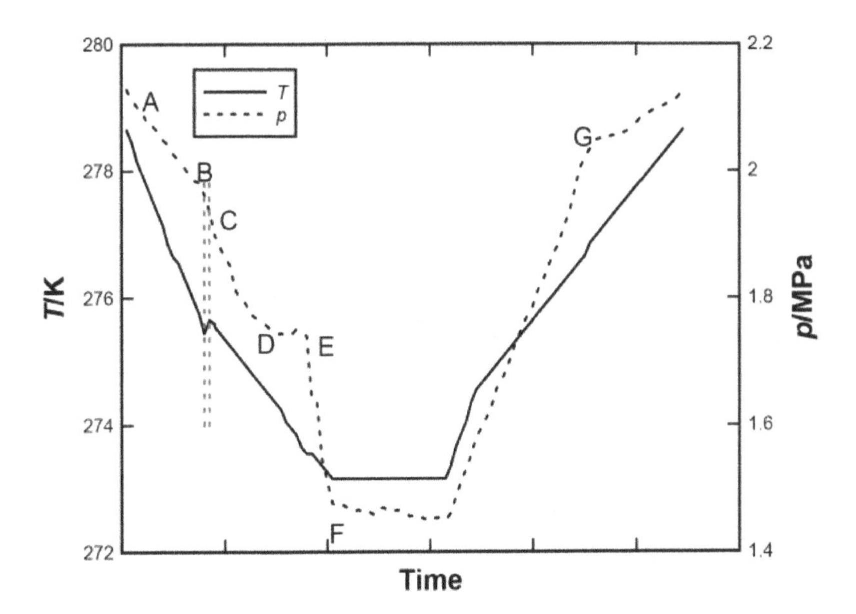

FIGURE 5: Typical curves of pressure and temperature changes with time for CO_2 hydrate in the mixed glass beads and silica gel: the broken line represents the value of pressure and the solid line indicates the temperature in the vessel.

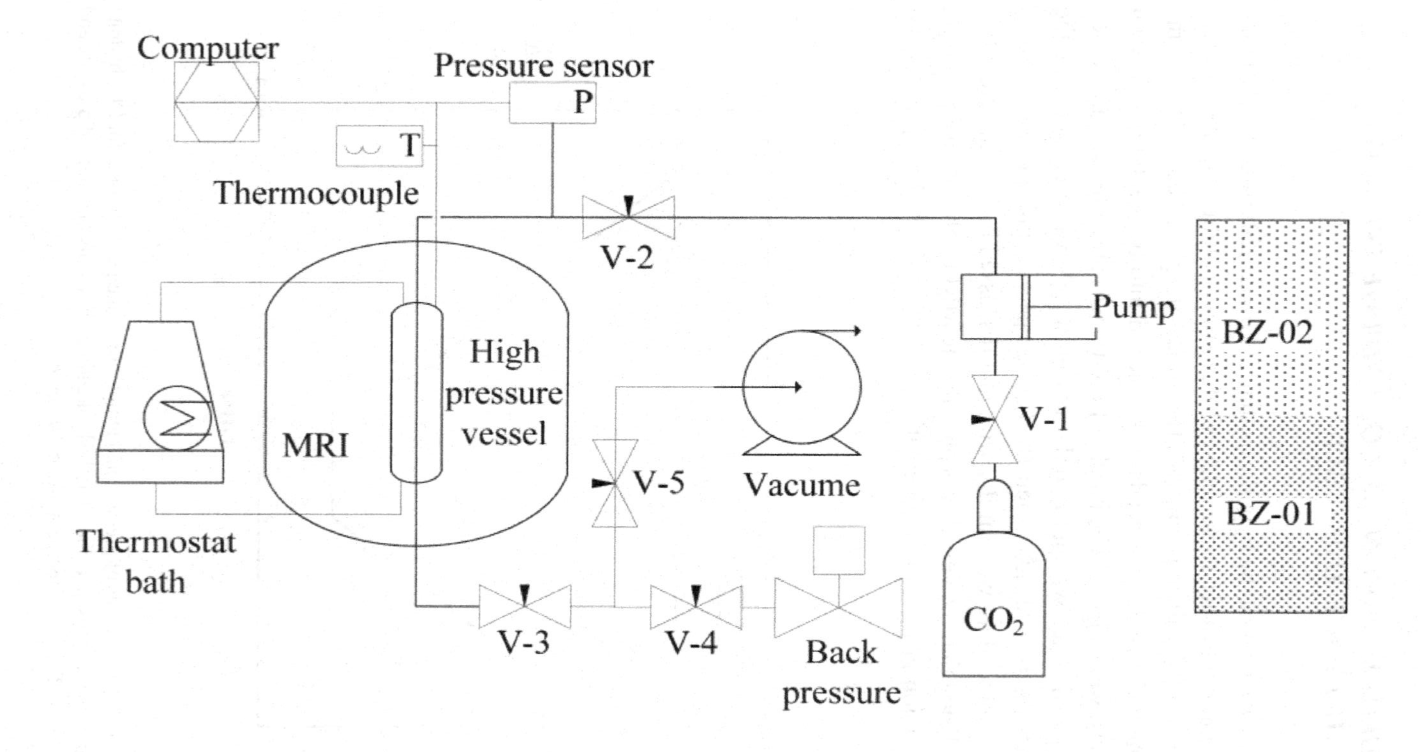

FIGURE 6: Schematic diagram of the MRI apparatus and the distribution of glass beads inthe high pressure vessel.

the distribution of glass beads in the high pressure vessel were shown in Figure 6. The experimental apparatus consisted of a high-pressure vessel, MRI system, data acquisition system, high-pressure pumps, and a low-temperature cooling system. The vessel is made of polyimide which is a non-magnetic material, and its design pressure is 12 MPa. The effective size for packed glass beads is Φ15 × 200 mm. The MRI (Varian) was operated at a resonance of 400 MHz, 9.4 Tesla, to measure hydrogen. The high-precision thermostat bath (F25-ME, produced by Julabo Labortechnik GmbH, Germany) filled with fluorocarbon (FC-40, supplied by 3M Company, USA) was used to control temperature precisely. 1H-MRI produces images of hydrogen contained in liquids, but does not image hydrogen contained in solids such as ice crystals or the CO_2 hydrates because of their much shorter transverse relaxation times. The detailed information for the MRI experiments was provided in previous work [27].

Images were obtained by using a fast spin-echo sequence, and the field of view was set to 40 × 40× 40 mm. Once the water existed in the vessel as liquid, the water distribution zone was bright in images obtained by MRI. When the water converted to hydrate (solid phase), the MRI cannot detect the water signals and the image becomes dark, so the sites that changed to dark in the image mean the formation of hydrate. Considering the randomicity of induction time, the experiment was carried out twice, the results are shown in Figure 7, and all of them indicate the CO_2 hydrate formed firstly in BZ-02 (the bigger glass beads) under these experimental conditions.

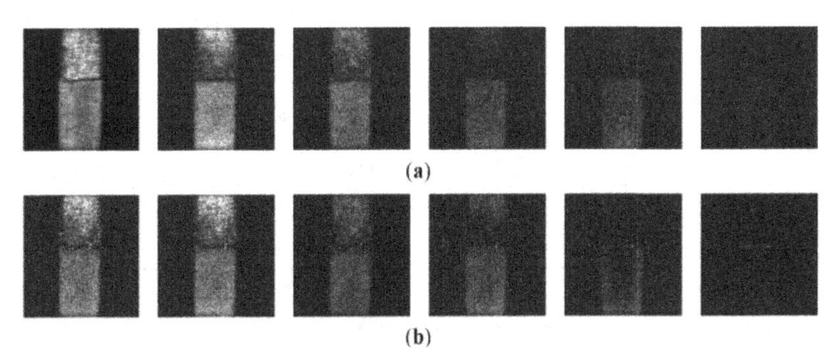

FIGURE 7: CO_2 hydrate formation process in BZ-01 and $BZ\text{-}0_2$ using MRI: (a) the first cycle; (b) the second cycle.

3.3.3 EFFECTS OF PORES SIZE ON CO_2 HYDRATE EQUILIBRIUM CONDITION

The presence of porous medium affected the phase equilibrium conditions of the CO_2-water-hydrate system. As shown in Figure 8, the hydrate equilibrium curve displayed a movement in the p-T diagram with the presence of glass beads and silica gel. The experimental data of CO_2 hydrate in bulk water obtained by Deaton et al., and the experimental data of CO_2 hydrate in porous medium with nominal 5.0 nm radii obtained by Smith et al. were quoted to compare with our data. The comparison clearly showed that the presence of porous medium with small-diameter pores affected the CO_2 hydrate equilibrium conditions when compared with that in bulk water. In other words, the capillary inhibition of porous medium makes the CO_2 hydrate equilibrium pressure increased at a certain temperature. The effects were mainly caused by the additional resistance effect of capillary surface tension, which leads to lower water activity and affects hydrate equilib-

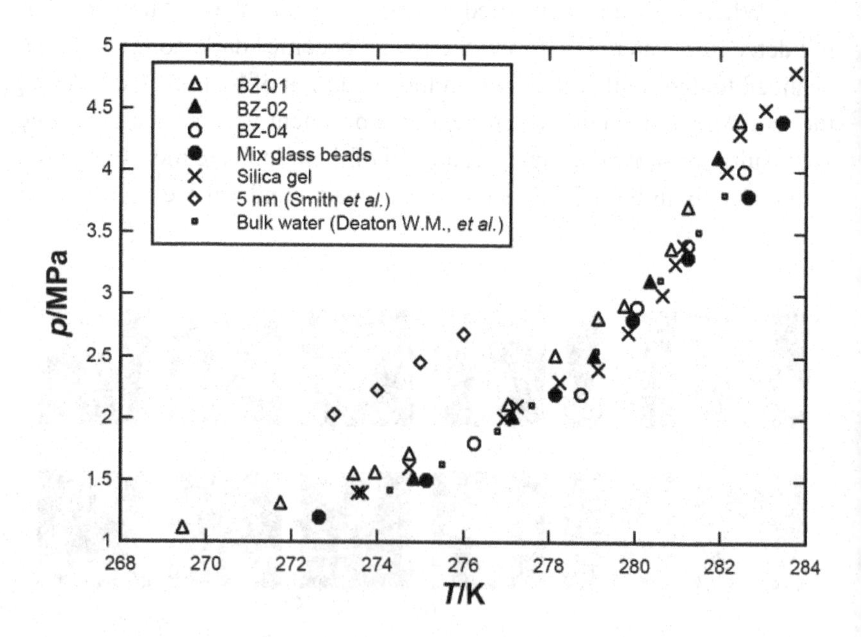

FIGURE 8: Measurement of equilibrium condition for CO_2 hydrate in different porous medium.

rium condition [28]. It is important to address the discrepancy of activity between bulk water and pore-water for understanding the hydrate equilibrium pressure increase in silica gel pores and glass beads pores reported in these study. By accounting for the effects of the pore size distribution on additional forces, we can conclude that the effect of the capillary force is to lower the activity of water in the pores [29]. Once the pore size was obtained, the equilibrium pressure of CO_2 hydrate in porous medium can be calculated. Based on the simulated theory, the increase of pore size caused the equilibrium pressure decrease as temperature was kept constant. When the pore size increased to some value, the effects of capillary force on hydrate equilibrium conditions becomes very small and can be ignored. The investigation of Turner et al. [23] indicated that the equilibrium temperature changes cannot be detected by the thermocouples in their equilibrium apparatus when the pores radius was larger than 600 Å.

3.3.4 PREDICTION OF HYDRATE EQUILIBRIUM CONDITION BY THE IMPROVED MODEL

The improved model of Song et al. [30] was used for predicting the equilibrium conditions for CO_2 hydrates in porous medium, which was based on the traditional model of van der Waals and Plateeuw [31]. In this model, the mechanical equilibrium of force between the interfaces in hydrate-liquid-vapor system was considered. The solubility of CO_2 gas was calculated using an empirical modification of Krichevsky–Kasarnovsky (K–K) equation [32]. This equation also includes the effect of pressure on the solubility. Having calculated the X_{gas}, X_w is calculated from:

$$X_W = 1 - X_{gas} = 1 - \frac{f_{gas}}{[\eta_3 \exp(p^{\eta_1} - \eta_2/RT)]} \tag{1}$$

where η_1, η_2, η_3 are equal to 0.3411, 22.33 and 899683.5 for CO_2, which can be found in the study of Nasrifar et al. [32].

Figure 9 shows the comparison between the experimental data and the prediction results. Considering the complexity of the system, the predictions show an acceptable agreement with the experimental data. The absolute average deviation of predicted temperature (Δ_{AADT}) and pressure (Δ_{AADP}) are defined as follows [33]:

$$\Delta_{AADT}=\left(\frac{1}{N_p}\right)\sum_{j=1}^{N_p}\left[\left|T_{cal}-T_{exp}\right|/T_{exp}\right]_j\times 100 \tag{2}$$

$$\Delta_{AADT}=\left(\frac{1}{N_p}\right)\sum_{j=1}^{N_p}\left[\left|P_{cal}-P_{exp}\right|/P_{exp}\right]_j\times 100 \tag{2}$$

where N_p denotes the number of data points.

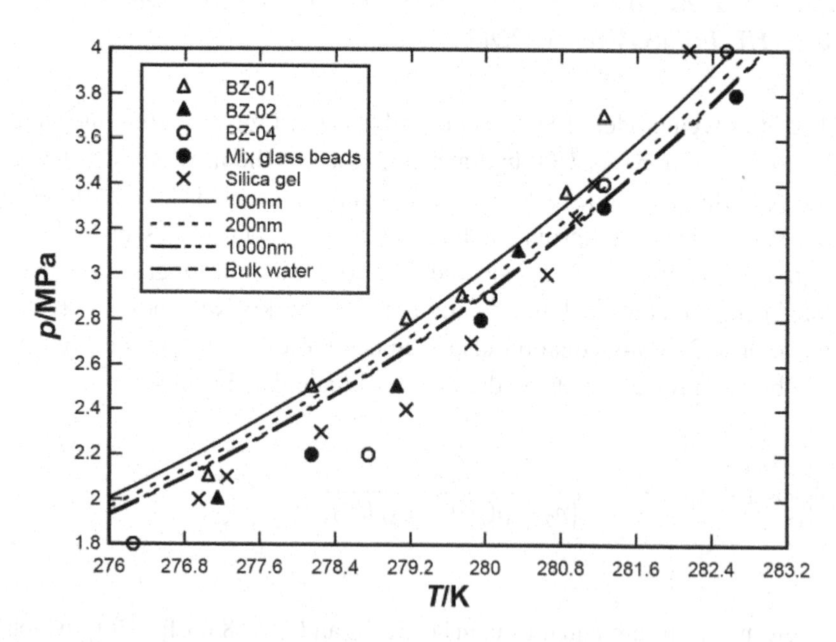

FIGURE 9: Comparison of measured and calculated equilibrium condition data for CO_2 hydrate in different porous media.

The error analysis of the predicted CO_2 hydrate equilibrium conditions for different porous media is shown in Table 2.

TABLE 2: Absolute average deviations of predicted CO_2 hydrate formation conditions.

Porous medium	T range/K	P range/MPa	Np	AADT (%)	AADP (%)
BZ-01	273.1–281.8	1.6–3.6	10	0.14	4.10
BZ-02	274.8–282.0	1.5–4.1	5	0.32	9.29
BZ-04 *	276.2–282.8	1.8–4.0	5	0.21	5.36
BZ-04 **	276.2–282.8	1.8–4.0	5	0.30	6.83
Mix glass beads *	275.2–282.7	1.5–3.8	5	0.20	8.78
Mix glass beads **	275.2–282.7	1.5–3.8	5	0.24	9.38
Silica gel *	274.8–282.2	1.6–4.0	10	0.20	5.82
Silica gel **	274.8–282.2	1.6–4.0	10	0.19	5.78

** comparison with calculation data of bulk water; ** comparison with calculation data of 1000 nm.*

The overall Δ_{AADT} for the improved model was 0.32%, respectively, and Δ_{AADP} were usually less 9.38%. There is a significant deviation between the measurements and the modeling results under both high pressure and low pressure conditions. However, this is not surprising due to the complexity of the system. The result that can be concluded without doubt is that the improved model gave good predictions for the CO_2 hydrate in porous media.

3.4 CONCLUSIONS

Experimental and calculated phase equilibrium conditions of carbon dioxide (CO_2) hydrate in porous media were investigated in this study. When CO_2 hydrate is formed in sediments with uniform pore size distribution, there is only one hydrate formation stage. When CO_2 hydrate is formed in sediments with non-uniform pore size distribution, there are two hydrate formation stages, and the MRI apparatus results with spin-echo sequence showed that the hydrate formed firstly in BZ-02 glass beads with the same pressure and temperature. The smaller the pore size corresponds to the lower hydrate equilibrium temperature at the same pressure. This was mainly caused by the additional resistance effect of capillary surface tension. An improved model, based on the traditional model of van der Waals and Plateeuw, was used to predict CO_2 hydrate equilibrium conditions, and the predictions showed good agreement with our experimental measurements.

REFERENCES

1. Ji, Y.H.; Ji, X.Y.; Feng, X.; Liu, C.; Lv, L.H.; Lu, X.H. Progress in the study on the phase equilibria of the CO_2-H_2O and CO_2-H_2O-NaCl systems. Chin. J. Chem. Eng. 2007, 15, 439–448.
2. Bachu, S. Sequestration of CO_2 in geological media: Criteria and approach for site selection in response to climate change. Energy Convers. Manag. 2000, 41, 953–970.
3. Unander, F. Energy Technology Perspectives: Scenarios and Strategies to 2050; Technical Report; International Energy Agency: Paris, France, 2008.
4. Sloan, E.D. Clathrate Hydrates of Natural Gases, 2nd ed.; Marcel Dekker Inc.: New York, NY, USA, 1998.
5. Yang, H.Q.; Xu, Z.H.; Fan, M.H.; Gupta, R.; Slimane, R.B.; Bland, A.E.; Wright, I. Progress in carbon dioxide separation and capture: A review. J. Environ. Sci. 2008, 20, 14–27.
6. Brewer, P.G.; Orr, F.M., Jr.; Friederich, G.; Kvenvolden, K.A.; Orange, D.L. Gas hydrate formation in the deep sea: In situ experiments with controlled release of methane, natural gas and carbon dioxide. Energy Fuels 1998, 12, 183–188.
7. Wendland, M.; Hasse, H.; Maurer, G. Experimental pressure-temperature data on three- and four-phase equilibria of fluid, hydrate and ice phases in the system carbon dioxide water. J. Chem. Eng. Data 1999, 44, 901–906.
8. Yang, S.O.; Yang, I.M.; Kim, Y.S.; Lee, C.S. Measurement and prediction of phase equilibria for water + CO_2 mixture in hydrate forming conditions. Fluid Phase Equilib. 2000, 175, 75–89.
9. Englezos, P.; Hall, S. Phase equilibrium data on carbon dioxide hydrate in the presence of electrolytes, water soluble polymers and montmorillonite. Can. J. Chem. Eng. 1994, 72, 887–893.
10. Breland, E.; Englezos, P. Equilibrium hydrate formation data for carbon dioxide in aqueous glycerol solutions. J. Chem. Eng. Data 1996, 41, 11–13.
11. Dholabhai, P.D.; Scott Parent, J.; Raj Bishnoi, P. Carbon dioxide hydrate equilibrium conditions in aqueous solutions containing electrolytes and methanol using a new apparatus. Ind. Eng. Chem. Res. 1996, 35, 819–823.
12. Dholabhai, P.D.; Scott Parent, J.; Raj Bishnoi, P. Equilibrium conditions for hydrate formation from binary mixtures of methane and carbon dioxide in the presence of electrolytes, methanol and ethylene glycol. Fluid Phase Equilib. 1997, 41,

13. Kang, S.P.; Chun, M.K.; Lee, H. Phase equilibria of methane and carbon dioxide hydrates in the aqueous $MgCl_2$ solutions. Fluid Phase Equilib. 1998, 147, 229–238.
14. Mohammadi, A.H.; Afzal, W.; Richon, D. Gas hydrates of methane, ethane, propane and carbon dioxide in the presence of single NaCl, KCl and $CaCl_2$ aqueous solutions: Experimental measurements and predictions of dissociation conditions. J. Chem. Thermodyn. 2008, 40, 1693–1697.
15. Handa, Y.P.; Stupin, D.Y. Thermodynamic properties and dissociation characteristics of methane and propane hydrates in 70-Å-radius silica gel pores. J. Phys. Chem. 1992, 96, 8599–8603.
16. Uchida, T.; Ebinuma, T.; Ishizaki, T. Dissociation condition measurements of methane hydrate in confined small pores of porous glass. J. Phys. Chem. B 1999, 103, 3659–3662.
17. Clennell, M.B.; Hovland, M.; Booth, J.S.; Henry, P.; Winters, W.J. Formation of natural gas hydrates in marine sediments 1: Conceptual model of gas hydrate growth conditioned by host sediment properties. J. Geophys. Res. 1999, 104, 22985–23003.
18. Henry, P.; Thomas, M.; Clennell, M.B. Formation of natural gas hydrates in marine sediments 2: Thermodynamic calculations of stability conditions in porous sediments. J. Geophys. Res. 1999, 104, 23005–23022.
19. Zatsepina, O.Y.; Buffet, B.A. Nucleation of CO_2 hydrate in porous medium. Fluid Phase Equilib. 2002, 200, 263–275.
20. Smith, D.; Wilder, J.; Seshadri, K. Thermodynamics of carbon dioxide hydrate formation in media with broad pore-size distributions. Environ. Sci. Technol. 2002, 36, 5192–5198.
21. Anderson, R.; Llamedo, M.; Tohidi, B.; Burgass, R.W. Experimental measurement of methane and carbon dioxide clathrate hydrate equilibria in mesoporous silica. J. Phys. Chem. B. 2003, 107, 3507–3514.
22. Kumar, A. Formation and Dissociation of Gas Hydrates in Porous Media. Master Thesis, University of Calgary, Calgary, Canada, 2005.
23. Turner, D.J.; Cherry, R.S.; Sloan, E.D. Sensitivity of methane hydrate phase equilibria to sediment pore size. Fluid Phase Equilib. 2005, 228–229, 505–510.
24. Aladko, E.Y.; Dyadin, Y.A.; Fenelonov, V.B.; Larionov, E.G.; Manakov, A.Y.; Mel'gunov, M.S.; Zhurko, F.V. Formation and decomposition of ethane, propane, and carbon dioxide hydrates in silica gel mesopores under high pressure. J. Phys. Chem. B 2006, 110, 19717–19725.
25. Song, Y.C.; Yang, M.J.; Liu, Y.; Li, Q.P. Influence of ions on phase equilibrium of methane hydrate. CIESC J. 2009, 60, 1362–1366.
26. Yang, M.J.; Song, Y.C.; Liu, Y.; Chen, Y.J.; Li, Q.P. Influence of pore size, salinity and gas composition upon the hydrate formation conditions. Chin. J. Chem. Eng. 2010, 18, 292–296.
27. Yang, M.J.; Song, Y.C.; Zhao, Y.C.; Liu, Y.; Jiang, L.L.; Li, Q.P. MRI measurements of CO_2 hydrate dissociation rate in a porous medium. Magn. Reson. Imaging 2011, 29, 1007–1013.
28. Wilder, J.W.; Seshadri, K.; Smith, D.H. Modeling hydrate formation in media with broad pore size distributions. Langmuir 2001, 17, 6729–6735.

29. Clarke, M.A.; Darvish, M.P.; Bishnoi, P.R. A method to predict equilibrium conditions of gas hydrate formation in porous media. Ind. Eng. Chem. Res. 1999, 38, 2485–2490.
30. Song, Y.C.; Yang, M.J.; Chen, Y.J.; Li, Q.P. An improved model for predicting hydrate phase equilibrium in marine sediment environment. J. Nat. Gas Chem. 2010, 19, 241–245.
31. Van der Waals, J.H.; Platteeuw, J.C. Clathrate solution. Adv. Chem. Phys. 1959, 2, 1–57.
32. Nasrifar, K. A model for prediction of gas hydrate formation conditions in aqueous solutions containing electrolytes and/or alcohol. J. Chem. Thermodyn. 2001, 33, 999–1014.
33. Mei, D.H.; Liao, J.; Yang, J.T.; Guo, T.M. Prediction of equilibrium hydrate formation conditions in electrolyte aqueous solutions. Acta Petrolei Sinica 1998, 14, 86–93.

PART II

GEOLOGICAL SEQUESTRATION OF CO_2

CHAPTER 4

GEOLOGICAL CARBON SEQUESTRATION: A NEW APPROACH FOR NEAR-SURFACE ASSURANCE MONITORING

LUCIAN WIELOPOLSKI

4.1 INTRODUCTION

Global warming and climate change are attributed to increases in the concentration of greenhouse gases (GHG) in the atmosphere, from anthropogenic emissions of CO_2, from the pre-industrial revolution level of about 260 ppm, to present day concentrations of about 391 ppm, viz., ~35% increase [1]. The main sources of GHG emissions are associated with burning fossil fuels, changing land usage, and cultivation of the soil. To combat global climate change will require a combination of approaches including improving energy efficiency and using alternative energy sources. Predictions of the increased use of energy globally during this century and continued reliance on fossil fuels point to a further rise in GHG emissions [2] with a concomitant one in atmospheric CO_2 concentrations. These consequences cannot be abated unless major changes are made in the way energy is produced and used; in particular, how carbon is managed [3,4]. Mitigating the forecast increase in fossil-fuel consumption includes pro-

---

This chapter was originally published under the Creative Commons Attribution License. Wielopolski L. Geological Carbon Sequestration: A New Approach for Near-Surface Assurance Monitoring. International Journal of Environmental Research and Public Health ***2011**,8 (2011). doi:10.3390/ ijerph8030818.*

ducing clean fuels, capturing industrially generated CO_2, and sequestering this CO_2 in deep geologic formations (carbon capture & sequestration (CCS)). The attractiveness of the CCS program stimulated significant investments by governments and the private sector to develop the necessary technologies, and to evaluate whether CO_2 control could be implemented safely and effectively to maintain the CO_2 in reservoirs. The United States Department of Energy (USDOE) prepared a roadmap for the CCS program [5]. The program's early planners recognized the potential risks of geological storage to humans and ecosystems that might arise from leaking injection wells, abandoned wells, across faults, and from ineffective confining layers. Hence, cost-effective, robust monitoring must be an integral part of and specifically designed for every individual CCS project.

Monitoring the status and the fate of a CO_2 plume from geological carbon sequestration (GCS) projects is mandatory as stipulated by the Environmental Protection Agency's (EPA's) permitting processes for underground injections [6-9]. The monitoring generally falls into two types; monitoring deep reservoirs to confirm their stability and integrity, and, monitoring above the reservoir, i.e., near-surface monitoring (NSM) of water, air, and soil to assure public health and environmental safety. The IPCC and the USDA reports outline these two domains, differing in their objectives and the instrumentation required for monitoring [10,11], as depicted schematically in Figure 1. In general, the IPCC guidelines [10] stipulate a 99% reservoir-retention capacity over a 1,000 year period. That, for a 200 Mt CO_2 reservoir, translates into a yearly acceptable leak of 2,000 t/year or ~5.5 t/day. Considering the surface area of a reservoir through which a gas could leak, its tortuous passage and dispersion on its movement from a depth of several thousand feet to the surface, we would expect very low fluxes of CO_2 to be evident at the surface. The exceptions might be leaks occurring near injection- and abandoned-wells, or known geological faults. Many of the well-established techniques of monitoring CO_2 in the atmosphere and in the near-surface areas were adopted directly for assessing leaks from geological carbon-sequestration sites in spite of their inadequate sensitivities and point measurements in space and time. Table 2 of the USDOE's report summarizes their basic characteristics and the challenges they pose for detecting low-level signals [11]. Thus, current instrumentation faces a double challenge of reducing the minimum detectable

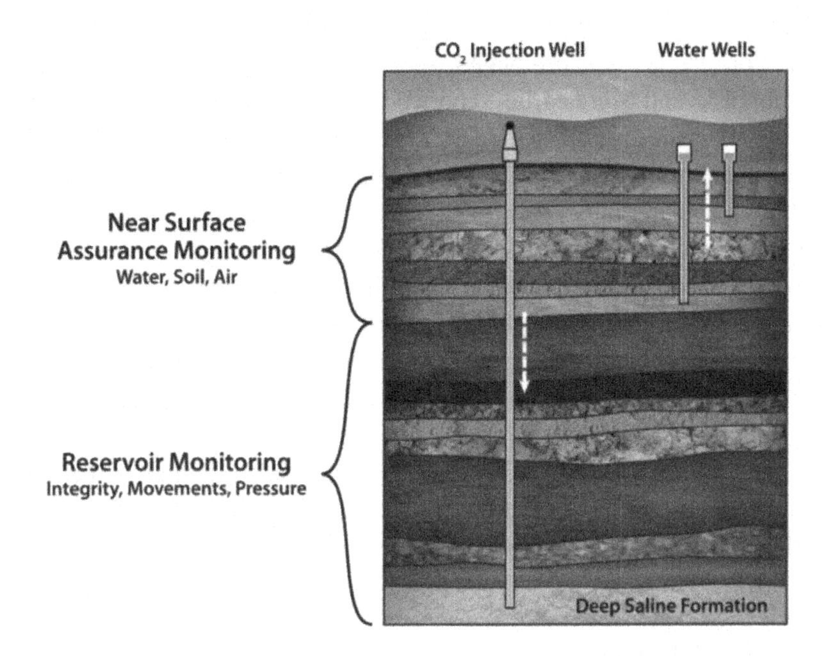

FIGURE 1: Scheme of two monitoring regions: One, near the surface for assurance monitoring; and, two, deep monitoring for evaluating the reservoir's integrity.

limit (MDL) with minimum detectable change (MDC), and distinguishing real changes from natural ones due to seasonal- and diurnal-variations in the field CO_2 fluxes. Point measurements might well be inadequate when the location of the leak is unknown, so that it probably is necessary to couple them with line- and area-integrated CO_2 measurements, or design sensor networks to cover the area [12,13].

To address the hurdles of the MDL, field natural variability and point measurements, a new approach that, rather than directly measuring the fluxes of seeping CO_2, measures a secondary quantity, namely total carbon in soil (TOC). Since the soil's CO_2 levels affect its pH and the activity of the plants' roots it contains, they influence the TOC levels. Hence, a slow CO_2 seepage will increase cumulatively the soil CO_2 content inversely impacting TOC. Lower noise and reduced natural variability surrounding the TOC, lowering the MDL levels is enabled. Measurements of TOC offer a temporal- and spatial-integration of the impact of prolonged low seepage of CO_2. Time integration is accomplished by measuring the cumulative effect on the TOC of prolonged exposure to changes in soil CO_2 [14]; Wielopolski and Mitra earlier reported such a decrease in TOC [15]. Others detailed the overall degradation of vegetation caused by CO_2 leaks from underground CO_2 springs in Mammoth Mountain, California, and in Latera caldera, Italy [16,17]. This paper emphasizes the benefits of the error reduction of the proposed new system and of using unique scanning capacity of the inelastic neutron scattering (INS) system for spatially integrated monitoring. Thus, the hypothesis tested is that a CO_2 leak would impact the vegetation and result in a near surface carbon suppression; like in the vicinity of natural CO_2 vents; and the objectives are to demonstrate the validity of the hypothesis and suitability of the INS to measure these changes. INS system is briefly described and the reduction in the error propagation and lowering of the MDL and MDC are outlined. Theoretically, both can be reduced to reasonably low levels.

4.2 SITE AND SETUP

4.2.1 SITE

The applicability of INS for monitoring GCS was demonstrated at the zero emission research and technology (ZERT) facility located on a former agricultural plot at the western edge of the Montana State University-Bozeman campus, Bozeman, Montana, USA. This facility was established for testing and tuning instrumentation for studying near-surface CO_2 transport and detection under controlled conditions. The site, located at an elevation of 1,495 m, is covered with vegetation consisting primarily of alfalfa (*Medicago sativa*), yellow blossom sweet clover (*Meliotus officinalis*), dandelion (*Taraxacum officinale)*, Canada thistle (*Cirdium arvense*), and a variety of grasses (family *Poaceae*). The field is typical of the Bozeman area, with alluvial sandy gravel deposits overlain by a few meters of silts and clays with a blanket of topsoil. There are two distinct soil horizons; a topsoil, some 0.2 to 1.2 m thick, of organic silt, clay, and some sand,

and an underlying deposit of sandy gravel extending down to about 5 m. Carbon-dioxide was introduced through a 100 m long horizontal well installed between 1 and 2.5 m deep, and injected at a rate of 0.3 tons per day for twenty eight days; Spangler et al., give more detailed information on the site and injection system [18,19]. Figure 2 shows the site with the CO_2 storage tank, and the transport line to a control hut that regulates and monitors the flow through the horizontal well. The hot spots indicate regions of high CO_2 flow that degraded the vegetation.

4.2.2 INS SYSTEM

The INS method is based on spectroscopy of gamma rays induced by fast (14 MeV) neutrons interacting with the elements present in soil via inelastic neutron scattering and thermal neutron capture processes. The INS

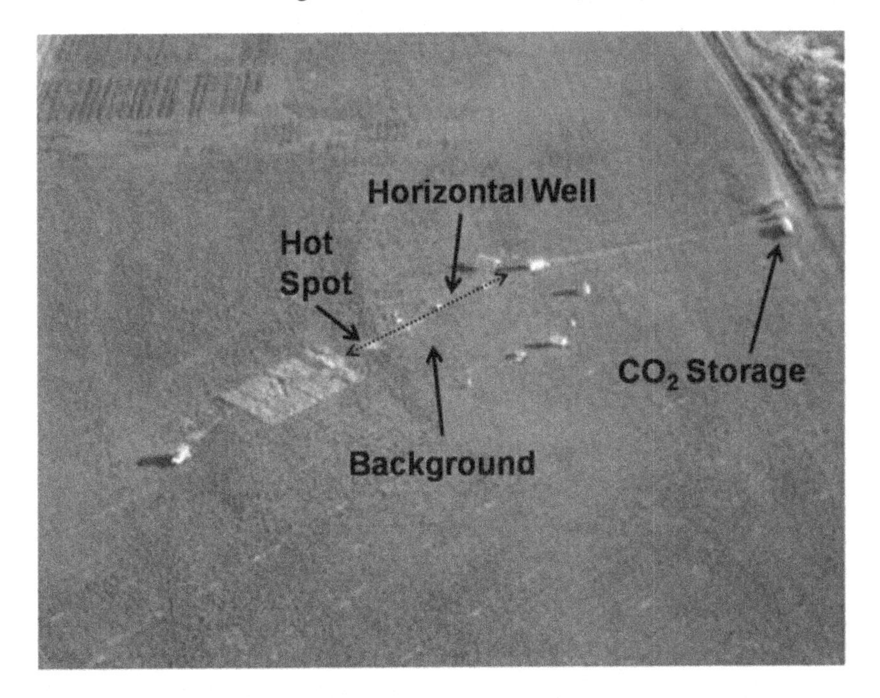

FIGURE 2: Site of the ZERT facility showing: a CO_2 storage tank, a flow control hat, and the location of the horizontal well. It also shows the measurement sites over a hot spot and the background region.

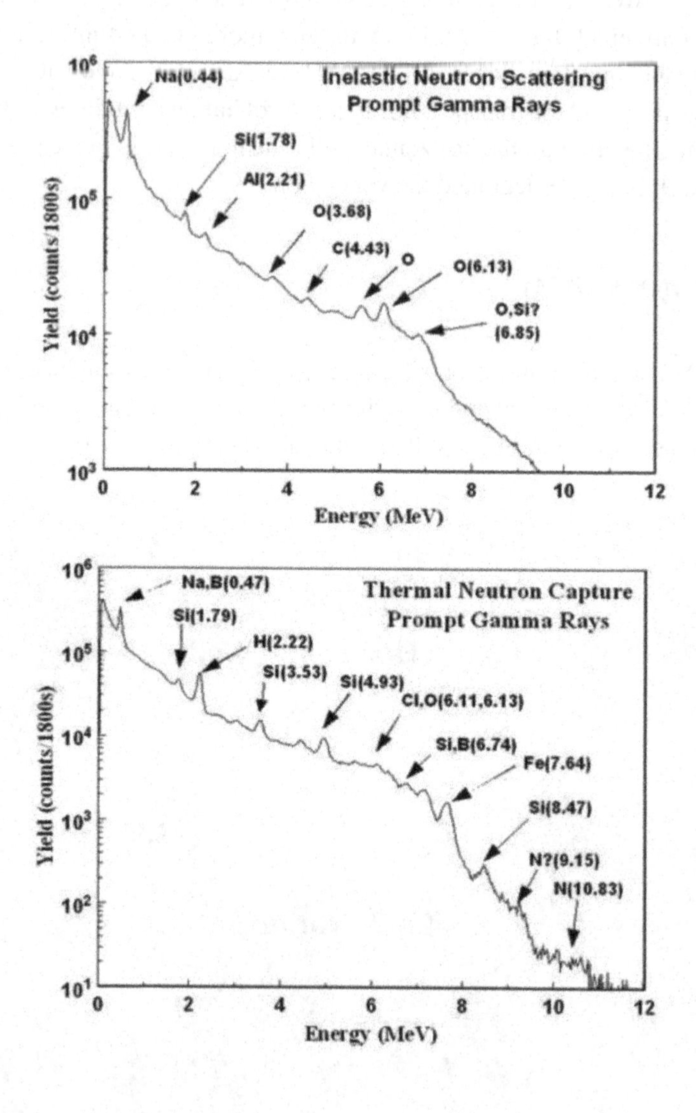

FIGURE 3: Typical gamma-ray spectra induced by inelastic neutron scattering and thermal neutron capture reactions.

system consists of a neutron generator (NG) that is turned off at the end of the data acquisition, detection and spectroscopy systems, and a power supply, all of which are mounted on a cart about 30 cm above the ground, thus enabling use in stationary or scanning modes of operation. Analysis and calibration of the characteristic elemental gamma-ray spectra resulting from inelastic neutron scatterings and thermal neutron captures (Figure 3) provide quantitative information on elemental concentrations in soil. The INS system interrogates large soil volume of about 0.3 m^3 to an effective depth of ~30 cm, as detailed by Wielopolski et al. [15,20]. The linear correlation between INS signal counts and carbon concentration was demonstrated in synthetic soils [21] and in natural fields using soil chemical analysis [22,23]. Thus, the net number of counts in the carbon peak can be expressed in terms of surface carbon concentration (g C/cm^2) using the slope of a regression line. Similarly, INS system's signal resulting from scanning capabilities, a key feature for spatial averaging, is converted to carbon content using the same calibration line. This is pertinent for detecting low level signals over large areas where the actual location of the leak is unknown. Uniquely, the error and MDL in the INS system can be lowered by extending the counting time or increasing the system's sensitivity, i.e., by increasing the number of detectors. These features are demonstrated in the following section on spectral analysis.

The soil carbon measurements at the ZERT facility were taken by placing the INS system above a "hot spot", marked in Figure 2 that was impacted by CO_2 leakage from the horizontal well. These measurements were compared with those taken away from the horizontal well.

4.3 SPECTRAL ANALYSIS

Statistics of nuclear counting follows a binomial distribution, which for a large number of counts $N > 12$ can be approximated by a normal distribution with a mean value, N, and standard deviation (SD) the square-root of N (sqrt(N)) [24]. By extension, in nuclear spectroscopy, the gamma-ray events in the detector are represented by the number of counts falling into contiguous energy intervals (channels). Figure 4 depicts a partial spectrum with expanded energy intervals where interest lies with the number of counts in the energy interval 'ab' embracing a carbon peak. The total number of counts in that energy interval T_t following T minutes of counting time is due to unknown incident signal counting rate S_r times T, and the background counting rate B_r times T. Thus, $T_t = S_rT + B_rT$ in which B_rT is the area of a trapezoid 'abcd' marked in Figure 4. Conversely, the net number of counts associated with an element (E) of interest, S_rT, is given by the difference $T_t - B_rT$. The INS's net counts are converted to conventional units of areal density (g E/m^2) by dividing the net signal by the sensitivity of the system, s, defined as the number of counts acquired during a counting period T, S_rT, per gram element per unit area; k is proportionality constant with matching units of g E/m^2. Thus $s = S_rT/k$, which also is the slope of the regression line that correlates INS yield versus the soil's carbon concentration. The experimentally determined quantities B_r, S_r and s represent the key performance parameters of an INS system from which other parameters are derived. Using the general uncertainty estimator of a function f(x,y,z…) given, to a first approximation, by Equation 1 [25],

$$\sigma_f^2 = \sum_i \left[\sigma_i^2 \left(\frac{\delta f}{\delta x}\right)_i^2\right] \quad (1)$$

It is possible to derive the SD of S_rT as $\sigma_S = \sqrt{(T_{tot} + B_rT)} = \sqrt{((S_r + 2B_r)T)}$. The minimum detection limit (MDL) is defined as the number of counts above the background that differs from the background by a given confidence level; for example for a 99% confidence level the peak must contain three standard deviation counts above the background, and thus we can write:

$$MDL = 3 \times \sqrt{(Br \times T)} \text{ (counts)} \quad (2)$$

Further, the relative SD for a signal at the MDL level, RSDMDL, is given by σ_{MDL}/MDL_c, Equation 3,

$$RSDMDL = \sqrt{[2/9 + 1/3sqrt(B_rT)]} \quad (3)$$

The RSDMDL, plotted in Figure 5, is bound between 0.745 for $B_rT = 1$ and approaches asymptotically 0.471 for $B_rT \rightarrow \infty$, B_r or T can be changed independently.

The elemental density corresponding to the number of counts given in Equation 2 is obtained by dividing Equation 2 by s, thus,

$$MDL_E = (3 \times k/S_r) \times \sqrt{}(B_r/T)/s \text{ (g E/cm}^2) \quad (4)$$

Similarly, the minimum detectable change (MDC) defined as a change of three standard deviations in the signal level error, we can write,

$$MDC = 3 \times \sqrt{}((S_r + 2B_r) \times T) \text{ (counts)} \quad (5)$$

and, in terms of elemental concentration,

$$MDC_E = (3 \times k/S_r) \times \sqrt{}((S_r + 2B_r)/T)/s \text{ (g E/cm}^2) \quad (6)$$

From Equations 4 and 6, it is apparent that increasing the counting time reduces the MDL_E and the MDC_E. Similarly, increasing the sensitivity of s or Sr, the signal counting-rate, by increasing the number of detectors also will lower the MDL_E and MDC_E. Finally, reducing the background counting-rate by improving the shielding of the system also will lower MDL_E and MDC_E. These features are graphed in Figure 6.

4.4 RESULTS

Soil carbon measurements were taken over two 28-day injections episodes, in 2008 and in 2009. The soil carbon levels were measured above a HS pre- and post-injection and away from the horizontal well. No chemical analysis of soil samples were performed in order not to disturb the soil

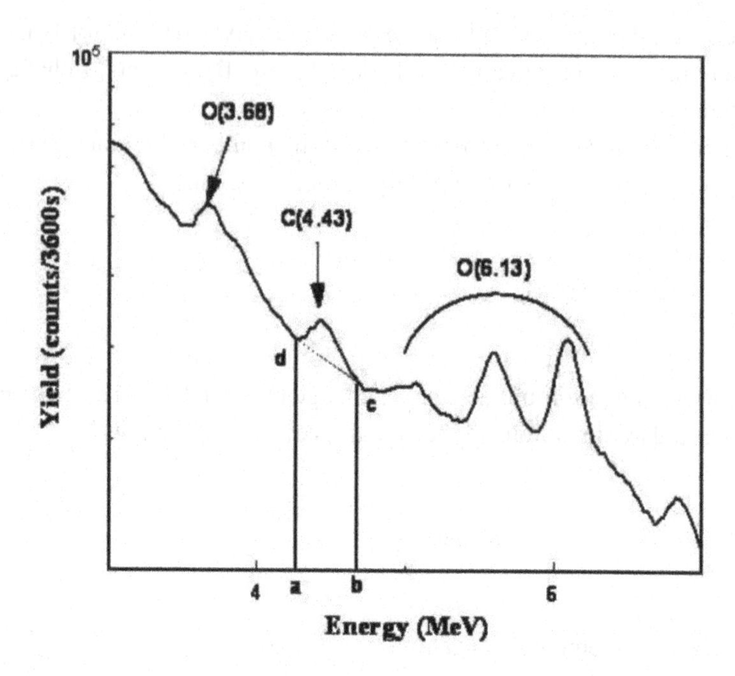

FIGURE 4: Partial gamma-ray spectrum in which an energy interval "ab," located under the carbon photopeak, marks the boundary of the total counts, Tt, and a background counts enclosed by the trapezoid's area "abcd."

CO_2 flow conditions. The net carbon yields, taken over one hour show a drop in soil carbon levels above a hot spot while simultaneously demonstrating no changes in silicon, oxygen and other elements in the background or above the HS; Table 1 shows the net counts in silicon (Si), oxygen (O) and carbon peaks [15]. To plot the graphs given in Equations 2, 4, and 6 the background count-rate, B_r, was averaged over the two injection episodes, Table 2. The lower background in 2009 is attributed to the malfunctioning of one of the three detectors, thus reducing the background by about a third. Correcting for this anomaly in 2009, the estimated mean background rate, Br, was about 50,000 counts/min, and the sensitivity, s, was approximately 1,500 counts/min/(kg C/m^2). Using these values the relative SD of a signal at the level of the detection limit given by Equation 3 is plotted versus time (Figure 5). Using the same values for B_r and s, the MDL_E and MDC_E, were calculated using Equations 4 and 6, respectively, and plotted in Figure 6. Quadrupling the number of detectors quadruples the signal and the background reducing the MDL_E and MDC_E by a factor of two. This is shown by the graph MDL_E-4Det in Figure 6.

TABLE 1: Analyses of the Si, O, and C peaks of the INS spectra measured during the 2008 and 2009 injection periods. Measurements were taken at a hot spot (HS), and the background (B) was determined in 2008 off the horizontal well; in 2009 it was determined off the well and at a pre-injection HS.

	2008—During Injection					
	Hot Spot (HS)			Background (B)		
	Si	O	C	Si	O	C
N	8	8	7	12	12	11
Mean	1,031,332	622,914	47,137	1,008,545	636,929	53,704
STD Deviations	24,858	28,328	3,610	16,530	48,235	4,731
STD Deviations (%)	2.4	4.5	7.4	1.6	7.5	8.8
STD Error (%)	0.8	1.6	2.8	0.5	2.2	2.7
$\Delta(1 - HS/B) \times 100$	2.3	−2.2	−14.0	---	---	---
	2009—Pre-Injection					
	Hot Spot			Background		
	Si	O	C	Si	O	C
N	9	9	8	5	5	5
Mean	787,977	650,746	79,728	759,986	665,833	81,228
STD Deviations	15,066	13,714	4,850	5,860	5,811	3,916
STD Deviations (%)	1.9	2.1	6.1	0.8	0.9	4.8
STD Error (%)	0.6	0.7	2.2	0.4	0.4	2.1
$\Delta(1 - HS/B) \times 100$	3.7	−2.3	−1.9	---	---	---
	2009—Post-Injection					
	Hot Spot			Background		
	Si	O	C	Si	O	C
N	9	9	8	3	3	3
Mean	842,562	628,521	78,850	812,448	635,948	84,718
STD Deviations	19,889	7,117	6,079	5,751	8,725	4,566
STD Deviations (%)	2.4	1.1	7.6	0.7	1.4	5.4
STD Error (%)	0.8	0.4	2.7	0.4	0.8	3.1
$\Delta(1 - HS/B) \times 100$	3.7	−1.2	−6.9	---	---	---

TABLE 2: Mean background counts during 2008 and 2009, and combined over two years; n is the number of measurements, SDEV is the standard deviation, and CV is the coefficient of variation (SDEV/sqrt(n)).

Year	2008	2009	Combined
n	20	27	47
Mean	3,338,759	2,102,343	3,232,342
SDEV (%)	43,119 (1.29)	15,384 (0.73)	65,455 (2.03)
CV (%)	9,642 (0.29)	2,961 (0.14)	9,548 (0.30)

4.5 DISCUSSION

Ideally no underground leakage of CO_2 should be occurring from underground reservoirs regardless of their size. However, practically, some very low leaks in the order of 0.01% over the expected life-time of a reservoir may be acceptable. The dispersion of the leaks over the reservoir's surface area and their dilution during migration toward the surface would result in very low changes in the surface fluxes. These amounts are below the detection limits of the current instrumentation that was tuned at test facilities operating with higher fluxes. Furthermore, current instrumentation provides point measurements in time and space. At potential leak sites, this instrumentation is being used near injection- and old abandoned-wells, and possibly along known faults. The concerns with MDLs and with covering the entire area above the reservoir, which may amount to hundreds of square-kilometers, continually are addressed by developing new improved instrumentation. One new approach is to monitor secondary parameters that are affected by CO_2 fluxes or, alternatively, combining a few modalities to improve the signal-to-noise ratio. Examples of secondary quantities include the quality of the drinking water, reflectance spectroscopy of the vegetation above-ground, and impact on the species forming the vegetation. However, noise levels and natural fluctuations continue to pose problems.

Monitoring carbon in soil, using an INS system, is yet another indirect method to detect possible leaks from deep reservoirs. The viability of INS was demonstrated by detecting a drop in the soil's carbon levels following fumigation with CO_2. The uniqueness of INS approach offers time integration of a cumulative effect of a low leak that slowly influences

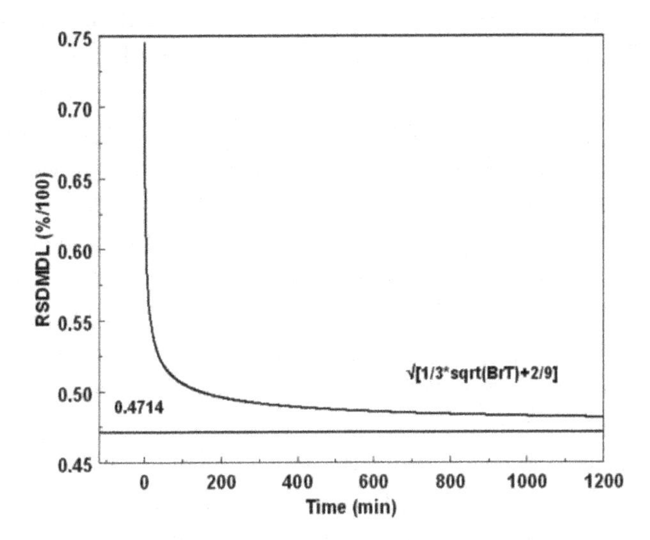

FIGURE 5: Relative standard deviation of a minimum detection limit signal (RSDMDL) based on Equation 3; it is bounded at 75%, and asymptotically approaches 47.14% for long counting times.

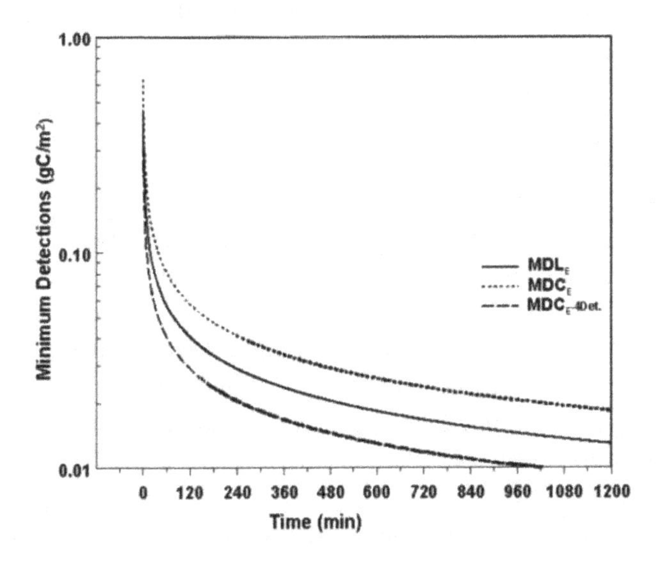

FIGURE 6: Minimum detection limit, MDL, and minimum detection change, MDC, respectively, based on Equations 4 and 6 versus counting time T. Increasing the sensitivity also impacts the MDC.

the vegetation and near-surface pH levels that, in turn, alter the carbon level. The non-destructive measurements made by INS enable us to acquire sequential readings in exactly the same spot. Its sensitivity is further enhanced by the ability to measure large volumes of soil when operating in static- and scanning-modes; in principle, this enables coverage of the entire area above the reservoir, thus providing spatial averaging of the signal from the entire site. These features are well suited for monitoring possible changes in the soil carbon for potential leaks in any location. In addition, a very unique feature of INS is that we can reduce errors and lower the detection limits by extending the counting time, increasing the sensitivity of the system, or lowering the background, thus enhancing the capacity of INS to detect potential CO_2 leaks.

The elemental peaks shown in Table 1 do not exhibit the same drop in 2009 as does the background in Table 2. The reason for this is not completely clear. It is speculated that, since the background radiation is more multidirectional than the specific peaks that originate in the soil, this may have to do with geometric factors depending on which detector malfunctioned, viz., the middle one or one of the side detectors. More experiments are needed to clarify this difference in response, as are others to determine the threshold values at which CO_2 fluxes begin to affect the vegetation and near-surface carbon storage.

4.6 SUMMARY

The hypothesis that leaking CO_2 suppresses the near surface carbon was validated and suitability of the INS system to measure these changes in soil was demonstrated. INS is a unique addition to the arsenal of tools for monitoring geological carbon sequestration. This new approach using INS offers the possibility of temporal-spatial integration, thus enhancing the capability for detecting low-level leaks. In addition, the paper detailed how the measurement error, MDL_E and MDC_E, can be reduced by extending the counting time and increasing the system's sensitivity. INS alone or in combination with other system will improve monitoring capabilities and enhance the success of the CCS programs. It would be highly desirable to perform controlled experiments in which soil CO_2 levels are doubled and record the threshold levels impacting the vegetation and TOC. These would have to be performed with different soil types.

REFERENCES

1. Tans, P. Trends in Atmospheric Carbon Dioxide—Mauna Loa; US Department of Commerce/ National Oceanic and Atmospheric Administration: Boulder, CO, USA. Available online: http://www.esrl.noaa.gov/gmd/ccgg/trends/ (accessed on 9 March 2011).
2. International Energy Outlook 2007; Energy Information Administration, US Department of Energy (EIA): Washington, DC, USA, May 2007. Available online: http://www.eia.doe.gov/ oiaf/ieo/ (accessed on 9 March 2011). .
3. Socolow, R.; Hotinski, R.; Greenblatt, J.B.; Pacala, P. Solving the climate problem—Technologies available to curb CO_2 emissions. Environment 2004, 46, 8-19.
4. Greenblatt, J.B.; Sarmiento, J.L. Variability and climate feedback mechanisms in ocean uptake of CO_2. In The Global Carbon Cycle: Integrating Humans, Climate, and the Natural World; Field, C.B., Raupach, M.R., Eds.; Island Press: Washington, DC, USA, 2004.
5. Carbon Sequestration Technology Roadmap and Program Plan; United States Department of Energy/Office of Fossil Energy/National Energy Technology Laboratory (USDOE/FE/NETL): Pittsburgh, PA, USA, 2007. Available online: http://www.netl.doe.gov (accessed on 9 March 2011).
6. Using the Class V Experimental Technology Well Classification for Pilot GS Projects; UIC Program Guidance (UICPG #83); Environmental Protection Agency (EPA): Washington, DC, USA, 2007.
7. Core Energy, LLC Class V UIC Injection Permit; Permit No. MI-137-5X$_2$5-0001; Environmental Protection Agency (EPA): Washington, DC, USA, draft version issued July 2008.
8. Underground Injection Control Program (UIC); Environmental Protection Agency (EPA): Washington, DC, USA, 12 February 2008. Available online: http://www.epa.gov/safewater/ uic/index.html (accessed on 8 June 2008).
9. Geologic Sequestration of Carbon Dioxide; Environmental Protection Agency (EPA): Washington, DC, USA, 2008. Available online: http://www.epa.gov/safewater/uic/ wells_sequestration.html (accessed on 9 March 2011).
10. International Panel on Climate Change (IPCC). IPCC Special Report on Carbon Doxide Capture and Storage; Metz, B., Davidson, O., de Coninck, H., Loos, M., Meyer, L., Eds.; Cambridge University Press: Cambridge, UK, 2005; p. 195.
11. Best Practices for: Monitoring, Verification, and Accounting of CO_2 Stored in Deep Geologic Formation; DOE/NETL-311/081508; United States Department of Energy/Office of Fossil Energy/National Energy Technology Laboratory (USDOE/FE/NETL): Pittsburgh, PA, USA, 2009.
12. Lewicki, J.L.; Hilley, G.E.; Laura Dobeck, L.; Spangler, L. Dynamics of CO_2 fluxes and concentrations during a shallow subsurface CO_2 release. Environ. Earth Sci. 2010, 60, 285-297.

13. Saripalli, P.; Amonette, J.; Rutz, F.; Gupta, N. Deign of sensor networks for long term monitoring of geological sequestration. Energy Convers. Manage. 2006, 47, 1968-1974.
14. West, J.M.; Pearce, J.M.; Coombs, P.; Ford, J.R.; Scheib, C.; Colls, J.J.; Smith, K.L.; Steven, M.D. The impact of controlled injection of CO_2 on the soil ecosystem and chemistry of an English lowland pasture. Energy Procedia 2009, 1, 1863-1870.
15. Wielopolski, L.; Mitra, S. Near-surface soil carbon detection for monitoring CO_2 seepage from a geological reservoir. Environ. Earth Sci. 2010, 60, 307-31_2.
16. Anderson, D.E.; Farrar, C.D. Eddy covariance measurement of CO_2 flux to the atmosphere from the area of high volcanogenic emissions, Mammoth Mountain, California. Chem. Geol. 2001, 177, 31-42.
17. Annunziatellis, A.; Beaubien, S.E.; Bigi, S.E.; Ciotoli, S.; Coltella, G.; Lombardi, M. Gas migration along fault systems and through the vadose zone in the Latera caldera (central Italy) implications for CO_2 geological storage. Int. J. Greenh. Gas Control 2008, 2, 253-372.
18. Spangler, L.H.; Dobeck, L.M.; Repasky, K.S.; Nehrir, A.; Humphries, S.; Barr, J.; Keith, C.; Shaw, J.; Rouse, J.; Cunningham, A.; et al. A controlled field pilot for testing near surface CO_2 detection techniques and transport models. Energy Procedia 2009, 1, 2143-2150.
19. Spangler, L.H.; Dobeck, L.M.; Repasky, K.S.; Nehrir, A.R.; Humphries, S.D.; Barr, J.L.; Keith, C.J.; Shaw, J.A.; Rouse, J.H.; Cunningham, A.B.; et al. A shallow subsurface controlled release facility in Bozeman, Montana, USA, for testing near surface CO_2 detection techniques and transport models. Environ. Earth Sci. 2010, 60, 227-239.
20. Wielopolski, L.; Hendrey, G.; Johnsen, K.; Mitra, S.; Prior, S.A.; Rogers, H.H.; Torber, H.A. Nondestructive system for analyzing carbon in the soil. Soil Sci. Soc. Am. J. 2008, 72, 1269-1277.
21. Wielopolski, L.; Mitra, S.; Hendrey, G.; Orion, I.; Prior, S.; Rogers, H.; Runion, B.; Torbert, A. Non-destructive Soil Carbon Analyzer (ND-SCA); BNL Report No.72200-2004; Brookhaven National Lab: Upton, NY, USA, 2004.
22. Wielopolski, L.; Johnston, K.; Zhang, Y. Comparison of soil analysis methods based on samples withdrawn from different volumes: Correlations versus Calibrations. Soil Sci. Soc. Am. J. 2010, 74, 812-819.
23. Wielopolski, L.; Chatterjee, A.; Mitra, S.; Lal, R. In Situ Determination of Soil Carbon Pool by Inelastic Neutron Scattering: Comparison with Dry Combustion. Geoderma 2010, 160, 394-399.
24. Evans, R.D. The Atomic Nucleus; McGraw-Hill Book Company: New York, NY, USA, 1955; p. 746.
25. Bevington, F.P.; Robinson, D.K. Data Reduction and Error Analysis for the Physical Sciences; McGraw-Hill Book Company: New York, NY, USA, 1969; p. 56.

CHAPTER 5

ENZYMATIC CARBON DIOXIDE CAPTURE

ALAIN C. PIERRE

5.1 INTRODUCTION

One of the main problems our world is presently facing, concerns the capture of anthropic carbon dioxide rejected in the atmosphere by human activities. This gas is considered as one of the main atmospheric components responsible for a greenhouse effect and an increase of the earth atmosphere temperature [1, 2], with many unwanted consequences, including the development of infectious diseases [3]. According to a report by the International Panel on Climate Change (IPCC) on the earth climate evolution, dating from 2007, the release of this gas in the atmosphere has increased by 80% from 1970 to 2004 and it accounted for 76.7% of the "Greenhouse Effects Gases" in 2004 [4]. An international agreement termed the "Kyoto Protocol," established by the United Nations Framework Convention on Climate Change, was initially signed in 1997 by 37 countries in order to reduce greenhouse gas (GHG) emissions [5]. This treaty was enforced in 2005 and the number of countries who ratified the convention increased to 191 in 2011. The target was to reduce the CO_2 emission by an amount depending on the country by comparison with a defined basis (8% in Europe, 7% in USA), over the five-year period 2008–2012.

This chapter was originally published under the Creative Commons Attribution License. Pierre AC. Enzymatic Carbon Dioxide Capture. ISRN Chemical Engineering ***2012*** *(2012). http://dx.doi.org/10.5402/2012/753687.*

Several methods are being developed or studied for this purpose [6, 7] and progress is being followed by the International Energy Agency (IEA) of the Organization for Economic Co-operation and Development (OECD) [8]. A general review was also published in a book chapter by Muradov [9]. Amongst them, one group of technology is proposing to use enzymes of the carbonic anhydrase type. The specificity of these enzymes is to catalyze the reversible transformation of neutral aqueous CO_2 molecules, termed $CO_2(aq)$ in this paper, to the ionic species H^+ and HCO_3^-. Very few reviews have specifically addressed these enzyme projects. To our knowledge, these comprise a recent publication by Shekh et al. [10] and a bibliography in a recent Ph.D. thesis by Favre [11]. However, the number of new research articles published has also significantly increased during the same time span, and the aim of the present paper is to present an up-to-date synthesis of this field.

5.2 PLACE OF ENZYME TECHNOLOGIES AMONGST THE MAIN CO_2 CAPTURE AND STORAGE (CCS) TECHNIQUES

Three major steps are being considered to tackle the anthropic CO_2 problem: the capture of this gas from the atmosphere, its transport to storage places, and its storage under various forms. These 3 steps are often gathered under the abbreviation "CCS," for "CO_2 Capture and Storage." Enzymes are concerned by the first step, that is, CO_2 capture, and also to some extent by the third one, to transform the captured CO_2 to carbonates for a safe storage, or possibly to more valuable products.

The main techniques developed to capture the CO_2 from industrial fumes can be classified as "postcombustion," "oxycombustion" or "precombustion" methods [12]. The enzymatic capture techniques can be classified within the first group of methods, where CO_2 is withdrawn from the industrial fumes produced by the combustion of hydrocarbons. Within this group, different CO_2 capture and storage techniques are actually in competition and they were reviewed in the 2005 report from the Intergovernmental Panel on Climate Change (IPCC) [13]. They comprise amine scrubbing, membrane separation, wet and dry mineral carbonation, pressure storage, and adsorption on solids or in liquids.

Typically, industrial fumes contain from 10% to 20% CO_2, mixed with nitrogen as the major component plus some lower percent of O_2 and H_2O vapor and a variety of other pollutants, in particular sulfur compounds. The processes most extensively studied rest on reversible carbonation reaction with amines. For instance, when an aqueous monoethylamine (MEA) solution is used, some ammonium carbamate partly hydrolyzed to a carbonate is produced [14]. To recover the CO_2 from the carbamate, it is then necessary to increase the temperature in order to displace the carbonation equilibrium towards CO_2 release. In the enzyme techniques, the amines are replaced by an aqueous solution of enzyme of the carbonic anhydrase family. As previously mentioned, the latter proteins can catalyze the reversible transformation of CO_2(aq) neutral species to ionic HCO_3^- species, provided adequate conditions can be satisfied as further detailed in the present paper.

By comparison, in the "oxycombustion" methods, combustion of the hydrocarbon is achieved in pure O_2 or in a mixture of O_2, H_2O vapor and CO_2. Consequently, the fumes are mostly constituted of H_2O vapor and CO_2, from which CO_2 can simply be separated if H_2O is condensed to the liquid state, by cooling [13]. At last in the "pre-combustion" methods, the fuel used is first converted to a mixture of CO_2 and H_2, often termed "syngas" [15].

In most cases, the recovered CO_2 can then be compressed to liquid CO_2 under a moderate pressure (e.g., 2 MPa at −20°C), to be transported by ships or trains. It can also be transported by pipeline, usually when brought to the supercritical fluid state, (temperature >31°C, pressure >7.4 MPa) [15].

The main storage methods which are being tested consist in injecting the captured CO_2 at great geological depth, at least 800 m, where it can hopefully remain for a time as long as possible [13, 15, 16]. The main geological sites considered for such storage comprise exhausted oil fields, unexploited coal seams where CO_2 could possibly react with the coal to produce some methane, and deep underground salinas which are actually evenly dispersed and abundant on Earth, so that they could offer a storage volume of the order of 10 times that from the other geological sites [13, 15–17]. Injection of the CO_2 in the ocean, at a depth beyond 1000 m where dense solid CO_2 hydrates could form, is also being considered [13].

A number of fundamental research studies also addressed the adsorption of CO_2 on solids, mainly basic solids. Carbonic anhydrase enzymes are concerned by a storage technique of CO_2 as solid carbonates. Such storage is often considered to be of lesser importance, because it would require abundant and cheap basic cation sources (Ca^{2+}, Mg^{2+}, Na^+, etc.) to be economically applicable. However, a number of basic scientific publications have addressed the use of carbonic anhydrase for this purpose, and they are reviewed in the last section of this paper.

CO_2 could also be used as a substrate to synthesize valuable chemicals, as reviewed by Sakakura et al. [18]. In particular, combined with a dehydrogenase, CA enzymes could be used to transform the captured CO_2 to methanol by a fully enzymatic process [19]. Besides, other biological techniques are also in progress such as the use of marine algae to perform a photocatalytic transformation of CO_2 to biofuels [20–24]. However, these subjects are outside the scope of the present paper.

5.3 THE PHYSICAL CHEMISTRY OF CO_2 CAPTURE IN AQUEOUS MEDIA

The general mechanism of CO_2 capture in aqueous media and its separation from other gases, can be decomposed in the 5 following steps [25].

1. Dissolution of the CO_2 gas molecules in water on the CO_2 capture side, at the gas/aqueous medium interface, according to the Henry's equilibrium [26–28]. As a result, neutral aqueous CO_2(aq) molecules are introduced in the aqueous film in direct contact with the gas.
2. Reversible conversion by deprotonation of the neutral CO_2(aq) species, usually termed hydration, to form anionic bicarbonate species HCO_3^-, according to a chemical equilibrium which is pH dependent.
3. Transport of both the neutral and anionic aqueous CO_2 species, from the CO_2 capture side towards the CO_2 release side, by molecular diffusion inside the aqueous medium and/or by forced fluid circulation.

4. Reverse conversion of the anionic HCO_3^- species to the neutral CO_2(aq) ones, according to the same chemical equilibrium as in step 2.
5. Evaporation of the CO_2(aq) in the gas to liberate CO_2 gas species, on the CO_2 release side, according to the same Henry's equilibrium as in step 1.

Regarding steps 1 and 5, the Henry's chemical equilibrium can be written as

$$CO_2(g) + H_2O \leftrightarrows CO_2(aq) \quad k_H \tag{1}$$

The equilibrium constant k_H of (1) is known as the Henry's constant, and it is usually written as in (2) known as the Henry's law:

$$\mathrm{N(CO_2(aq))} = \frac{\mathrm{P(CO_2(g))}}{k_H} \tag{2}$$

According to this law, the molar fraction N(CO_2(aq)) of the CO_2(aq) species in the aqueous film, in equilibrium with a gas phase with which it is in direct contact, is proportional to the partial pressure P(CO_2(g)) in this gas. This equilibrium equation concerns both the capture side and the release side. After conversion of the molar fraction N(CO_2(aq)) to the molar concentration [CO_2(aq)] in water, (2) can be transformed to

$$[\mathrm{CO_2(aq)}] = \frac{\mathrm{P(CO_2(g))}}{(0.018k_H)} \tag{3}$$

The exact nature of these neutral CO_2(aq) species is controversial. It is generally admitted that they essentially comprise CO_2 molecules more or less loosely solvated by H_2O molecules to which they can be linked by fluctuating hydrogen bonds [26, 29, 30]. One of these neutral molecular species is the carbonic acid molecule H_2CO_3, which could actually be synthesized in a virtually pure state in special conditions, from an exact stoichiometric molecular ratio $N(CO_2)/N(H_2O) \approx 1$ [31]. However, these H_2CO_3 molecules are metastable and they become very unstable in the presence of a slight excess of water. Hence they remain present in very low molar ratio solutions (<3/1000) in CO_2 saturated water at 25°C, by comparison with the simply solvated $CO_2(aq)$ species [28, 31–34].

The Henry's equilibrium is a direct consequence of simple molecular collisions at the interface between the gas phase and the liquid phase, which do not involve chemical reactions. Hence, regarding the first layers of liquid water molecules in direct contact with the gas, it is implicitly considered that this equilibrium is very rapidly established and maintained, independently of further diffusion or transformations of the neutral $CO_2(aq)$ species [30]. Consequently, for a given $P(CO_2(aq))$ partial pressure, the concentration $[CO_2(aq)]$ in the aqueous strata in direct contact with this gas can be reasonably considered as being constant.

On the other hand, equilibrium with a thicker water layer, such as needed for instance to experimentally determine the Henry's constant, is much slower. The reason is this requires a diffusion of both the neutral and anionic CO_2 species, from the aqueous strata in direct contact with the gas towards the whole liquid volume. Fortunately also, when the CO_2 Henry's constant is determined in pure water (no electrolyte added), the neutral $CO_2(aq)$ species are by very large dominating over the anionic ones, as summarized further on. Consequently, (3) practically concerns the neutral species only. At last, to favor the dissolution of $CO_2(aq)$ species on the capture side, as well as the release of CO_2 gas on the release side, the exchange surface between the gas phases and the aqueous medium must also be designed to be as high as possible. This point is very important to design efficient CO_2 "scrubbers."

The solubility of CO_2 in pure water under a partial pressure $P(CO_2(aq))$ ranging from 0.1 MPa (1 atm) to 100 MPa was reviewed in 2003 by Diamond and Akinfiev [26]. For lower $P(CO_2(g))$ partial pressures more in line with CO_2 capture from industrial fumes, it was reviewed in 1991 by Carroll et al. [27] and by Crovetto [28]. For instance, according to Crovetto:

$$\text{Ln}(k_H) = \frac{4.800 + 3934.40}{T} - \frac{941290}{T} \tag{4}$$

where the Henry constant k_H is expressed in bar (1 bar = 105 Pa) and the temperature T in Kelvin. As an example, for distilled water saturated in CO_2 under a partial pressure $P(CO_2(g)) = 0.1$ MPa (= 1 bar ≈ 1 atm), this equation indicates concentrations $[CO_2(aq)] \approx 33.7$ mmol L^{-1} at 25°C and 76.5 mmol L^{-1} at 0°C. The temperature is therefore an important parameter, since the $CO_2(aq)$ concentration in water increases significantly when the temperature decreases.

Regarding steps 2 and 4, the first deprotonation equilibrium or so-called hydration of $CO_2(aq)$ species to form bicarbonate anions HCO_3^-, can be written [35]:

$$CO_2\,(aq) + H_2O \leftrightarrows H^+ + HCO_3^-$$

$$K_{a1} = 10^{-6.35} = 4.47 \times 10^{-7} \text{ at } 25°C \tag{5}$$

According to (5), the pH rapidly falls below 7 as soon as CO_2 is dissolved in distilled water at an initial pH 7, as this is indeed the case to determine the Henry's constant in pure water. On the other hand, if the pH can be maintained at a value > $pK_{a1} = 6.35$ with the help of a buffer, the formation of ionic HCO_3^- species is favored, although the concentration of neutral $CO_2(aq)$ species remain fixed at the gas liquid interface by the Henry's law. Overall, because the HCO_3^- anions are much more soluble in water than the neutral $CO_2(aq)$ species, a much larger total CO_2 concentration can be dissolved in aqueous solution. This result is at the base of the idea to use a catalyst to capture CO_2 in aqueous media, where the catalyst role is simply to accelerate the formation of HCO_3^- anions.

The kinetic mechanism underlying (5) largely depends on the nature of the catalyst used and carbonic anhydrase enzymes only constitute one type of catalyst. Without any catalyst, hence at an acidic equilibrium pH, the forward reaction to produce HCO_3^- anions from $CO_2(aq)$ species is first order with a rate constant $\approx 0.15\ s^{-1}$. The reverse reaction is faster, with a rate constant $\approx 50\ s^{-1}$ [33, 34]. Possibly, it could be considered that the HCO_3^- molecule constitutes the transition state. In basic conditions where OH^- anions are abundant, the main mechanism involves a direct attack of these anions on the $CO_2(aq)$ species. Hence any base is a catalyst of the CO_2 capture and competes with the carbonic anhydrase enzyme. The enzyme catalytic mechanism involves its active site, as briefly summarized further on. Overall, as any catalyst, the enzyme only modifies the kinetics rate of both the forward and reverse reactions, not the thermodynamic equilibrium.

It is important to note that the Henry exchange mechanism operates for all gas components present on the capture side, including O_2 and N_2. However, with the latter species, no formation of highly soluble anions such as HCO_3^- occurs. Hence the overall concentration of these components in water, and their further transport rate towards the release side, remains much lower than that of CO_2, provided the pH is such that HCO_3^- species are abundant. Unfortunately, this is not the case of other pollutant species such as SO_2, which can also produce very soluble anions such as .

Step 3 of a CO_2 capture system, which is the transport of the CO_2 aqueous species from the capture side to the release side, applies both to the neutral $CO_2(aq)$ species and HCO_3^- anions. The liquid medium transport itself can be forced, with the help of circulating pumps as in one type of process under development presented in the next section, or it can be spontaneous by simple molecular diffusion. In both cases, the CO_2 transport must be fast enough to not be the rate limiting step. This implies a fast circulating pump system in the former case, or a very short diffusion distance such as for instance across thin water films in the latter case. In the CO_2 capture systems based on such thin aqueous films, transfer of the $CO_2(aq)$ and HCO_3^- species from the capture face towards the release side by diffusion of these species is illustrated Figure 1, which gathers the 5 previous steps. Besides, the diffusion coefficients of $CO_2(aq)$, HCO_3^- and other gas species such as N_2 and O_2, are of the same order of magnitude, because their molecular weights are relatively close to each other. Hence, as previously mentioned, it is indeed necessary to maximize the HCO_3^-

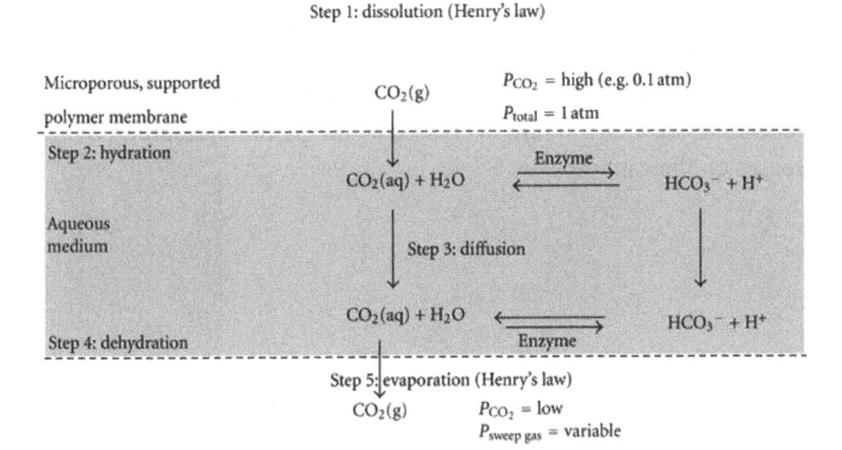

FIGURE 1: Illustration of the CO_2 transfer mechanism inside a thin liquid membrane [140].

concentration on the capture side, with the help of a catalyst and in the appropriate pH range, so as to increase the overall separation selectivity in CO_2 relative to other species.

In the case of a thin aqueous membrane, the overall CO_2 transfer rate across the membrane is described by (6) where Φ (CO_2), expressed in mol s^{-1} m^{-2} is the CO_2 flux density carried per second across 1 m^2 of liquid membrane; P(CO_2(capture)) and P(CO_2(release)) are the CO_2(gas) partial pressures, in Pascal, on both sides of the liquid membrane and P defines the membrane permeance, measured in mol $s^{-1}m_{membrane}^{-2}$ Pa^{-1}:

$$\Phi(CO_2) = P\,[P(CO_2(capture)) - P(CO_2\,(release))] \quad (6)$$

Another direct consequence of the CO_2 capture in aqueous media concerns the influence of the partial pressure P(CO_2(capture)) on the capture side. According to Henry's law (3), the [CO_2(aq)] concentration dissolved in water increases with the CO_2 partial pressure P(CO_2(g)) in the gas in contact with the aqueous medium. In turn, the [H^+] and [HCO_3^-] equilibrium concentrations increase with P(CO_2(g)), as a consequence of the hydration equilibrium reactions in (5), unless an increasingly stronger and faster reacting buffer can be added to maintain a pH > pK_{a1}. But this becomes increasingly difficult to achieve as $P(CO_2(g))$ increases. Accordingly, experimental results on thin water films gathered by Bao et Trachtenberg [36] and reproduced in Figure 2, confirm that the enzyme efficiency to accelerate the CO_2 capture decreases as the partial pressure $P(CO_2(g))$ increases on the capture side.

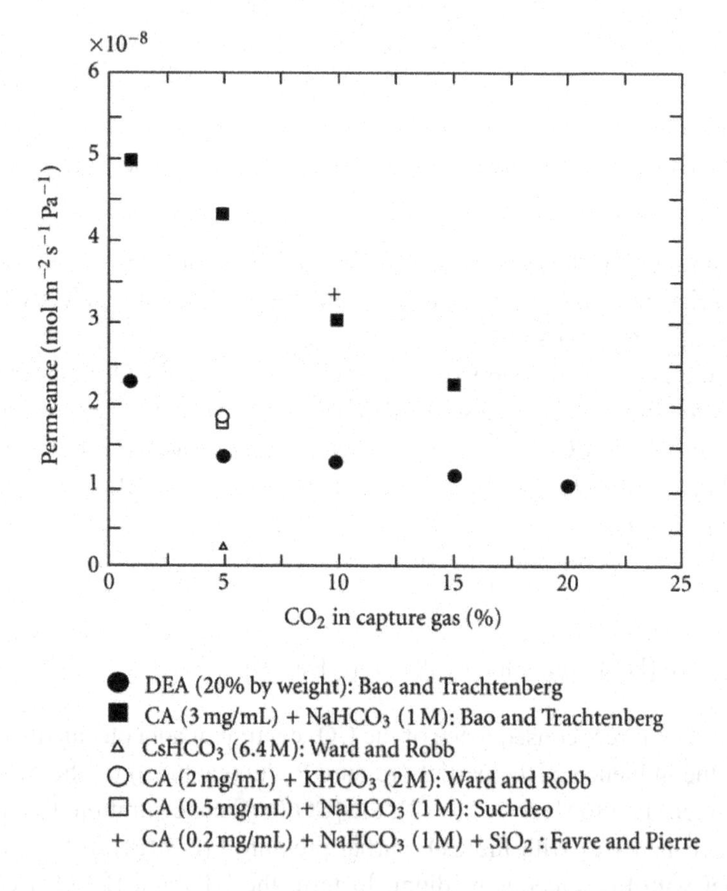

FIGURE 2: CO_2 permeance as a function of the CO_2 percent in the capture gas after the results of Bao et Trachtenberg [36], completed by those of Ward and Robb [107], Suchdeo and Schultz [108], and Favre and Pierre [140]. Adapted from Bao and Trachtenberg [36].

5.4 ENZYMATIC CO_2 CAPTURE

Carbonic anhydrase enzymes are labelled as belonging to the group E.C.4.2.1.1 by the International Enzyme Commission, in agreement with the International Union of Pure and Applied Chemistry (IUPAC). These numbers indicate that they belong to class 4, the class of lyases which gathers the enzymes able to catalyze a reaction of addition on a substrate carrying a double bond (such as O=C=O), to sub-class 2 corresponding to the creation of a single C–O bondby addition of an oxygen atom (Carbon-oxygen lyase), with an oxygen atom brought by an aqua group (Hydro-lyase, first number 1), being the product, or substrate (inverse reaction) amongst of a list of possible substrates (carbonate dehydratase-second number 1) [37]. Enzymes of this group are actually present in the 3 classes of the living world: prokaryotes, archaea, and eukaryotes [38, 39].

Historically, the first carbonic anhydrase (CA) was discovered in 1933 by Meldrum and Roughton when studying the factors responsible for a rapid transition of the bicarbonate anions HCO_3^- from erythrocytes towards the lung capillary [40]. In 1939, CA of plant origin were shown to be different from the previously known CA [41]. In 1940, Keilin and Mann purified AC extracts from bovine erythrocytes and they showed that the CA contained a Zn atom in their active site [42–44]. In 1963, Veitch and Blankenship discovered AC enzymes in prokaryotes [45] and the first purified AC extracted from such a source was achieved in 1972 from Neisseria Sicca [46]. The first genetic sequence of a purified AC of prokaryote origin (bacteria *Escherichia coli*) was established in the 1990's [47] and this metalloenzyme of molecular weight of 24 KDa was the first β class CA [48], while the previous ones were classified in the α class.

In αCA, the active catalytic center is built about a Zn atom in tetrahedral coordination with 3 histidine residues, plus 1 water molecule [49, 50]. For instance, as illustrated in Figure 3 for a human αCAII, this active center is localized in the cavity of a protein comprising a polypeptidic chain of 260 amino acids of molar mass 29 kg mol^{-1} (29 kDalton) [34].

βCA predominate in plants and algae. Their main difference with αAC is that they have an oligomeric quaternary structure composed of 2 to 6 monomers, which are each roughly similar to a full αCA. In each monomer,

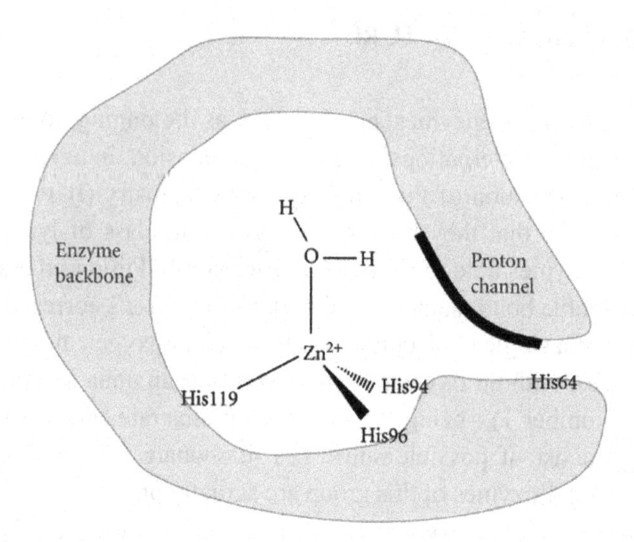

FIGURE 3: Schematic structure and active center of human αACII [34].

the Zn atom is moreover coordinated to 2 cysteines, 1 histidine, and to 1 aspartate by the intermediate of its carboxylate termination. The first CA from an Archaeon was isolated and purified in 1994 by Albert and Ferry from *Methanosarcina thermophila* [51]. It showed a different amino sequence by comparison with the previous α and β CA, hence it was placed in a new γ-CA class. Contrary to the previous CA, their Zn atom is coordinated in a penta mode to 3 histidines and 2 water molecules.

During the following years, much progress was achieved to discover new AC varieties and to understand the catalytic mechanism of these enzymes, particularly regarding the AC of human origin [52–54]. While α and β CA were found to predominate in Eukaryotes, γ CA were mostly present in Archaea [39, 46]. A new CA of molar mass 27 KDa was extracted from *Thalassiosira weissflogii* [55] and it showed a different amino acid sequence, compared to α, β and γ CA. Hence it opened a new δ CA class [39]. To this one must add another type of CA purified and sequenced from the shell of *Halothiobacillus neapolitanus*. This enzyme can also be found in the shells of Marine cynobacteria *Prochlorococcus* and *Synechococcus.* Its molecular weight of 57.3 KDa and its tertiary structure showed two domains, similar to the β-CA, except that only one of these domains had a Zn binding site. This CA converts HCO_3^- to CO_2 inside shells where the CO_2 is incorporated within the biomass by the enzyme Ribulose Bis phosphate Carboxylase (Rubisco) [56]. It was placed it in a new ε-CA class. δ and ε CA are present in eukaryotic algae and phytoplankton. Their Zn is coordinated 2 cysteines, 1 histidine, and 1 water molecule. At last, a ζ-CA class, comprising CA of molar weight 69 KDa containing a cadmium atom in their active site in place of a zinc, was isolated from the marine diatom Thalassiosira weissflogii. [57].

Globally, the AC enzymes based on Zn atoms are classified in 5 groups labelled α, β, γ, δ, and ε which all have in common to catalyze for the interconversion of HCO_3^- and CO_2 [34, 5_2, 58, 59]. However, in more details regarding the mammals αAC, 4 sub-classes can distinguished [53, 58, 60].

1. Cytosolic αCA, present in the cytoplasm of cells, themselves comprising several sub-sub-groups labelled CA-I, -II, -III, -VII, and –XIII.
2. Mitochondrial αCA present in cells mitochondria (groups CA-VA and -VB).
3. Secreted αCA present in saliva and milk (group CA-VI).
4. Membrane binding αCA (groups CA-IV, -IX, -XII, -XIV, and -XV).

To these one must add 3 "acatalytic" CA isoforms with unclear functions (CA-VIII, -X, and -XI). Overall, 16 different forms of isozymes (or isoenzymes), which are different forms of an enzyme type coexisting in a same living organism, could be identified in the mammals, of which 10 in humans.

The panel of CA enzymes available to capture CO_2 is indeed large and it keeps increasing. For instance recently, Ramanan et al. reported the isolation, purification, and sequencing of CA from the Enterobacter bacteria *Citrobacter freundii* and *Bacillus subtilis*. [61]. Progress regarding CA enzymes also concerns their extraction and purification techniques. Da Costa et al. compared 2 different purification techniques of bovine CA (BCA): one by extraction with the organic solvents chloroform and ethanol, and the other by ammonium sulfate precipitation [62]. In a CO_2 hydration assay, the first technique provided the highest enzyme activity, for a recovery of

98% and a purification factor of 104-fold. Im Kim et al. compared the hydration activity of a cheaper recombinant α-type CA from *Neisseria gonorrhoeae* (NCA) which they highly expressed in *Escherichia coli*, with a more expensive commercial BCA. The activity of both CA was found to be equivalent. Even the nonpurified NCA showed a significant activity, which opens the route to less expensive enzymatic CO_2 capture processes [63]. On the other hand, Trachtenberg patented a new γ-carbonic anhydrase enzyme which could operate in the temperature range of 40–85°C [64]. Regarding this last point, an interesting geological discovery must also be mentioned. Along the mid-ocean ridge system where tectonic plates are moving away from each other, sea water penetrates the fissures of the volcanic bed and is heated by the magma. This heated sea water rises to the surface and, although this environment seems very hostile, many microorganisms happen to prosper. Amongst them, some microorganisms have developed efficient CO_2 assimilation processes [65]. In a quite different domain, an artificial, bifunctional enzyme containing both a CA moiety from *Neisseria gonorrhoeae* and a cellulose binding domain (CBD) from *Clostridium thermocellum* was synthesized. This new biocatalyst opens the route to the development of new immobilized enzyme CO_2 capture systems [66]. Besides, the synthesis of biomimetic analogs of CA enzymes is also being investigated. After immobilization on a support, these catalytic complexes could be used to design "biomimetic" CO_2 cap-

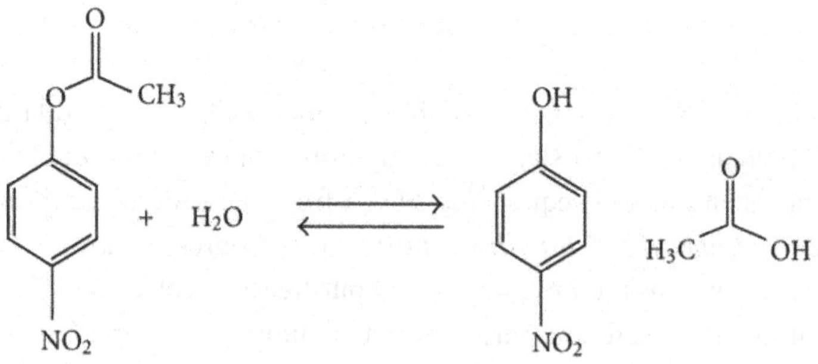

FIGURE 4: Hydrolysis reaction of para-nitrophenylacetate (p-NPA) to para-nitrophenol (p-NP) [35].

ture systems more robust than the true enzymatic ones [67, 68]. Overall, new more efficient and cheaper enzymatic systems for CO_2 capture may reasonably be expected to progressively appear in the future.

Carbonic anhydrase enzymes are known to catalyze 2 different types of equilibrium reactions [35]. First, as a "hydrase," they catalyze the equilibrium hydration and dehydration reactions of $CO_2(aq)$, previously presented (5). Secondly, as an "esterase," they hydrolyze substrates such as the para-nitrophenylacetate (p-NPA) to para-nitrophenol (p-NP) according to Figure 4.

The capture of CO_2 is concerned by the hydrase activity. The catalytic properties and mechanism of CA enzymes in CO_2 hydration were the subject of many papers which are only partially reviewed here. Overall, the experimental techniques used to measure this activity were generally derived from an electrochemical method first designed by Wilbur and Anderson [69]. These scientists studied 3 types of techniques, based on manometry, colorimetry, and electrochemistry respectively. The manometric techniques rest on a measure of the gas pressure, in a CO_2 containing atmosphere in equilibrium with an enzyme solution in a buffer. The colorimetric technique rests on a measure of the time for a change in color, when a color indicator is mixed in the enzyme solution. Actually, the latter technique was first investigated by Brinkman in 1933 [70] and later successively modified by Meldrum and Roughton [40], Philpot and Philpot [71], and finally Wilbur and Anderson [69]. The electrochemical technique rests on a measure of the pH decreasing rate with a pH electrode, during CO_2 capture. This is often done at low temperature (e.g., ≈4°C) [69] in CO_2 saturated water to which a buffer at a pH slightly above 8 is added, altogether with a variable mass of enzyme m_{enz} (a few mg). The pH decreasing rate with time, ($d[H^+]/dt$) is usually determined in a pH range about pH 7.

Because $CO_2(aq)$ hydration also occurs without any enzyme, it is necessary to subtract the non-enzymatic contribution from the data obtained with the enzyme. This operation really gives the "added" contribution to CO_2 hydration, due to the enzyme. For instance, let t_0 and t_{enz} designate the times measured when the pH decreases from 7.5 to 6.5, respectively, without enzyme and with a mass m_{enz} of CA. The activity per mg of enzyme, really a "relative added activity" v_r, can be expressed inUnits per mg ($U\,mg^{-1}$) according to Wilbur equation (7), where v_{enz} is the pH decreasing rate with the enzyme and v_0 the pH decreasing rate without any enzyme:

$$v_r(\mathrm{U\ mg^{-1}}) = \frac{v_{enz} - v_0}{v_0 m_{enz}} - \frac{t_0 - t_{enz}}{t_{enz} \cdot m_{enz}} \tag{7}$$

In most papers, v_r is often simply termed v.

Biochemists traditionally measure the enzyme concentration in preparations in "units" ("U"), where 1 U is defined as the quantity of enzyme which catalyzes the transformation of 1 μmol of substrate in 1 minute in conditions which must be specified (substrate nature and concentration, temperature, liquid medium, pH). However, regarding CO_2(aq) hydration, the "units" defined through (7) are not of a same nature. They rather designate an "added relative activity" which largely depends on the buffer used, because the non-enzymatic contribution significantly depends on this buffer. The latter brings OH^- anions which are catalyst competitors to the enzyme, as previously mentioned. Hence (7) cannot be used to indicate the "enzyme concentration" of a CA preparation. For this purpose, the hydrolysis of para-nitrophenyleacetate (p-NPA) into para-nitrophenol (p-NP) according to Figure 4 has to be used instead. This reaction is typically followed by measuring the UV-visible absorption due to p-NP, at a wavelength of 400 nm and 25°C, after mixing an aqueous CA solution at pH 7.5 with a p-NPA solution in acetonitrile. Because p-NPA also undergoes self dissociation, the rate of the self dissociation measured in the same conditions without enzyme must be subtracted from the data obtained with the enzyme [72].

On the other hand, it is necessary to keep using Wilbur's equation (7) when the aim is to study the factors affecting CO_2 hydration. Different scientific groups have applied this equation with variations regarding the enzyme concentration, the CO_2 saturated water volume and the buffer nature and molarity. Overall, the hydration kinetics of CO_2(aq) by CA enzymes was extensively studied [44, 49, 50, 73–76]. However, the oldest data were often affected by significant error magnitudes, as reported by Bondetal. [30], Mirjafari et al. [77] or Ozdemir [72]. In the most recent developments, CO_2 saturated water and an enzyme solution in a buffer are rapidly mixed in astop-flowcell. The pH time evolution is then followed by recording the visible light absorption at a characteristic wavelength of a pH dependant color indicator [44, 74].

It must also be mentioned that other CO_2-capture assaying methods have also been developed for many decarboxylating enzymes. In particular, some methods involve a radiometric measurement of the trapped $^{14}CO_2$ by scintillation counting, which can be performed in capped tubes or in the µL wells of titration plates [78].

The simplest model of enzyme kinetics applied to CA enzymes is the very classical Henri-Michealis-Menten model [79]. The chemical reactions underlying this model can be summarized by (8) in the case of CO_2 hydration, where $CA.CO_2(aq)$ stands for a so-called enzyme-substrate Michaelis complex:

$$CA + CO_2(aq) \underset{k_{-1}}{\overset{k_1}{\leftrightarrows}} CA \cdot CO_2(aq) \underset{k_{-2}}{\overset{k_{cat}}{\leftrightarrows}} CA + HCO_3^- + H^+ \tag{8}$$

In 1913, next to a work by Henri [80, 81], Michaelis and Menten [82] considered that at the beginning of product formation (presently HCO_3^-), the second inverse reaction corresponding to the kinetic constant k_{-2} was very slow and could be neglected, so that the initial formation rate v_0 of this product could be written as:

$$v_0 = \frac{d[\mathrm{HCO_3^-}]}{dt} = k_{cat}[\mathrm{CA{\cdot}CO_2(aq)}] \tag{9}$$

This mechanism also implies that the species CA, $CO_2(aq)$, and $CA.CO_2(aq)$ were in quasi-thermodynamic equilibrium, described by a thermodynamic equilibrium constant K_m termed the Michaelis constant defined by

$$K_m = \frac{k_{-1}}{k_1} = \frac{[CA][CO_2(aq)]}{[CA \cdot CO_2(aq)]} \tag{10}$$

Overall the following Michaelis-Menten rate equation (11) previously established by Henri [80, 81] is applied to CO_2(aq) hydration:

$$v_0 = \frac{v_{max}[CO_2(aq)]}{K_m + [CO_2(aq)]} \tag{11}$$

In this equation, v_{max} is the maximum initial formation rate of the product HCO_3^-, obtained when the total enzyme concentration $[CA]_t$ is engaged in a Michaelis complex CA.CO_2(aq).

Consider

$$v_{max} = k_{cat}[CA]_t \tag{12}$$

Amongst the many publications related to this simple model, a number of them report some values for k_{cat}, K_m, k_{cat}/K_m, and possibly v_{max}. An example of such kinetic constants are for instance gathered in Table 1 for the most efficient human CA (isoenzyme CA II), regarding both the forward and reverse reactions, for which CO_2(aq) and HCO_3^- are, respectively, the substrates. Table 1 shows that the enzymatic turnover number of the forward hydration reaction is high, $k_{cat} \approx 10^6$ s^{-1}, as well as the ratio of this turnover to the Michaelis constant $k_{cat}/K_m^{CO2} \approx 8.33 \times 10^7$ s^{-1} M^{-1}. These high numbers are at the origin of the idea to investigate the enzymatic capture of CO_2.

However, according to (5), a proton H^+ is also exchanged during CO_2(aq) hydration. This is illustrated in Figure 5 regarding the active site of a αCA enzyme, for the most simple mechanism model. The latter model involves 4 successive steps [49, 50, 75] as follows.

1. The enzyme ligands close to the enzyme active site induce a polarization of the O–H bond in the H_2O molecule coordinated to the Zn atom. This facilitates the deprotonation of this aqua ligand which is transformed to an OH− ligand, while the lost proton is captured by another histidine residue close to the active site.
2. The oxygen atom of this OH- ligand performs a nucleophilic attack onto the C atom of an incoming CO_2 molecule.
3. As a result of this nucleophilic attack, an HCO_3^- anion is bound to the Zn coordination center.
4. This HCO_3^- ligand is labile and in turn it can be rapidly exchanged for an H_2O ligand, which regenerates the enzyme active site in its initial state.

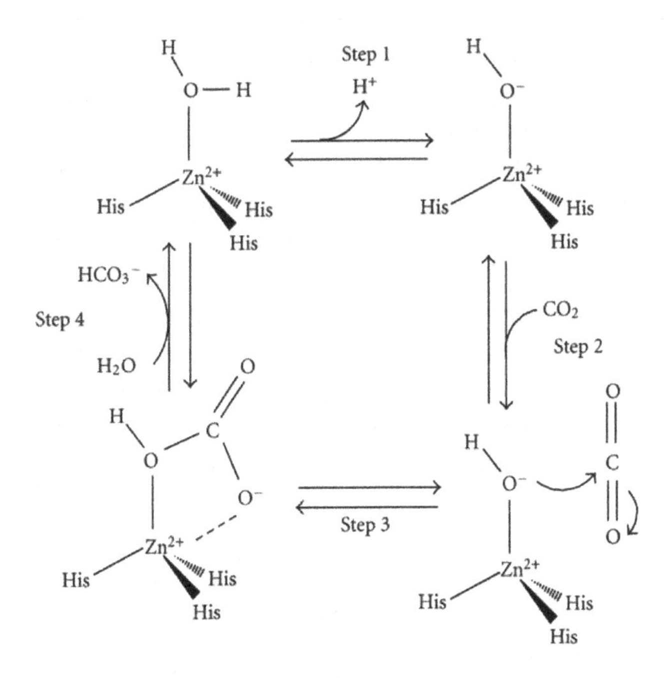

FIGURE 5: Simplified catalytic mechanism of an αCA enzyme [49].

TABLE 1: Michaelis-Menten kinetic constants for the forward and reverse equilibrium reactions involved in CO_2(aq) hydration, according to references [73, 76].

Substrate	$K_{cat}(s^{-1})$	K_m(M)	K_{cat}/K_m $(s^{-1}M^{-1})$
CO_2(aq)	10^6	0.012	8.33×10^7
HCO_3^-	4×10^5	0.026	1.54×10^7

However, the Michaelis-Menten rate equation model implies that this proton exchange is not rate limiting, while many studies later showed that the nature of a buffer mixed in the enzymatic solution could significantly modify the CO_2 hydration kinetics. Indeed, as previously mentioned, the base form B of a buffer couple (B/BH+) is in competition with the enzyme to exchange a proton, hence to catalyze the hydration reactions in (5). Hence, other models were developed to specifically address this point. The various kinetic models of CO_2 hydration catalyzed by a CA of human source (HCAII) were reviewed and analyzed in detail by Larachi [74]. His result was that the kinetic model most consistent with the reliable experimental data published by different investigators, was a model termed pseudo random Quad Quad Iso Ping Pong. This model involves a proton transfer competition between, on one hand the enzyme active site and the CO_2/HCO_3^- couple, on the other hand between the later couple and the B/BH+ buffer couple. The term Ping-Pong indicates that the HCO_3^- anions must first leave the enzyme active site, before they can interact with the buffer. The termPseudo indicates that the enzyme mechanism requires at least 2 different enzyme-substrate complexes, since a proton must be transported from inside the enzyme active site, towards the active site entrance, before it can leave the enzyme and react with the buffer. The 2 different complexes of the protonated enzyme, designated by E_W and $_HE$ in Figure 6 indeed constitute 2 different isomers of this enzyme. In the first one (E_W) the proton belongs to the H_2O ligand inside the enzyme active site. In the second one ($_HE$) this proton is located near the entrance of the active site cavity, although it still belongs to the enzyme, while the H_2O active site ligand is transformed to OH-. Hence, the most important modification brought by the pseudo random Quad Quad Iso Ping Pong model to the simplified model presented in Figure 5, concerns previous step 1. As illustrated in Figure 6, it involves a transformation of the enzyme from the E_W isomer to the $_HE$ one (Step 1(a)). In this process, the proton H^+ withdrawn from the H_2O active site ligand in the E_W isomer, is transported along a "proton tunnel" to the external entrance of the channel leading to the active site. This tunnel is constituted by a series of H_2O molecules covering lateral sites of the channel. At the end of this transfer, the proton remains linked to the enzyme via the imidazole ring of a histidine residue (His64) located on the external surface of the enzyme, a conformation which corresponds to the isomer. In a further step 1(b), this $_HE$ enzyme proton is finally exchanged with the buffer B, to produce BH+.

Step 3 of the simplified model presented in Figure 5 is itself subdivided in 2 sub-steps. First, a CO_2 molecule links via a single bond the O atom of the OH− enzyme ligand (step 3(a) in Figure 7 leading to a HO–CO_2 ligand). Secondly, this single bond transforms to a chelate bond via 2 oxygen atoms, with transfer of the H atom to the CO3 group terminal position (Step 3(b) in Figure 7 leading to a ligand). Step 4 is itself unchanged as it involves the replacement of an HCO_3^- ligand by H_2O, while the product is liberated. The active site with a full H_2O ligand corresponds to the previous EW enzyme isomer.

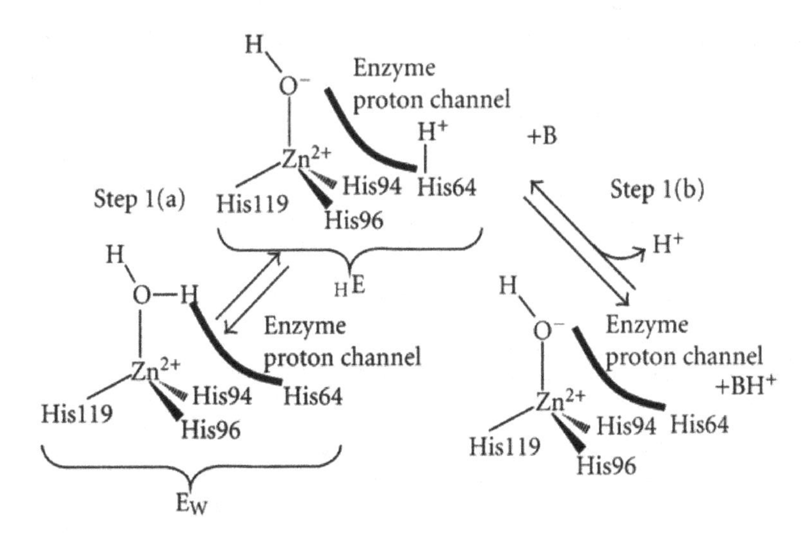

FIGURE 6: Steps 1(a) and 1(b) of the "Random Quad Quad Pseudo Iso Ping Pong" CO_2 hydration model, catalyzed by human αCA (HαCA), adapted from Larachi [74] and Rowlett and Silverman [44].

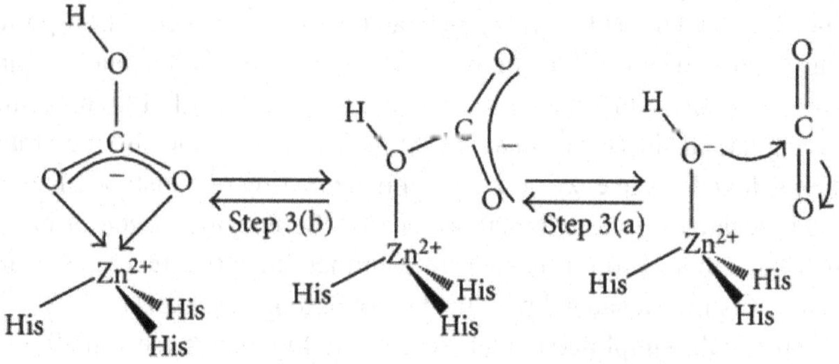

FIGURE 7: Step 3 of the "Random Quad Quad Pseudo Iso Ping Pong" CO_2 hydration model, catalyzed by human CA (HCA), adapted from Larachi [74] and Rowlett and Silverman [44]

A reverse protonation of the liberated HCO_3^- to form back some CO_2(aq) is also possible outside the enzyme, depending on the pH and the buffer couple B/BH+ used, in particular when the concentration of HCO_3^- is high. Moreover, an inhibition of the enzyme active site by the products HCO_3^- or CO_2, respectively, for the forward and reverse CO_2 hydration reactions, must be taken into account when any of these compounds are present in excess in the solution.

The full set of kinetic equations describing this model is very complex and would require a large development to present them. The readers interested in this point are recommended to refer to the publication by Larachi [74].

A more simple and more approximate equation (13) applicable to CO_2 capture in conditions when the substrate concentration is low enough to ignore inhibition reactions, was proposed by Rowlett and Silverman [44]. It only takes into account the Michaelis constant K_m^{CO2} and the corresponding turnover number k_{cat}, as well as a constant k_4 which describes the proton transfer kinetics between the enzyme isomer $_HE$ and a buffer B,

$$\frac{[E]_t}{\vartheta} = \frac{1}{K_4[B]} + \frac{1}{k_{cat}}\left(1 + \frac{K_m^{CO_2}}{[CO_2]}\right) \tag{13}$$

When inhibition by HCO_3^- and/or CO_2 must be taken into account, this equation can be transformed to(14)

$$\frac{[E]_t}{\vartheta} = \frac{k_{cat}[B]}{K_m^B\left[1 + \left([HCO_3^-]/K_i^{HCO_3^-}\right)\left(1 + K_i^{CO_2}/[CO_2]\right) + [B]\left(1 + K_m^{CO_2}/[CO_2]\right)\right]} \tag{14}$$

with K_m^B = Buffer Michaelis constant, K_i^{CO2} = inhibition constant by CO_2, K_i^{HCO3-} = inhibition constant by HCO_3^-.

The previous effective buffer constant is given by:

$$k_4 = \frac{k_{cat}}{K_m^B} \tag{15}$$

An apparent HCO_3^- inhibition constant can also be defined by

$$K_{i,app}^{HCO_3^-} = \frac{K_i^{HCO_3^-}}{1 + K_i^{CO_2}/[CO_2]} \tag{16}$$

Some kinetic constants for a few buffers are gathered in Table 2. Overall, the influence of the pH on the CA activity in CO_2(aq) hydration is therefore complex. However, according to data on k_{cat} provided by Berg et al. [34] and reported in Figure 8, the human CA reaches its maximum activity at pH > 8, hence when a buffer is added and the base buffer B can compete with the enzyme. These results are consistent with those of Ramanan et al., who reported that the enzyme from *Bacillus subtilis* was stable in the pH range 7.0 to 11.0, with a maximum activity in the pH range 8 to pH 8.3 [61, 85]. The stability of CA actually depends on the enzyme source, as shown in a study on CA from *Pseudomonas fragi, Micrococcus lylae, Micrococcus luteus* 2 and commercial bovine CA (BCA) in a pH range 8.0–9.0 and temperature range 35–45° [86]. In the latter study, the stability was the highest for the *Micrococcus luteus* 2 CA.

TABLE 2: Magnitude of the kinetic constants in (13) to (16) for a few buffers, according to reference [44].

Buffer	pH	pK_a	K_m^B (mM)	k_4 (M^{-1} s^{-1})	$K_{i,app}^{HCO3-}$ (mM)
Triethanolamine	8.4	7.8	1.2	5.4×10^8	10 ± 1
HEPES*	8.2	7.5	3.5	2.1×10^8	10 ± 1
4-methylimidazole	8.4	7.8	1.3	1.2×10^9	5.4 ± 0.3
1-methylimidazole	8.0	7.2	2.4	2.9×10^8	6.5 ± 0.4

The influence of temperature on the human and bovine CA activity in CO_2(aq) hydration was studied by Ghannam et al. in a pH = 9 buffer [87]. The data on k_{cat} and K_m^{CO2} are respectively reported in Figures 9 and 10 for these 2 enzymes. For both enzymes, k_{cat} and K_m^{CO2} increased as the temperature increased in the range from ≈5°C to ≈30°C. HCAII was somewhat more active than BCA (higher k_{cat} for HCA), and it showed a better affinity for CO_2 (lower K_m).

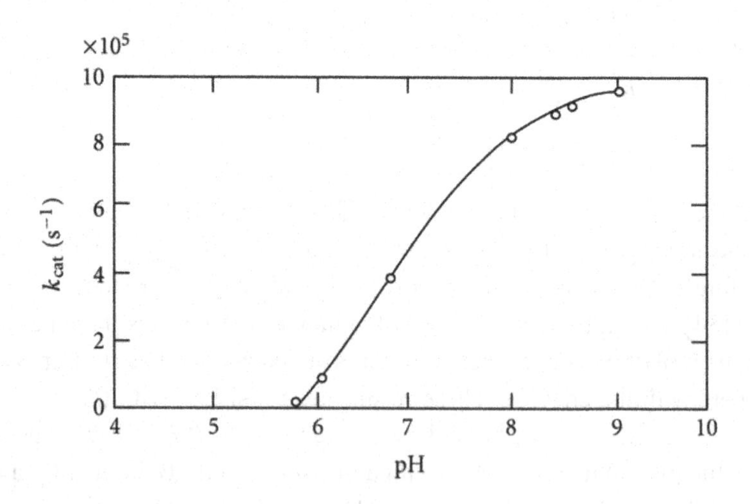

FIGURE 8: Effect of the pH on a human CA, in CO_2(aq) hydration. Adapted from [34].

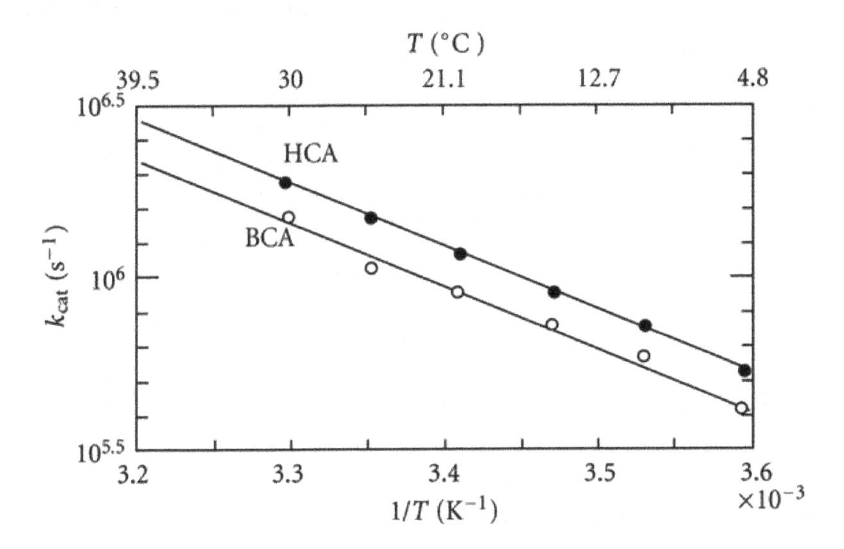

FIGURE 9: Influence of the temperature on k_{cat} for human CA (HCA) and bovine CA (BCA). Adapted from [87].

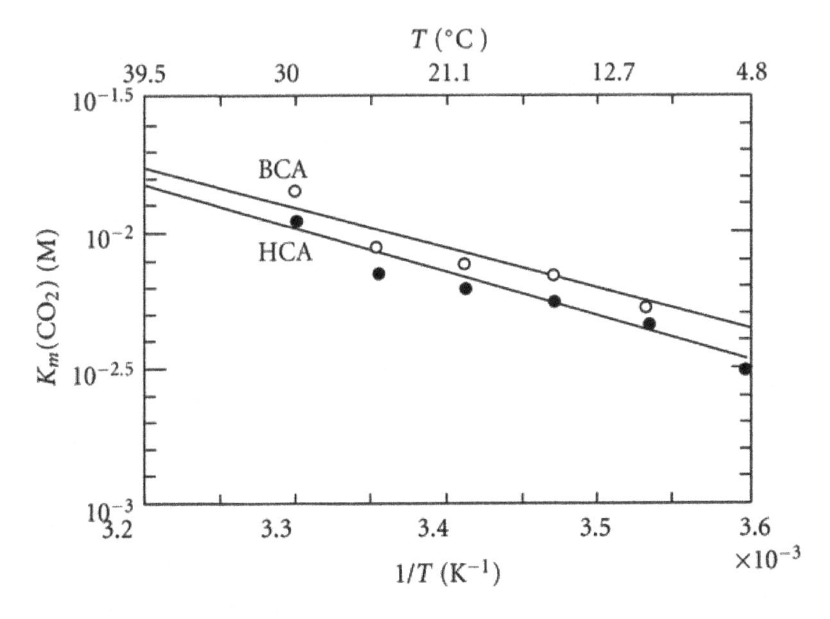

FIGURE 10: Influence of the temperature on K_m^{CO2} for human CA (HCA) and bovine CA (BCA). Adapted from [87].

However, an increased denaturation of CA enzymes occurs when the temperature increases because the enzyme conformation is progressively altered. Hence the lifetime when they remain active is shortened. This particular aspect was studied in the hydrolysis reaction of para-nitrophenylacetate [88, 89] by BCA enzyme, in which the thermal denaturation kinetic could be described by an Arrhenius type equation (17):

$$\frac{-d[E]}{dt} = k_d[E] \tag{17}$$

In this equation, [E] designates the active enzyme concentration and k_d a kinetic denaturation constant. The latter constant itself followed an Arrhenius type law as a function of the temperature T, according to (18) where E_d is a denaturation energy

$$k_d = A_d e^{(-Ed/RT)} \tag{18}$$

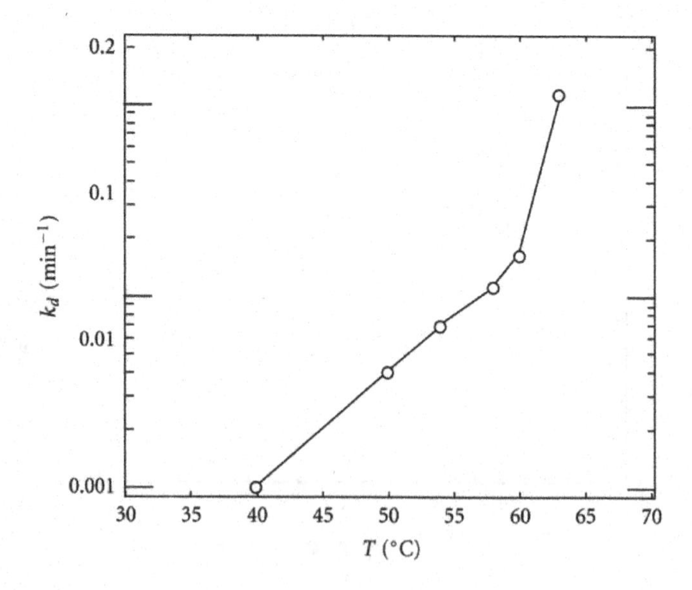

FIGURE 11: Denaturation kinetic constants of bovine erythrocyte CA (BCA) as a function of the temperature in the hydrolysis reaction of para-nitrophenylacetate, according to [88].

Some numerical values of k_d were determined by Kanbar and Ozdemir and they are reported Figure 11 [88]. Practically, after 120 min at 40°C, 50°C or 60°C, the residual enzyme activity was respectively 90%, 70% and $_2$0% of its initial activity.

At last, various chemical species inhibit the CA activity, in particular some present in industrial flue gases from which CO_2 must be captured. A study was carried out by Ramanan et al. on CA from *Bacillus subtilis* [61, 85]. The anions Cl^-, HCO_3^-, and CO_3^{2-} and the metal cations Pb^{2+} and Hg^{2+} were found to significantly inhibit the CA activity, while Ca^{2+} and Mn^{2+} were weak inhibitors and Co^{2+}, Cu^{2+} and Fe^{3+} were found to enhance this activity. Regarding SO_4^{2-}, a major pollutant in industrial flue gases, they found this anion activated CA, contrary to the results of Bond et al. [30]. The inhibition by Cl^-, SO_4^{2-}, NO_3^-, HCO_3^- and the cations As^{3+}, Ca^{2+}, Mg^{2+}, Hg^{2+}, Mn^{2+}, Cd^{2+}, Cu^{2+}, Zn^{2+}, Co^{2+}, Pb^{2+}, Fe^{2+}, Ni^{2+}, Se^{2+}, Na^+, and K^+ was also studied on CA from *Pseudomonas fragi, Micrococcus lylae, Micrococcus luteus* 2, and BCA by Sharma and Bhattacharya [86]. The level of inhibition was found to depend on the ion and the enzyme. It was significantly higher for BCA and *M. luteus* 2 CA, in particular by the anions.

5.5 ENZYMATIC CO_2 CAPTURE SCRUBBERS

In this paper, CO_2 scrubbers designate systems to separate CO_2(g) from other gaseous components. In 2008, Lacroix and Larachi reviewed the different types of CA enzymatic scrubbers in development [90]. These comprised membrane contactors using free CA solutions, to release gaseous CO_2 as well as to precipitate calcium carbonate, contactors using immobilized CA, namely counter-current and cross-con-current packed columns, and contactors using either free or particle-immobilized CA. These authors also examined a list of possible CA enzymes to capture CO_2 and to potentially produce useful organic compounds.

Overall, three enzymatic CO_2 capture techniques are being industrially developed, to which other scientific research publications must be added. In a first process developed by the company "CO_2 Solution Inc." and schematically illustrated in Figure 12, the enzyme is immobilized on a solid support, itself packed in a bed reactor [30, 91–95]. An aqueous solution is sprayed through a nozzle at the top of the reactor. It washes a counter flow of the gas containing the CO_2 to be captured, itself injected at the lower end of the reactor. Capture of the CO_2 occurs when the opposite flows of aqueous solution and gas from which CO_2 must be scrubbed out, percolate through a supported enzyme bed. A second reactor in which the former aqueous CO_2 solution is sprayed, for instance in a carrier gas or in a partial vacuum, makes it possible to recover the CO_2 gas. In reactors of this type, the exchange mechanisms between the liquid and the gas in the presence of the CA enzyme, are critical. In order to favor these exchanges, Fradette et

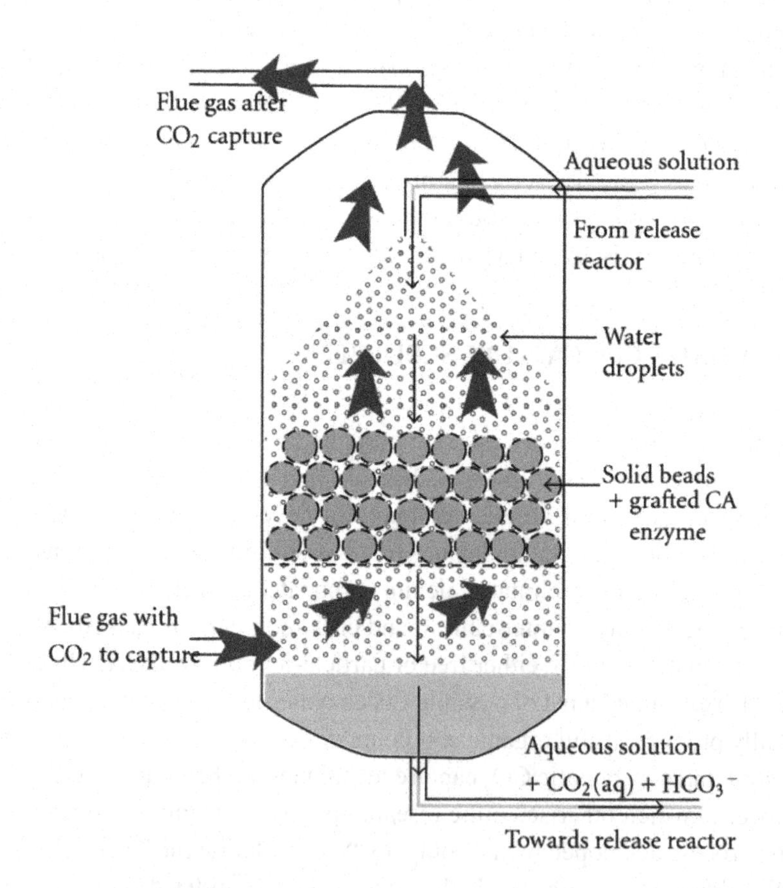

FIGURE 12: Schematic illustration of a CO_2 capture unit by gas washing, of the CO_2 Solution Inc. Company. Adapted from [93]

al. deposited a patent according to which the liquid layer flows as droplets on the enzyme immobilized on elongated supports, in such a way that the CO_2 hydration reaction occurs within the flowing liquid [96]. According to the authors, a prototype was tested in 2004 in an aluminum foundry of the Alcoa Inc. company, during a non-stop one month period. It made possible to capture 80% of CO_2 from the industrial fumes [30]. The enzyme permitted to reduce the reactor size by comparison with the same process without any enzyme. The process was also found to be more economical then a process based on CO_2 capture by an amine solution, which required to heat the amine solution in order to recover the CO_2 [14, 93]. In a variant of such a process, Bhattacharya et al. immobilized the enzyme by covalent grafting on silica coated porous steel and water was sprayed down through the flue gas. The best results were obtained with an enzyme support pore size of ≈2 μm and an enzyme load of 2 mg mL^{-1} [97]. Besides, it was also shown that CA enzymes could promote the absorption kinetics of CO_2 in potassium carbonate or aqueous amine solvents [91, 98–101].

The gas-liquid and liquid-solid mass transfer exchange mechanisms were examined in details by Iliuta and Larachi, for Robinson-Mahoney and packed beads reactors [102]. These authors showed that, for immobilized enzyme, these mechanisms could significantly alter the CO_2 hydration kinetics. The most remarkable increase in CO_2 removal was obtained by integrating immobilized-enzyme absorption with ion-exchange resin microparticles to remove excess enzyme inhibiting HCO_3^- anions [103]. As a consequence, they developed a 3-phase reactor, comprising HCAII enzyme, immobilized in the longitudinal channels washcoat of a post combustion monolith, in which an aqueous slurry containing resin exchange beads was flowed [10_2]. Other researchers proposed to enhance the exchange mechanisms limiting the CO_2 hydration rate, by ultrasonic techniques [104].

In a second type of process initially developed by the National Aeronautics and Space Administration (NASA) to purify the ambient atmosphere of confined inhabited cabins, the CO_2 is captured through thin aqueous films in which some CA is dissolved [25, 105]. The CO_2 concentration of such atmospheres is low, typically of the order of 0.1% or less. A schematic illustration of the membrane sandwich involved is presented in Figure 13. The core of the liquid membrane comprises a thin (e.g., 330 μm thick) layer of enzymatic solution in an aqueous phosphate buffer, squeezed in between 2 microporous hydrophobic polypropylene membranes, themselves retained by thin metal grids to insure the liquid membrane thickness and rigidity. The CO_2 from the atmosphere to purify, spontaneously dissolves inside the liquid membrane on one face of the membrane. It diffuses across the liquid membrane and evaporates out the other liquid membrane on the opposite face, either in vacuum or in a carrier gas. Analysis of the capture and release gases with a mass spectrometer showed that the enzyme permitted a selective diffusion of CO_2, in ratio of 1400 to 1by comparison with N_2 and 866 to 1 by comparison with O_2. As previously discussed, this result is due to the fact the N_2 and O_2 can only dissolve as neutral molecules; hence their solubility in water is limited by comparison with CO_2. These selectivities were superior to those achieved with a 20% (by weight) diethaloamine (DEA) solution, respectively, 442 to 1 and 270 to 1. By comparison with the same liquid membrane without enzyme, the CA permitted to decrease the overall resistance to CO_2 transport through the membrane, by 71%. Besides, the diffusion of HCO_3^- and H^+ ions across a liquid membrane could possibly be accelerated by an electrochemical process [106]. For a gas containing 0.1% CO_2, Cowan

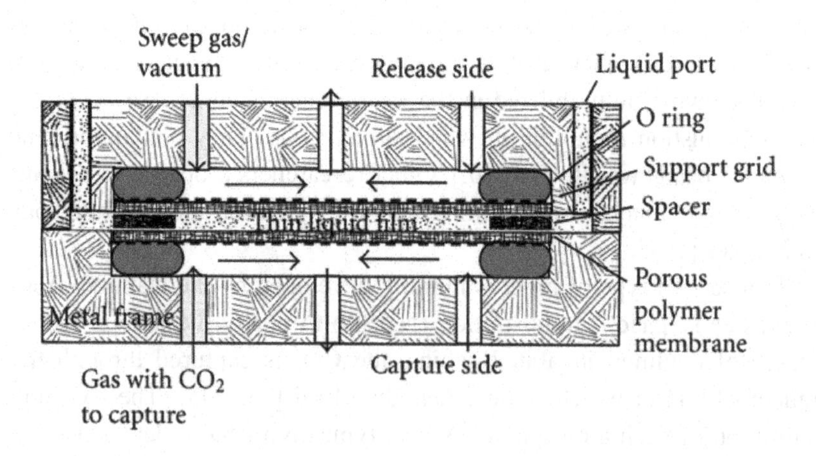

FIGURE 13: Schematic illustra of a thin liquid membrane system such as developed by NASA to capture CO_2. Adapted from [25].

et al. showed that a liquid membrane with CA enzyme was very stable, even in the presence of dry feed, and had a CO_2 permeance of 4.71×10^{-8} mol $m^{-2} Pa^{-1} sec^{-1}$ at ambient temperature and pressure [105]. Ward and Robb were the first to apply simple diffusion liquid enzyme membranes to gases containing 5% CO_2, in which the enzyme was dissolved in cesium or potassium bicarbonate solutions [107]. Suchdeo and Schultz used enzyme solutions in sodium bicarbonate [108]. Matsuyama and colleagues extended such enzymatic liquid films to gases containing up to 15% de CO_2, more representative of industrial fumes [109, 110].

Technical difficulties may appear due to the drying of aqueous film during long time operation. Humidifiers such as based on polysulfone were proposed to humidify the capture and release gases [105]. However, to better solve this problem, Trachtenberg et al. adapted the technique to networks of hollow microporous fibers in which the flue gas and the release gases could flow [36, 111]. Next to this progress, the Carbozyme company developed a technology which is schematically illustrated in Figure 14, based on such hollow microporous propylene microfibers, separated by control separators made with thin oxide powders, the whole system bathing in a excess aqueous enzyme solution. The enzyme was directly immobilized on the external faces of the microfibers. Water vapor under moderate vacuum (15 kPa) was used as the sweeping gas at a low flow rate in the release microfibers. The CO_2 content in the sweeping gas reached ≈95%, for a flue gas containing ≈15% de CO_2. No significant loss of enzyme activity was observed during a 5 days continuous run [11$_2$], and a conservative run time of $_2$500 hours was selected before needing to change the enzyme [111]. Several refinements were later designed, regarding the microfiber network geometry, the nature of hollow microfibers and the variety of CA used. This system was found to be efficient for a flue gas containing from 0.05 to 40% CO_2, at a temperature ranging from 15 to 85°C with a particular γCA isozyme [39, 11$_2$]. Trachtenberg et al. also confirmed that the system permeance, as well as the selectivity of CO_2 transfer with respect to O_2, N_2, and Ar, decreased when the CO_2 content in the flue gas increased [36, 11$_2$]. Hollow fiber membrane reactors were also designed and modeled with success by Zhang et al. Their set-up comprised some CA immobilized in nanocomposite hydrogel/hydrotalcite thin films, used as thin layers to separate the fibers [113–115]. With 0.1% (v/v) of CO_2 in

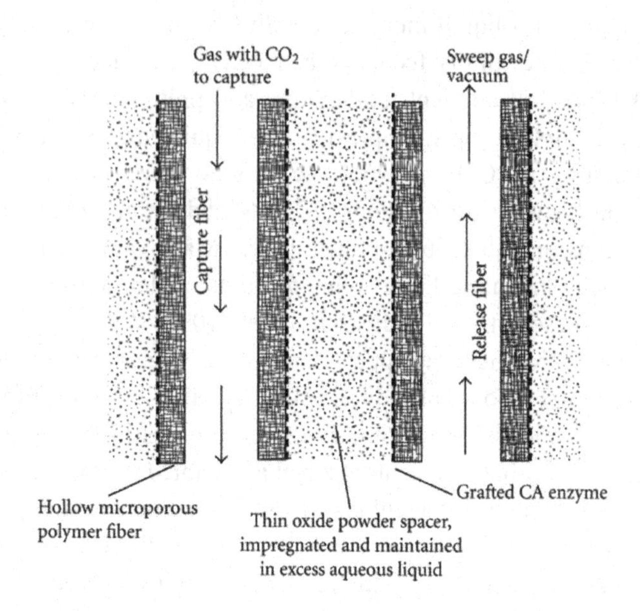

FIGURE 14: Schematic illustration of a membrane CO_2 capture system basd on microporous microfibers, such as developed by the Carbozyme Inc. company. Adapted from [111].

the flue gas, a permeance of mol m^{-2} s^{-1} Pa^{-1} was achieved for a selectivity of CO_2 of 820 over N_2 and 330 over O_2, and a stable performance during a 30 h run [114].

Some patents were also deposited by the Novozymes company. The latter proposed to combine various CO_2 capture and release units, such as those developed by the CO_2 Solution or Carbozyme Companies, interconnected by fluid circulation pipes [83]. Some results obtained with hollow microfiber units containing 0.03 mg mL^{-1} αCA extracted from the bacteria "Bacillus Clausii KSM-K16," dissolved in a 1 M pH 8 sodium bicarbonate solution and applied to a flue gas containing 15% CO_2, are gathered in Table 3.

The CO_2 scrubbers presented above mostly use immobilized CA. One reason is that commercial CA are costly and their immobilization on a support permits to use them for a longer time before losing them by leaching, hence to decrease the operational cost. This explains that a number of researchers addressed the problem of CA immobilization.

TABLE 3: Percent CO_2 in the release gas after a 75 min. capture test, with or without αCA enzyme, according to [83]. The flue gas composition was 15% CO_2 and 85% N_2.

Liquid	% de CO_2 in the release gas after 75 min test	Initial liquid pH	Final liquid pH
Water, no enzyme	0	---	---
$NaHCO_3$, 1 M, no enzyme	3.3	8	8.3
$NaHCO_3$, 1 M + 0.03 mg mL^{-1} enzyme	10	8.1	8,8

Actually, the immobilization of CA on solid supports is not recent. In 1988, Crumbliss et al. published a paper on the immobilization of BCA on silica beads and graphite rods [116]. The enzyme was covalently linked after activation of the graphite with amide bonds, while glutaraldehyde was used as an intermediate between the enzyme and the beads. The BCA surface coverage on the silica beads was reported to be superior to previously reported data on silica beads and polyacrylamide gels and comparable to that on other organic matrix supports. In 2001, Bond et al. immobilized BCA by adsorption on chitosane and alginate beads, which they applied in a sequestration process of CO_2 as stable solid carbonates [30, 117, 118]. In 2003, Hosseinkhani and Nemat-Gorgani adsorbed partially unfolded CA on hydrophobic alkyl substituted sepharose 4B supports. The octyl substituted support provided the best thermal stability and highest $k_{cat}/K_{m(app)}$, which was attributed to an irreversible thermal inactivation of the enzyme by interaction with the alkyl support groups [119]. Dilmore et al. studied the adsorption of CA on wet polyacrylamidegel beads functionalized with amino groups facilitating the adsorption of CO_2 [120]. Adsorption on chitosan and alginate supports was often used, such as for CA from *Pseudomonas fragi, Micrococcus lylae, Micrococcus luteus* 2, and *Bacillus pumilus* [121–123]. The immobilized enzymes showed improved storage stability and retained up to 50% of the initial activity after 30 days [123]. The immobilization of CA from *Bacillus pumilus* on chitosan beads was also studied by Wanjari et al. [124]. In p-NPA hydrolysis, they determined that K_m and v_{max} were 2.36 mM and 0.54 μmole min^{-1} mL^{-1}, respectively, for the immobilized CA, versus 0.87 mM and 0.93 μmole min^{-1} mL^{-1}, respectively, for the free CA [124]. For CA immobilized on ordered mesoporous aluminosilicate, K_m, v_{max}, and k_{cat} were 0.158 mM,

2.307 μmole min^{-1} mL^{-1}, and 1.9 s^{-1}, respectively [125]. Besides, whole cells of *Bacillus pumilus* were immobilized on different chitosan and sodium alginate based materials, which also improved their esterase activity by comparison with the free cells [121]. For the same *Bacillus pumilus* CA immobilized on chitosan activated alumina-carbon composite beads, the K_m and v_{max} values were 10.35 mM and 0.99 μmole min^{-1} mL^{-1} [126]. Overall, in p-NPA hydrolysis, it was found that K_m often decreased after immobilization, which denoted a greater affinity of the CA for the substrate, while v_{max} increased.

The main drawback of adsorption techniques is the enzyme progressively desorbs and is leached out during repeated tests. To solve this problem, the enzyme can be covalently grafted on a support, and a few techniques have been experimented in this direction. Bhattacharya et al. immobilized CA on iron particles coated with γ aminopropyltriethoxysilane, by grafting via dicarbocarbodi-imide (DCC) bonds or via dicarboxy bonds after conversion of the support surface groups with succinic anhydride. Cyanogen bromide coupling on an intermediate thin glass coating was also applied. Immobilization was also carried out by CA copolymerization with gluteraldehyde in methacrylic acid polymer beads. All these methods were reported to provide excellent results regarding the activity (98% activity retention) and leaching, in particular the DCC and dicarboxy coupling methods [97]. Belzil and Parent grafted human CA on nylon 6.6 Raschig rings [127]. The best activity in CO_2 hydration from a gas containing 20% CO_2 at 1°C was obtained by grafting CA from an enzyme solution contained 0.5 mg mL^{-1} CA. 73% of the enzyme was actually grafted on the support and 45% of this enzyme was active. However the relative hydration yield for the immobilized enzyme, defined by (19) itself derived from (7), was only 20% compared to 85% for the free enzyme:

$$\text{Yield} = v_r(\text{U mg}^{-1})m_{enz} = \frac{v_{enz} - v_0}{v_0} \tag{19}$$

Zhang et al. investigated the covalent grafting of CA on a hybrid-Poly (acrylic acid-co-acrylamide)/hydrotalcite nanocomposite termed "PAA-AAm/HT" [113, 114]. Hydrotalcite is a basic inorganic material of composition $M_g6Al_2(CO_3)(OH)_{16}\cdot4(H_2O)$ [128, 129] and coupling was achieved by the intermediate of N-hydroxysuccinimide (NHS) and DCC. Up to 4.6 mg of enzyme per gram of support could be grafted and 76.8% of the initial enzyme activity could be retained after immobilization. A covalent coupling method was also developed to graft CA enzymes onto silica nanoparticles made by spray pyrolysis [130]. These immobilized enzymes exhibited a significantly improved thermal stability compared to the free counterpart. Lee et al. studied the immobilization of CA by single or multiple attachments to polymers, themselves deposited onto Fe_3O_4 magnetic aggregates [131]. Yadav et al. immobilized CA on silylated chitosan beads, to precipitate $CaCO_3$ [13$_2$]. They observed that the immobilized CA had a longer storage stability than the free enzyme and retained 50% of its initial activity up to 30 days. They also developed core-shell single enzyme nanoparticles (SEN-CA), by covering the CA surface with a thin layer of chitosan, which showed an improved stability by comparison with the free enzyme [133].

Vinoba et al. compared BCA immobilized on SBA-15 by various techniques comprising covalent attachment (BCA-CA), adsorption (BCA-ADS), and cross-linked enzyme aggregation (BCA-CLEA). They found all were promising reusable catalysts [134]. In the hydrolysis reaction of para-nitrophenyl acetate (p-NPA), the k_{cat}/K_m values were 740.05, 660.62, and 680.11 $M^{-1}s^{-1}$, respectively, by comparison with 873.76 $M^{-1}s^{-1}$ for free BCA. In the hydration of CO_2, the k_{cat} values were 0.58, 0.36, 0.78 s^{-1} by comparison with 0.79 s^{-1} for free BCA, respectively, indicating that BCA-CLEA showed a comparatively higher hydration rate than the other immobilized CA, although it remained lower than the free CA [135]. Vinoba et al. also immobilized human carbonic anhydrase (HCA) via electrostatic interactions on silver nanoparticles confined in amine-functionalized mesoporous SBA-15 [136]. The latter retained ≈87% of its initial activity after 30 days. Similarly, they immobilized HCA on Au nanoparticles assembled over amine/thiol-functionalized, mesoporous SBA-15 [137]. Depending on the grafting agent K_m ranged fom 22.35 to 27.75 mM and k_{cat}/K_m from 1514.09 to 1612.25 $M^{-1}s^{-1}$ in p-NPA hydrolysis. With HCA simply covalently immobilized on SBA-15 via various amines, the values ranged from 7182 to 7569 $M^{-1}s^{-1}$ [138].

Besides adsorption and covalent grafting, enzymes can also be efficiently entrapped within porous supports. Such entrapment can be done within polyurethane foams, next to an initial protocol developed by Wood et al. in 1982 [139]. This technique was extended with success to CA enzymes for CO_2 capture, by Kanbar and Ozdemir [88] hydrolysis reaction of para-nitrophenylacetate (p-NPA) to para-nitrophenol (p-NP) and by Ozdemir [72]. Polyurethane immobilized CA could be used without any activity loss in aqueous media for 7 successive CO_2 capture tests and the optimum operational temperature was in a range from 35°C to 45°C. Simple entrapment in the open pores of a porous polymeric membrane is possible by immersion of the membrane in an enzyme solution, such as done by Favre and Pierre [140] with BCA, for a thin membrane system. In a 1 M $NaHCO_3$ solution at initial pH ≈ 8, the existence of an optimum enzyme concentration of 0.2 mg mL^{-1} was observed. The permeance was quite comparable with the data gathered by BaO and Trachtenberg and reported in Figure 2. The $NaHCO_3$ solution helped to maintain a high [HCO_3^-] concentration by displacing (5) towards a higher pH, while the electrical neutrality was insured by the Na^+ cations. As a further development hybrid sol-gel membranes were moreover made by impregnation of the previous polymeric membranes with a SiO_2 sol made from tetramethylorthosilicate (TMOS) [140]. After SiO_2 gelation and drying, the hybrid polymer-SiO_2 membranes were then impregnated with an enzyme aqueous solution in a buffer. It was observed that SiO_2 moderately increased the membrane permeance. Based on these results, a moderate catalytic action of silica in the capture of CO_2 could be proposed. It relied on the fact that the isoelectric point (i.e.p) of SiO_2 is low: i.e; p. ≈ 2.5 à 3 [141]. At pH > i.e.p, as this is the case in CO_2 capture, SiO_2 carries an excess of ≡SiO− negative surface charges. Hence the equilibrium between ≡SiOH and ≡SiO− surface sites in SiO_2, is similar to that between −Zn-OH_2 and −ZnOH− in the active enzyme site, so that a catalytic mechanism similar to that of the enzyme could be proposed. However this possible catalytic effect remained moderate and could be impeded by a necessary diffusion inside of CO_2(aq) inside the gel fine pore texture, depending on the membrane architecture.

In a variation of the process, nylon membranes was impregnated with a silica sol in which the enzyme was dissolved. SiO_2 gelation then directly occurred during the CO_2 transfer [11]. Leaks were rapidly induced by shrinkage of the gel about the nylon the fibers, followed by rapid drying, but these leaks occurred more rapidly with the CA enzyme than without it. This result led to investigation the action of CA enzyme and of CO_2(aq) on the gelation of silica sol made from TMOS [142]. It was found that both additives accelerated the gelation of SiO_2, although in a different way. While CO_2(aq) acted as an acidic gelation catalyst, leading to a so-called "polymeric" SiO_2 gel characterized by a very high specific surface area and small mesopores mixed with micropores [143], the CA enzyme acted as a basic catalyst resulting in a more "colloidal" gel, characterized by a lower specific surface area, mesopores of bigger size, and a reduced contraction during drying. These results were consistent with several other reports on the capability of other enzymes or proteins to catalyze the formation of silica from a liquid precursor. Such effects were indeed observed with polypeptides [144, 145], silicatein [146], lysozyme [147, 148], papaïn and trypsin [149], and a lipase from *Burkholderia cepacia* [150]. Monolithic SiO_2 gels were synthesized by Frampton et al. by hydrolysis of tetraethoxysilane (TEOS), catalyzed by α-chymotrypsin or trypsin and by hydrolysis of phenyltrimethoxysilane (PhTMOS) catalyzed by pepsin, in a time scale where gels were not obtained without any enzyme [149].

Because SiO_2 is slightly acidic, the deposition of a basic solid was also attempted in nylon and hybrid nylon-SiO_2 membranes. This was done by impregnating these membranes with a $CaCl_2 \cdot 2H_2O$ solution containing 0.2 mg mL^{-1} enzyme at pH $\approx$ 10.5, so as to deposit some $CaCO_3$ in situ during CO_2 capture. In the nylon membranes, the best permeance was obtained with a $CaCl_2 \cdot 2H_2O$ concentration of 0.146 M and was equivalent to that obtained with a 1 M $NaHCO_3$ solution at pH $\approx$ 8. SEM micrographs showed that small calcite crystals had deposited on the nylon fibers. On the other hand, in hybrid nylon-SiO_2 membranes, the permeance was slightly lower than that of similar hybrid nylon-SiO_2 membranes impregnated with in a 1 M $NaHCO_3$ solution. SEM micrographs and EDX microanalysis showed that besides calcite, some calcium silicate was also formed. Because SiO_2 is slightly soluble at pH 10.5, it is therefore possible that coprecipitation of the solubilized SiO_2 with the calcium from $CaCl_2$ may have produced this calcium silicate.

5.6 ENZYMATIC CO_2 STORAGE AS SOLID CARBONATES

The storage of CO_2 as solid carbonates requires the carbonation of basic cations dissolved or in contact with an aqueous medium. The most noteworthy candidates are alkaline or alkaline earth cations available in hydroxides such $Ca(OH)_2$ or oxides such as MgO and CaO [13, 30, 61, 77, 85, 117, 151–153]. Natural minerals such as wollastonite (CaSiO3) [154], serpentine ($Mg_3Si_2O_5(OH)_4$) [155], and olivine (Mg_2SiO_4) [156] provide such cations. More interestingly, metallurgical slags, lignite ashes, or chemical brine rejects such as from the oil industry [157] contain a significant concentration of such cations and could be used for carbonation. Besides, materials such as concrete do contain a high concentration of such cations and are abundantly used in civil engineering. Actually, the incorporation of CA enzyme in cement compositions was attempted with success, with the aim of designing civil engineering materials able to capture the CO_2 from the air and directly sequester it as solid carbonates within porous coatings of building walls [158].

The products of these carbonations are stable solid carbonates, for exemple, $MgCO_3$ and $CaCO_3$ which are themselves the major components of naturallimestone. Hence they present no environmental inconvenience for long term safe disposal [13, 61, 85], and they could possibly be reused in civil engineer constructions.

A first important drawback is that, considering the molar mass of the silicates mentioned above, from 1.6 to 3.7 tons of silicate source and from 2.6 to 4.7 tons of products would have to be handled, per ton of CO_2 to be stored. Hence, huge geological deposits would have to be mined when the cation source has to be extracted as a solid from such sites.

A second drawback is that the pH decreases when the carbonation reaction proceeds, due to the fact that the first deprotonation (5) as well as the second deprotonation equilibrium of CO_2(aq) to form the carbonate anions CO_3^{2-} according to (20) [159], both decrease the pH:

$$HCO_3^- \leftrightarrows H^+ + CO_3^{2-} \quad K_{a2} = 10^{-10.33} \tag{20}$$

Given the magnitude of K_{a2}, CO_3^{2-} anions predominate at pH > 10.5. Hence, it is necessary to maintain a high pH to induce the precipitation of a solid carbonate. Indeed, precipitation of a solid carbonate, such as $CaCO_3$, is itself the result of a dissolution/precipitation equilibrium as described in (21), where the solubility product $K_s = [Ca^{2+}][CO_3^{2-}]$ relates the ion concentrations in the liquid solution in equilibrium with the solid being precipitated:

$$Ca^{2+} + CO_3^{2-} \leftrightarrows CaCO_3\,(s) \quad 1/K_S \tag{21}$$

According to (21), CO_3^{2-} anions are continuously withdrawn from the liquid solution during precipitation of the solid phase. Hence, new CO_3^{2-} anions must be continuously supplied in the solution in order for precipitation to keep proceeding. This supply is achieved in accordance with (20), which simultaneously brings supplementary H^+ cations, so that the pH keeps decreasing unless a buffer is continuously supplied. If this is not done, the HCO_3^- anions again predominate when the pH reaches a value $<pK_{a2}$ (20), so that precipitation stops, and the previously precipitated solid carbonate may even redissolve. This explains that in laboratory batch studies with a given initial buffer concentration, the final mass of $CaCO_3$ precipitated was relatively the same with or without enzyme. Only the initial rate to reach this final mass changed. Depending on the CO_2 capture system used, it was actually shown that the mass of $CaCO_3$ precipitated could only be used as an indicator of the CO_2 capture rate at the beginning of precipitation [140]. This problem can be solved by carefully monitoring the continuous rate of addition of basic ashes or brines relative to the flow rate of CO_2 to sequester [157].

Moreover, the deposit of $CaCO_3$ from natural brine solutions supersaturated in both bicarbonate anions and Ca^{2+} cations, was extensively studied by geochemists, for instance Dreybrodt et al. [160]. In this case the deposit of $CaCO_3$ occurs when the solution is placed in contact with an atmosphere where the $P(CO_2(g))$ partial pressure is lower than that corresponding to equilibrium with the brine solution. When this occurs, some CO_2 is not captured but released from the brine by reverse dehydration of HCO_3^-, according to the equilibrium equation (5). The situation is equivalent to the release of CO_2 on the release side of a thin liquid membrane. However, simultaneously, a proton H^+ is captured, which increases the pH and in turn displaces the equilibrium reaction in (20), in favour of the formation of CO_3^{2-} anions. The latter in turn induce the precipitation of $CaCO_3$. Overall, one mole of $CO_2(g)$ is released per mole of deposited $CaCO_3$. If this was applied to the captured CO_2, half of this CO_2 would be released in the air. Obviously, such a situation must be avoided when the aim is to capture and fully sequester $CO_2(g)$. For this purpose, the $P(CO_2(g))$ partial pressure in the gas in contact with the brine, must not be lower than the equilibrium partial pressure corresponding to the brine $[HCO_3^-]$ concentration.

TABLE 4: Solubility product as a function of the temperature , for various cristallographic forms of $CaCO_3$, according to Gal et al. [84].

Mineral Name	Crystallographic structure	$-\log K_S$ (at 25° C)	$-\log K_S$ as a function of T (K) or θ (°C)
Amorphous	Amorphous	6.40	$= 6.1987 + 0.005336\theta + 0.001096\theta^2$
Ikaïte	Monoclinic	6.62	$= 1696/T + 0.9336$
Vatérite	Hexagonal	7.91	$= 172.1295 + 0.07793T - 3074.688/T - 71.595\log T$
Aragonite	Orthorhombic	8.34	$= 171.9773 + 0.07793T - 2903.293/T - 71.595\log T$
Calcite	Rhombohedral	8.48	$= 171.9065 + 0.07793T - 2839.319/T - 71.595\log T$

From the kinetic point of view, the second deprotonation of $CO_2(aq)$ (20) is much faster than the first one (5) and it does not a priori necessitate to use a catalyst. However, because a CA enzyme catalyzes the formation of HCO_3^- from $CO_2(aq)$ (5), which in turn displaces successively the equilibrium reactions in (20) to form more CO_3^{2-} anions, and the precipitation of $CaCO_3$ (21), this $CaCO_3$ precipitation is indirectly catalyzed by the enzyme. This is particularly true when carbonation is directly made in the medium where CO_2 is captured, where the first and second deprotonation steps of $CO_2(aq)$ occur in the same medium.

At a given temperature, the solubility limite K_s in (21) is a thermodynamic constant. However, its value depends on the solid phase which first nucleates, as indicated in Table 4 [84]. In this table, the final solid phase most often obtained is the stable thermodynamic phase calcite. The other crystalline forms are metastable, although they may easily nucleate and grow first, before calcite, depending on the conditions. When this occurs, they eventually redissolve in a second stage to reprecipitate as calcite, as illustrated further on. Overall, these phases are not very soluble at high temperature, but their solubility increases as the temperature decreases.

The precipitation kinetic of $CaCO_3$ in a $CaCl_2$ solution, or of other alkaline earth cation carbonates, was studied by several authors, in particular Pocker and Bjorkquist [29], Bond et al., [30, 117] and Druckenmiller and Maroto-Valer [161]. Various techniques were used to follow such precipitation, in particular the precipitation onset time according to turbidity data, the $[Ca^{2+}]$ concentration before and after precipitation, the total "inorganic carbon" concentration in solution, and the pH decreasing rate in industrial brines containing Ca^{2+} cations. The enzyme itself could directly be present in the precipitation medium, or used to first catalyze the formation of HCO_3^- anions in a pH range from 8.55 to 8.7, while the Ca^{2+} containing brine was added in a second step. Ramanan et al. [61, 85] compared $CaCO_3$ precipitation with the enzymes from *Citrobacter freundii* and *Bacillus subtilis*. They showed that the crude enzymes were much less active than the purified ones. Li et al. showed that CA of microbial origin and bovine CA both accelerated the precipitation of $CaCO_3$ and favoured the formation of the calcite phase [162]. An acceleration of $CaCO_3$ precipitation was also observed by Da Costa et al. on bovine CA (BCA) extracted by $_2$ different techniques [62] and by Kim et al. with a cheaper recombinant αCA from *Neisseria gonorrhoeae* (NCA) [63]. With purified CA from Pseudomonas fragi, immobilized by adsorption on chitosan, Anjaba et al. observed a more than 2 fold increase in calcite $CaCO_3$ sequestration by comparison with the free enzyme, in 5 minutes precipitation tests [122, 123]. With *Bacillus pumilus* CA adsorbed on chitosan beads, Wanjari et al. [124] showed that the precipitation of $CaCO_3$ was also accelerated by comparison with the free CA. Mirjafari et al. determined the mass of $CaCO_3$ precipitated in an aqueous $CaCl_2.2H_2O$ solution containing a buffer, when some CO_2 saturated water was added to the solution [77]. They also followed the evolution of the liquid turbidity as a function of time. Sharma and Bhattacharya compared CA extracted from *Pseudomonas fragi, Micrococcus lylae*, and *Micrococcus luteus* 2 with commercial bovine CA (BCA) [86]. The 3 bacterial CA exhibited enhanced CO_2 sequestration compared to the commercial BCA.

With CA immobilized on silylated chitosan beads, Yadav et al. determined apparent Michaelis constants K_m and v_{max} for the precipitation of $CaCO_3$ [132]. For this purpose, they quanticized the $CaCO_3$ by gas chromatography after decomposing it with HCl to release the captured CO_2(g). They found that K_m was higher for the immobilized enzyme than for the free enzyme (respectively 4.547 mM and 1.211 mM), while v_{max} was relatively unchanged (1.018 and 1.211 mmol min^{-1} mg^{-1}). The CO_2 sequestration capacity was found to be best enhanced with CA immobilized on core-shell CA-chitosan nanoparticles [133]. Kim et al. showed that the precipitation rate of $CaCO_3$ was about 3-fold faster with BCA and a CA enzyme extracted from oyster shell, than without enzyme [163]. In CO_2 capture and sequestration as $CaCO_3$ in 2 successive steps, BCA immobilized by various techniques on SBA-15 supports [134] displayed a similar $CaCO_3$ precipitation capability [135]. With human carbonic anhydrase (HCA) immobilized via electrostatic interactions on silver nanoparticles, themselves confined to amine-functionalized mesoporous SBA-15, they reported a CO_2 capture rate ~25 fold higher than that of free HCA after 30 cycles [136]. HCA was also immobilized on Au nanoparticles assembled over amine/thiol-functionalized mesoporous SBA-15 and the $CaCO_3$ final mass precipitated per test was similar to that of free HCA. However, the immobilized BCA retained its activity after 20 days storage at 25° and 20 recycling [137]. With HCA simply covalently immobilized on SBA-15 via various amines, the immobilized HCA efficiency in CO_2 hydration was 36 times greater than free HCA, and 75% of initial enzymatic activity was retained after 40 cycles [138]. Favre et al. investigated the deposition kinetics of $CaCO_3$ in a mixture of CO_2 saturated water, a buffer at different pH and an aqueous $CaCl_2 \cdot 2H_2O$ solution, for different enzyme concentration [164]. Overall, it was shown that the enzyme could drastically increase the apparent precipitation rate of $CaCO_3$(s), during the first minute. However, a maximum in this precipitation rate was observed, for an enzyme mass ≤ 0.3 mg mL^{-1}. This result could be explained by a faster increase of the formation rate of HCO_3^- and H^+ ions during the first deprotonation step, when the enzyme concentration increased. Hence, the pH decreased more rapidly to a low value, unfavorable to the formation of CO_3^{2-} ions. In turn, this stopped the precipitation of $CaCO_3$(s) at an earlier time. Globally, if the pH was not maintained at a high enough value (e.g., 10) the total mass of $CaCO_3$(s) precipitated did not depend on the presence of enzyme. The enzyme only modified the time to reach equilibrium where $CaCO_3$(s) precipitation stopped. The final mass of $CaCO_3$(s) precipitated only depended on the initial buffer nature, pH and concentration. The precipitates were also analyzed by X-ray diffraction [164]. At 20°C and initial pH 8.4 or 9.4, the only phase observed was calcite when no enzyme was added. At initial pH 10.5, vaterite was predominant. On the other hand, still at pH = 10.5,

FIGURE 15: SEM micrographs of $CaCO_3$ solid particles precipitated from $CaCl_2$ at initial pH 10.5 and 20°C: (a) with CA enzyme; (b) without enzyme. Adapted from [164].

the enzyme favored the conversion of vaterite to calcite. The same observation was made at 5°C, although this conversion to calcite was uncomplete at this temperature. Scanning electron micrographs of these $CaCO_3$ deposits are shown in Figure 15. Both $CaCO_3$ phases have a hexagonal structure [165, 166], but vaterite is more complex than calcite. Solid vaterite particles displayed a porous spherical shape made by aggregation of nano crystallites, while calcite particles are characterized by a well defined rhombohedral shape with marked facets.

It was known that, in some cases, enzymes may possibly catalyze the formation of a given solid phase [142]. However, more common mechanisms are likely. Indeed, a spontaneous change in the crystallographic form and/or particle shape is often observed when solid particles nucleate and grow in liquid media [167]. Shape transitions without any crystallographic phase change can be due to a change in the concentration magnitude of the oligomers responsible for precipitation. These oligomers are formed from the chemical precursor of the solid, in the present case $CaCl_2 \cdot 2H_2O$. Moreover, the first particles which nucleate do not necessarily correspond to the most stable thermodynamic phase. They really depend on the nature of those precursor oligomers which first reached a critical supersaturation for nucleation, and which may correspond to a metastable phase. On the other hand, the most stable thermodynamic phase (presently calcite) is likely to form in a second slower step, by dissolution of the first metastable solid phase, followed by re-precipitation to calcite. In the study of Favre et al., the first $CaCO_3$ phase nucleating at pH 10.5 was vaterite [164]. But this phase is also more soluble than calcite [84] so that it could redissolve and reprecipitate to calcite, at a lower rate. Consequently, metastable vaterite could more likely be observed during the first stage of precipitation, in particular when the overall precipitation kinetics was slow, hence at 5°C rather than at 20°C and/or when no enzyme was present.

5.7 CONCLUSION

Carbonic anhydrase are amongst the most well known enzymes, since they operate in most living organisms, including human beings where they play an important role. Their catalytic mechanism in the hydration of CO_2(aq) molecules has been extensively studied, and the summary presented in this review has stressed out the fact that this was a complex mechanism, requiring the use of pH buffers with which the enzyme was in direct competition. Nonetheless the well understood chemical physics laws underlying the capture of CO_2 in aqueous medium have permitted to develop several types of efficient CO_2 capture reactors. In particular, hollow microfibers reactors seem very promising for applications to industrial fumes. To improve their applicability, significant progress on several points may also be expected. These concern the cost of these enzymes, their catalytic activity, their stability in time and their resistance to pollutants such as sulfur compounds. Indeed, the large variety of carbonic anhydrase enzymes available in the living organisms, with a very different resistance capability to the operational parameters involved, may permit to anticipate that their application to capture CO_2 will increasingly emerge as an efficient, environmental friendly technique, applicable with very moderate energy consumption, without requiring any heating.

REFERENCES

1. C. D. Keeling, T. P. Whorf, M. Wahlen, and J. Van Der Plicht, "Interannual extremes in the rate of rise of atmospheric carbon dioxide since 1980," Nature, vol. 375, no. 6533, pp. 666–670, 1995.
2. M. R. Raupach, G. Marland, P. Ciais et al., "Global and regional drivers of accelerating CO2 emissions," Proceedings of the National Academy of Sciences of the United States of America, vol. 104, no. 24, pp. 10288–10293, 2007.
3. A. J. McMichael, R. E. Woodruff, and S. Hales, "Climate change and human health: present and future risks," The Lancet, vol. 367, no. 9513, pp. 859–869, 2006.
4. Core Writing Team, IPCC, "Climate change 2007: synthesis report," in Contribution of Working Groups I, II and III to the Fourth Assessment Report of the Intergovernmental Panel on Climate Change, R. K. Pachauri and A. Reisinger, Eds., p. 104, IPCC, Geneva, Switzerland, 2007.
5. United Nations Framework Convention on Climate Change (UNFCCC), Kyoto Protocol Reference Manuel on Accounting of Emissions and Assigned Amounts, 2008, http://unfccc.int/2860.php.
6. M. M. Maroto-Valer, D. J. Fauth, M. E. Kuchta, Y. Zhang, and J. M. Andrésen, "Activation of magnesium rich minerals as carbonation feedstock materials for CO2 sequestration," Fuel Processing Technology, vol. 86, no. 14-15, pp. 1627–1645, 2005.

7. M. Abu-Khader, "Recent progress in CO2 capture/sequestration: a review," Energy Sources, Part A, vol. 28, no. 14, pp. 1261–1279, 2006.
8. International Energy Agency, Policy Strategy for Carbon Capture and Storage, OECD/IEA, Paris, France, 2012, http://www.iea.org/publications/freepublications.
9. N. Z. Muradov, "Fossil fuel decarbonization: in the quest for clean and lasting fossil energy," in Carbon-Neutral Fuels and Energy Carriers, N. Z. Muradov and T. N. Veziroglu, Eds., pp. 667–786, CRC Press, Taylor and Francis, 2012.
10. A. Y. Shekh, K. Krishnamurthi, S. N. Mudliar et al., "Recent advancements in carbonic anhydrase driven processes for CO2 sequestration: minireview," Critical Reviews in Environmental Science and Technology, vol. 42, no. 14, pp. 1419–1440, 2012.
11. N. Favre, Captage enzymatique de dioxyde de carbone [Ph.D. thesis], Report no. 109-2011, Université Claude Bernard-Lyon 1, Lyon, France, 2011.
12. M. Ramdin, T. W. de Loos, and T. J. H. Vlugt, "State-of-the-Art of CO2 Capture with Ionic Liquids," Industrial & Engineering Chemistry Research, vol. 51, pp. 8149–8177, 2012.
13. Working Group III of the Intergovernmental Panel on Climate Change, IPCC, "IPCC special report on carbon dioxide capture and storage," B. O. Metz, H. Davidson, C. de Coninck, M. Loos, and L. A. Meyer, Eds., p. 442, Cambridge University Press, New York, NY, USA, 2005.
14. G. S. Goff and G. T. Rochelle, "Monoethanolamine degradation: O2 mass transfer effects under CO2 capture conditions," Industrial and Engineering Chemistry Research, vol. 43, no. 20, pp. 6400–6408, 2004.
15. A. Kishimoto, Y. Kansha, C. Fushimi, and A. Tsutsumi, "Exergy recuperative CO2 gas separation in pre-combustion capture," Clean Techn Environ Policy, vol. 14, pp. 465–474, 2012.
16. C. A. Griffith, D. A. Dzombak, and G. V. Lowry, "Physical and chemical characteristics of potential seal strata in regions considered for demonstrating geological saline CO2 sequestration," Environmental Earth Sciences, vol. 64, no. 4, pp. 925–948, 2011.
17. K. Z. House, B. Altundas, C. F. Harvey, and D. P. Schrag, "The immobility of CO2 in marine sediments beneath 1500 meters of water," ChemSusChem, vol. 3, no. 8, pp. 905–912, 2010.
18. T. Sakakura, J. C. Choi, and H. Yasuda, "Transformation of carbon dioxide," Chemical Reviews, vol. 107, no. 6, pp. 2365–2387, 2007.
19. Y. Amao and T. Watanabe, "Photochemical and enzymatic methanol synthesis from HCO3- by dehydrogenases using water-soluble zinc porphyrin in aqueous media," Applied Catalysis B, vol. 86, no. 3-4, pp. 109–113, 2009.
20. E. Ono and J. L. Cuello, "Feasibility assessment of microalgal carbon dioxide sequestration technology with photobioreactor and solar collector," Biosystems Engineering, vol. 95, no. 4, pp. 597–606, 2006.
21. P. Kaladharan, S. Veena, and E. Vivekanandan, "Carbon sequestration by a few marine algae: observation and projection," Journal of the Marine Biological Association of India, vol. 51, pp. 107–110, 2009.
22. R. Morgan, "Photobioreactor and method for processing polluted air," From PCT International Application, WO 2008097845 A1 20080814, 2008.

23. M. Aresta, A. Dibenedetto, and G. Barberio, "Utilization of macro-algae for enhanced CO2 fixation and biofuels production: development of a computing software for an LCA study," Fuel Processing Technology, vol. 86, no. 14-15, pp. 1679–1693, 2005.
24. K. Skjånes, P. Lindblad, and J. Muller, "BioCO2—a multidisciplinary, biological approach using solar energy to capture CO2 while producing H2 and high value products," Biomolecular Engineering, vol. 24, no. 4, pp. 405–413, 2007.
25. J. Ge, R. M. Cowan, C. Tu, M. L. McGregor, and M. C. Trachtenberg, "Enzyme-based CO2 capture for advanced life support," Life Support & Biosphere Science, vol. 8, no. 3-4, pp. 181–189, 2002.
26. L. W. Diamond and N. N. Akinfiev, "Solubility of CO2 in water from -1.5 to 100°C and from 0.1 to 100 MPa: evaluation of literature data and thermodynamic modelling," Fluid Phase Equilibria, vol. 208, no. 1-2, pp. 265–290, 2003.
27. J. J. Carroll, J. D. Slupsky, and A. E. Mather, "The solubility of carbon dioxide in water at low pressure," Journal of Physical and Chemical Reference Data, vol. 20, pp. 1201–1209, 1991.
28. R. Crovetto, "Evaluation of solubility data of the system CO2-H2O from 273 K to the critical point of water," Journal of Physical and Chemical Reference Data, vol. 20, pp. 575–589, 1991.
29. Y. Pocker and D. W. Bjorkquist, "Stopped-flow studies of carbon dioxide hydration and bicarbonate dehydration in H2O and D2O. Acid-base and metal ion catalysis," Journal of the American Chemical Society, vol. 99, no. 20, pp. 6537–6543, 1977.
30. G. M. Bond, J. Stringer, D. K. Brandvold, F. A. Simsek, M. G. Medina, and G. Egeland, "Development of integrated system for biomimetic CO2 sequestration using the enzyme carbonic anhydrase," Energy and Fuels, vol. 15, no. 2, pp. 309–316, 2001.
31. C. S. Tautermann, A. F. Voegele, T. Loerting et al., "Towards the experimental decomposition rate of carbonic acid (H2CO3) in aqueous solution," Chemistry A European Journal, vol. 8, pp. 66–73, 2002.
32. B. H. Gibbons and J. T. Edsall, "Rate of hydration of carbon dioxide and dehydration of carbonic acid at 30°C," The Journal of Biological Chemistry, vol. 238, pp. 3502–3507, 1963.
33. G. T. Tsao, "The effect of carbonic anhydrase on carbon dioxide absorption," Chemical Engineering Science, vol. 27, no. 8, pp. 1593–1600, 1972.
34. J. M. Berg, J. L. Tymoczko, and L. Stryer, Biochemistry, W.H. Freeman and Company, 5th edition, 2002.
35. S. Lindskog and J. E. Coleman, "The catalytic mechanism of carbonic anhydrase," Proceedings of the National Academy of Sciences of the United States of America, vol. 70, no. 9, pp. 2505–2508, 1973.
36. L. Bao and M. C. Trachtenberg, "Facilitated transport of CO2 across a liquid membrane: comparing enzyme, amine, and alkaline," Journal of Membrane Science, vol. 280, no. 1-2, pp. 330–334, 2006.
37. E. C. Webb, Enzyme Nomenclature, Academic Press, 1992, http://www.chem.qmw.ac.uk/iubmb.

38. K. S. Smith, C. Jakubzick, T. S. Whittam, and J. G. Ferry, "Carbonic anhydrase is an ancient enzyme widespread in prokaryotes," Proceedings of the National Academy of Sciences of the United States of America, vol. 96, no. 26, pp. 15184–15189, 1999.
39. B. C. Tripp, K. Smith, and J. G. Ferry, "Carbonic anhydrase: new insights for an ancient enzyme," The Journal of Biological Chemistry, vol. 276, no. 52, pp. 48615–48618, 2001.
40. N. Meldrum and F. J. W. Roughton, "Carbonic anhydrase: its preparation and properties," The Journal of Physiology, vol. 80, pp. 113–142, 1933.
41. A. C. Neish, "Studies on chloroplasts. Their chemical composition and the distribution of certain metabolites between the chloroplasts and remainder of the leaf," Biochemical Journal, vol. 33, pp. 300–3308, 1939.
42. D. Keilin and T. Mann, "Carbonic anhydrase," Nature, vol. 144, no. 3644, pp. 442–443, 1939.
43. D. Keilin and T. Mann, "Carbonic anhydrase. Purification and nature of the enzyme," Biochemical Journal, vol. 34, pp. 1163–1176, 1940.
44. R. S. Rowlett and D. N. Silverman, "Kinetics of the protonation of buffer and hydration of CO2 catalyzed by human carbonic anhydrase II," Journal of the American Chemical Society, vol. 104, no. 24, pp. 6737–6741, 1982.
45. F. P. Veitch and L. C. Blankenship, "Carbonic anhydrase in bacteria," Nature, vol. 197, no. 4862, pp. 76–77, 1963.
46. L. Adler, J. Brundell, S. O. Falkbring, and P. O. Nyman, "Carbonic anhydrase from Neisseria sicca, strain 6021 I. Bacterial growth and purification of the enzyme," Biochimica et Biophysica Acta, vol. 284, no. 1, pp. 298–310, 1972.
47. M. B. Guilloton, J. J. Korte, A. F. Lamblin, J. A. Fuchs, and P. M. Anderson, "Carbonic anhydrase in Escherichia coli. A product of the cyn operon," The Journal of Biological Chemistry, vol. 267, no. 6, pp. 3731–3734, 1992.
48. M. B. Guilloton, A. F. Lamblin, E. I. Kozliak et al., "A physiological role for cyanate-induced carbonic anhydrase in Escherichia coli," Journal of Bacteriology, vol. 175, no. 5, pp. 1443–1451, 1993.
49. J. F. Domsic and R. McKenna, "Sequestration of carbon dioxide by the hydrophobic pocket of the carbonic anhydrases," Biochimica et Biophysica Acta, vol. 1804, no. 2, pp. 326–331, 2010.
50. C. T. Supuran, "Carbonic anhydrase inhibitors: possible anticancer drugs with a novel mechanism of action," Drug Development Research, vol. 69, no. 6, pp. 297–303, 2008.
51. B. E. Alber and J. G. Ferry, "A carbonic anhydrase from the archaeon Methanosarcina thermophila," Proceedings of the National Academy of Sciences of the United States of America, vol. 91, no. 15, pp. 6909–6913, 1994.
52. S. Lindskog, "Structure and mechanism of Carbonic Anhydrase," Pharmacology and Therapeutics, vol. 74, no. 1, pp. 1–20, 1997.
53. E. E. Rickli, S. A. Ghazanfar, B. H. Gibbons, and J. T. Edsall, "Carbonic anhydrase from human erythrocytes," The Journal of Biological Chemistry, vol. 239, pp. 1065–1078, 1964.
54. G. P. Miscione, M. Stenta, D. Spinelli, E. Anders, and A. Bottoni, "New computational evidence for the catalytic mechanism of carbonic anhydrase," Theoretical Chemistry Accounts, vol. 118, no. 1, pp. 193–201, 2007.

55. T. W. Lane and F. M. M. Morel, "Regulation of carbonic anhydrase expression by zinc, cobalt, and carbon dioxide in the marine diatom Thalassiosira weissflogii," Plant Physiology, vol. 123, no. 1, pp. 345–352, 2000.
56. S. Heinhorst, E. B. Williams, F. Cai, C. D. Murin, J. M. Shively, and G. C. Cannon, "Characterization of the carboxysomal carbonic anhydrase CsoSCA from Halothiobacillus neapolitanus," Journal of Bacteriology, vol. 188, no. 23, pp. 8087–8094, 2006.
57. T. W. Lane and F. M. M. Morel, "A biological function for cadmium in marine diatoms," Proceedings of the National Academy of Sciences of the United States of America, vol. 97, no. 9, pp. 4627–4631, 2000.
58. C. T. Supuran and A. Scozzafava, "Carbonic anhydrases as targets for medicinal chemistry," Bioorganic and Medicinal Chemistry, vol. 15, no. 13, pp. 4336–4350, 2007.
59. M. R. Sawaya, G. C. Cannon, S. Heinhorst et al., "The structure of β-carbonic anhydrase from the carboxysomal shell reveals a distinct subclass with one active site for the price of two," The Journal of Biological Chemistry, vol. 281, no. 11, pp. 7546–7555, 2006.
60. S. Breton, "The cellular physiology of carbonic anhydrases," Journal of the Pancreas, vol. 2, no. 4, supplement, pp. 159–164, 2001.
61. R. Ramanan, K. Kannan, N. Vinayagamoorthy, K. M. Ramkumar, S. D. Sivanesan, and T. Chakrabarti, "Purification and characterization of a novel plant-type carbonic anhydrase from Bacillus subtilis," Biotechnology and Bioprocess Engineering, vol. 14, no. 1, pp. 32–37, 2009.
62. O. J. da Costa, L. Sala, G. P. Cerveira, and S. J. Kalil, "Purification of carbonic anhydrase from bovine erythrocytes and its application in the enzymatic capture of carbon dioxide," Chemosphere, vol. 88, no. 2, pp. 255–259, 2012.
63. I. G. Kim, B. H. Jo, D. G. Kang, C. S. Kim, Y. S. Choi, and H. J. Cha, "Biomineralization-based conversion of carbon dioxide to calcium carbonate using recombinant carbonic anhydrase," Chemosphere, vol. 87, no. 10, pp. 1091–1096, 2012.
64. M. C. Trachtenberg, "Novel enzyme compositions for removing carbon dioxide from a mixed gas," U.S. Patent Application Publication US 20080003662 A1 2008010, 2008.
65. Z. Minic and P. D. Thongbam, "The biological deep sea hydrothermal vent as a model to study carbon dioxide capturing enzymes," Marine Drugs, vol. 9, no. 5, pp. 719–738, 2011.
66. Z. Liu, P. Bartlow, R. M. Dilmore et al., "Production, purification, and characterization of a fusion protein of carbonic anhydrase from Neisseria gonorrhoeae and cellulose binding domain from Clostridium thermocellum," Biotechnology Progress, vol. 25, no. 1, pp. 68–74, 2009.
67. C. Thibaud-Erkey and H. Cordatos, "Post-combustion CO2 capture with biomimetic membrane," Preprints of Symposia—American Chemical Society, Division of Fuel Chemistry, vol. 56, no. 1, pp. 232–233, 2011.
68. H. J. Kulik, S. E. Wong, E. Y. Lau et al., "Computational design of carbonic anhydrase mimics for carbon dioxide capture," in Proceedings of the 240th ACS National Meeting, Boston, Mass, USA, August 2010.

69. K. M. Wilbur and N. G. Anderson, "Electrometric and colorimetric determination of carbonic anhydrase," The Journal of Biological Chemistry, vol. 176, no. 1, pp. 147–154, 1948.
70. R. Brinkman, "The occurrence of carbonic anhydrase in lower marine animals," The Journal of Physiology, vol. 80, pp. 171–173, 1933.
71. F. J. Philpot and J. S. L. Philpot, "CCCVI. A modified colorimetric estimation of carbonic anhydrase," Biochemical Journal, vol. 30, pp. 2191–2193, 1936.
72. E. Ozdemir, "Biomimetic CO2 sequestration: 1. Immobilization of carbonic anhydrase within polyurethane foam," Energy and Fuels, vol. 23, no. 11, pp. 5725–5730, 2009.
73. D. Voet, J. G. Voet, and G. Rousseau, Biochimie, De Boeck University, 2005.
74. F. Larachi, "Kinetic model for the reversible hydration of carbon dioxide catalyzed by human carbonic anhydrase II," Industrial and Engineering Chemistry Research, vol. 49, no. 19, pp. 9095–9104, 2010.
75. C. T. Supuran, "Carbonic anhydrases—a overview," Current Pharmaceutical Design, vol. 14, no. 7, pp. 603–614, 2008.
76. D. Whitford, Proteins: Structure and Function, John Wiley & Sons, 2005.
77. P. Mirjafari, K. Asghari, and N. Mahinpey, "Investigating the application of enzyme carbonic anhydrase for CO2 sequestration purposes," Industrial and Engineering Chemistry Research, vol. 46, no. 3, pp. 921–926, 2007.
78. J. H. Zhang, R. C. C. Qi, T. Chen et al., "Development of a carbon dioxide-capture assay in microtiter plate for aspartyl-β-hydroxylase," Analytical Biochemistry, vol. 271, no. 2, pp. 137–142, 1999.
79. I. H. Segel, Enzyme Kinetics, John Wiley & Sons, 1993.
80. V. Henri, "Théorie générale de l'action de quelques diastases," Comptes Rendus de l'Académie des Sciences, Paris, vol. 135, pp. 916–919, 1902.
81. V. Henri, Lois GénéraLes de L'Action des Diastases, Herman, Paris, France, 1903.
82. L. Michaelis and M. L. Menten, "Kinetics of invertase action," Biochemistry Zeitshrift, vol. 49, pp. 333–369, 1913.
83. P. Saunders, S. Salmon, M. Borchert, and L. P. Lessard, "Modular membrane reactor and process for carbon dioxide extraction," Novozymes, International Patent WO 2010/014774 A2, 2010.
84. J. Y. Gal, J. C. Bollinger, H. Tolosa, and N. Gache, "Calcium carbonate solubility: a reappraisal of scale formation and inhibition," Talanta, vol. 43, no. 9, pp. 1497–1509, 1996.
85. R. Ramanan, K. Kannan, S. D. Sivanesan et al., "Bio-sequestration of carbon dioxide using carbonic anhydrase enzyme purified from Citrobacter freundii," World Journal of Microbiology and Biotechnology, vol. 25, no. 6, pp. 981–987, 2009.
86. A. Sharma and A. Bhattacharya, "Enhanced biomimetic sequestration of CO2 into CaCO3 using purified carbonic anhydrase from indigenous bacterial strains," Journal of Molecular Catalysis B, vol. 67, no. 1-2, pp. 122–128, 2010.
87. A. F. Ghannam, W. Tsen, and R. S. Rowlett, "Activation parameters for the carbonic anhydrase II-catalyzed hydration of CO2," The Journal of Biological Chemistry, vol. 261, no. 3, pp. 1164–1169, 1986.

88. B. Kanbar and E. Ozdemir, "Thermal stability of carbonic anhydrase immobilized within polyurethane foam," Biotechnology Progress, vol. 26, no. 5, pp. 1474–1480, 2010.
89. R. A. Sheldon, "Enzyme immobilization: the quest for optimum performance," Advanced Synthesis and Catalysis, vol. 349, no. 8-9, pp. 1289–1307, 2007.
90. O. Lacroix and F. Larachi, "Scrubber designs for enzyme-mediated capture of CO2," Recent Patents on Chemical Engineering, vol. 1, no. 2, pp. 93–105, 2008.
91. S. Fradette and O. Ceperkovic, CO2 Solution Inc., "An improved carbon dioxide absorption solution", PCT International Application, WO 2006089423 A1 20060831, 2006.
92. T. Hamilton, "Capturing Carbon with enzymes," 2008, http://www.technologyreview.com/Energy/18217/.
93. CO2 Solution Company, Enzymatic power for carbon capture, "The process", 2012, http://www.co2solutions.com/en/the-process.
94. C. Parent and F. Dutil, CO2 Solution Inc., "Triphasic bioreactor and process for gas effluent treatment", International Patent WO 2004/007058 A1, 2004.
95. C. Parent, A. Barry, S. Fradette, and R. Lepine, CO2 Solution Inc., "Gas purification apparatus and process using biofiltration and enzymatic reactions" Canadian Patent CA 2554395 A1, 2006.
96. S. Fradette, A. Belzil, M. Dion, and R. Parent, "Enzymatic process and bioreactor using elongated structures for carbon dioxide capture treatments," PCT International Application WO 2011054107 A1 20110512, 2011.
97. S. Bhattacharya, M. Schiavone, S. Chakrabarti, and S. K. Bhattacharya, "CO2 hydration by immobilized carbonic anhydrase," Biotechnology and Applied Biochemistry, vol. 38, no. 2, pp. 111–117, 2003.
98. Reardon and P. John, "Advanced low energy enzyme catalyzed carbonate solvents for CO2 capture-exploring operating conditions," Preprints of Symposia—American Chemical Society, Division of Fuel Chemistry, vol. 56, no. 1, pp. 224–225, 2011.
99. M. Su and V. Haritos, "Use of enzyme catalysts in CO2 post combustion capture (PCC) processes," PCT International Application WO 2010045689 A1 20100429, 2010.
100. N. J. M. C. Penders-Van Elk, P. W. J. Derks, G. F. Versteeg, and S. Fradette, "Enzyme enhanced carbon dioxide capture and desorption processes," CT International Application WO 2012055035 A1 20120503, 2012.
101. N. J. M. C. Penders-van Elk, P. W. J. Derks, S. Fradette, and G. F. Versteeg, "Kinetics of absorption of carbon dioxide in aqueous MDEA solutions with carbonic anhydrase at 298 K," International Journal of Greenhouse Gas Control, vol. 9, pp. 385–392, 2012.
102. I. Iliuta and F. Larachi, "New scrubber concept for catalytic CO2 hydration by immobilized carbonic anhydrase II & in-situ inhibitor removal in three-phase monolith slurry reactor," Separation and Purification Technol, vol. 86, pp. 199–214, 2012.
103. F. Larachi, O. Lacroix, and B. P. A. Grandjean, "Carbon dioxide hydration by immobilized carbonic anhydrase in Robinson-Mahoney and packed-bed scrubbers," Chemical Engineering Science, vol. 73, pp. 99–115, 2012.
104. S. Salmon, J. Holmes, and P. Saunders, "Ultrasound-assisted regeneration for CO2 capture processes," IP.com Journal, vol. 10, no. 2B, article 12, 2010.

105. R. M. Cowan, J. J. Ge, Y. J. Qin, M. L. McGregor, and M. C. Trachtenberg, "CO2 capture by means of an enzyme-based reactor," Annals of the New York Academy of Sciences, vol. 984, pp. 453–469, 2003.
106. Y. J. Lin, S. Datta, M. P. Henry et al., "An electrochemical process for energy efficient capture of CO2 from coal flue gas," Preprints of Symposia—American Chemical Society, Division of Fuel Chemistry, vol. 56, no. 1, pp. 311–312, 2011.
107. W. J. Ward and W. L. Robb, "Carbon dioxide-oxygen separation: facilitated transport of carbon dioxide across a liquid film," Science, vol. 156, no. 3781, pp. 1481–1484, 1967.
108. S. R. Suchdeo and J. S. Schultz, "Mass transfer of CO2 across membranes: facilitation in the presence of bicarbonate ion and the enzyme carbonic anhydrase," Biochimica et Biophysica Acta, vol. 352, no. 3, pp. 412–440, 1974.
109. H. Matsuyama, M. Teramoto, and K. Iwai, "Development of a new functional cation-exchange membrane and its application to facilitated transport of CO2," Journal of Membrane Science, vol. 93, no. 3, pp. 237–244, 1994.
110. H. Matsuyama, A. Terada, T. Nakagawara, Y. Kitamura, and M. Teramoto, "Facilitated transport of CO2 through polyethylenimine/poly(vinyl alcohol) blend membrane," Journal of Membrane Science, vol. 163, no. 2, pp. 221–227, 1999.
111. M. C. Trachtenberg, R. M. Cowan, D. A. Smith et al., "Membrane-based, enzyme-facilitated, efficient carbon dioxide capture," Energy Procedia, vol. 1, pp. 353–360, 2009.
112. M. C. Trachtenberg, D. A. Smith, R. M. Cowan, and X. Wang, "Flue gas CO2 capture by means of a biomimetic facilitated transport membrane," in Proceedings of the AIChE Spring Annual Meeting, Houston, Tex, USA, 2007, http://www.carbozyme.us/pub_abstracts.shtml.
113. Y. T. Zhang, T. T. Zhi, L. Zhang, H. Huang, and H. L. Chen, "Immobilization of carbonic anhydrase by embedding and covalent coupling into nanocomposite hydrogel containing hydrotalcite," Polymer, vol. 50, no. 24, pp. 5693–5700, 2009.
114. Y. T. Zhang, L. Zhang, H. L. Chen, and H. M. Zhang, "Selective separation of low concentration CO2 using hydrogel immobilized CA enzyme based hollow fiber membrane reactors," Chemical Engineering Science, vol. 65, no. 10, pp. 3199–3207, 2010.
115. Y.-T. Zhang, X.-G. Dai, G.-H. Xu et al., "Modeling of CO2 mass transport across a hollow fiber membrane reactor filled with immobilized enzyme," AIChE Journal, vol. 58, pp. 2069–2077, 2012.
116. A. L. Crumbliss, K. L. McLachlan, J. P. O'Daly, and R. W. Henkens, "Preparation and activity of carbonic anhydrase immobilized on porous silica beads and graphite rods," Biotechnology and Bioengineering, vol. 31, no. 8, pp. 796–801, 1988.
117. N. Liu, G. M. Bond, A. Abel, B. J. McPherson, and J. Stringer, "Biomimetic sequestration of CO2 in carbonate form: role of produced waters and other brines," Fuel Processing Technology, vol. 86, no. 14-15, pp. 1615–1625, 2005.
118. B. Krajewska, "Application of chitin- and chitosan-based materials for enzyme immobilizations: a review," Enzyme and Microbial Technology, vol. 35, no. 2-3, pp. 126–139, 2004.
119. S. Hosseinkhani and M. Nemat-Gorgani, "Partial unfolding of carbonic anhydrase provides a method for its immobilization on hydrophobic adsorbents and protects it

against irreversible thermoinactivation," Enzyme and Microbial Technology, vol. 33, no. 2-3, pp. 179–184, 2003.
120. R. Dilmore, C. Griffith, Z. Liu et al., "Carbonic anhydrase-facilitated CO2 absorption with polyacrylamide buffering bead capture," International Journal of Greenhouse Gas Control, vol. 3, no. 4, pp. 401–410, 2009.
121. C. Prabhu, S. Wanjari, S. Gawande et al., "Immobilization of carbonic anhydrase enriched microorganism on biopolymer based materials," Journal of Molecular Catalysis B, vol. 60, no. 1-2, pp. 13–21, 2009.
122. A. Sharma, A. Bhattacharya, and A. Shrivastava, "Biomimetic CO2 sequestration using purified carbonic anhydrase from indigenous bacterial strains immobilized on biopolymeric materials," Enzyme and Microbial Technology, vol. 48, no. 4-5, pp. 416–426, 2011.
123. C. Prabhu, S. Wanjari, A. Puri et al., "Region-specific bacterial carbonic anhydrase for biomimetic sequestration of carbon dioxide," Energy and Fuels, vol. 25, no. 3, pp. 1327–1332, 2011.
124. S. Wanjari, C. Prabhu, R. Yadav, T. Satyanarayana, N. Labhsetwar, and S. Rayalu, "Immobilization of carbonic anhydrase on chitosan beads for enhanced carbonation reaction," Process Biochemistry, vol. 46, no. 4, pp. 1010–1018, 2011.
125. S. Wanjari, C. Prabhu, T. Satyanarayana, A. Vinu, and S. Rayalu, "Immobilization of carbonic anhydrase on mesoporous aluminosilicate for carbonation reaction," Microporous Mesoporous Mater, vol. 160, pp. 151–158, 2012.
126. C. Prabhu, A. Valechha, S. Wanjari et al., "Carbon composite beads for immobilization of carbonic anhydrase," Journal of Molecular Catalysis B, vol. 71, no. 1-2, pp. 71–78, 2011.
127. A. Belzil and C. Parent, "Qualification methods of chemical immobilizations of an enzyme on solid support," Biochemistry and Cell Biology, vol. 83, no. 1, pp. 70–77, 2005.
128. M. N. Bennani, D. Tichit, F. Figueras, and S. Abouarnadasse, "Synthèse et caractérisation d'hydrotalcites Mg-Al. Application à l'éldolisation de l'acétone," Journal of Chemical Physics, vol. 96, pp. 498–509, 1999.
129. "Hydrotalcite," http://webmineral.com/data/Hydrotalcite.shtml.
130. S. Zhang, Y. Lu, M. Rostab-Abadi, and A. Jones, "Immobilization of a rbonic anhydrase enzyme onto flame spray pyrolysis-based silica nanoparticles for promoting CO2 absorption into a carbonate solution for post-combustion CO2 capture," in Proceedings of the 243rd ACS National Meeting & Exposition, San Diego, Calif, USA, March 2012.
131. J. S. Lee, K. J. Schilling, and P. A. Johnson, "Fabrication of polymer ated magnetic Fe3O4 nanocomposites with magnetic stirring method for the immobilization of enzymes," in Proceedings of the 242nd ACS National Meeting & Exposition, Denver, Colo, USA, August 2011.
132. R. Yadav, S. Wanjari, C. Prabhu et al., "Immobilized carbonic anhydrase for the biomimetic carbonation reaction," Energy and Fuels, vol. 24, no. 11, pp. 6198–6207, 2010.
133. R. Yadav, T. Satyanarayanan, S. Kotwal, and S. Rayalu, "Enhanced carbonation reaction using chitosan-based carbonic anhydrase nanoparticles," Current Science, vol. 100, no. 4, pp. 520–524, 2011.

134. M. Vinoba, D. H. Kim, K. S. Lim, S. K. Jeong, S. W. Lee, and M. Alagar, "Biomimetic sequestration of CO2 and reformation to CaCO3 using bovine carbonic anhydrase immobilized on SBA-15," Energy and Fuels, vol. 25, no. 1, pp. 438–445, 2011.
135. M. Vinoba, M. Bhagiyalakshmi, S. K. Jeong, Y. I. Yoon, and S. C. Nam, "Immobilization of carbonic anhydrase on spherical SBA-15 for hydration and sequestration of CO2," Colloids and Surfaces, B, vol. 90, pp. 91–96, 2012.
136. M. Vinoba, M. Bhagiyalakshmi, S. K. Jeong, Y. I. Yoon, and S. C. Nam, "Carbonic anhydrase conjugated to nanosilver immobilized onto mesoporous SBA-15 for sequestration of CO2," Journal of Molecular Catalysis B, vol. 75, pp. 60–67, 2012.
137. M. Vinoba, K. S. Lim, S. K. Lee, S. K. Jeong, and M. Alagar, "Immobilization of human carbonic anhydrase on gold nanoparticles assembled onto amine/thiol-functionalized mesoporous SBA-15 for biomimetic sequestration of CO2," Langmuir, vol. 27, no. 10, pp. 6227–6234, 2011.
138. M. Vinoba, M. Bhagiyalakshmi, S. K. Jeong, Y. I. Yoon, and S. C. Nam, "Capture and sequestration of CO2 by human carbonic anhydrase covalently immobilized onto amine-functionalized SBA-15," The Journal of Physical Chemistry C, vol. 115, no. 41, pp. 20209–20216, 2011.
139. L. L. Wood, F. J. Hartdegen, and P. A. Hahn, "Enzymes bound to polyurethane," US Patent US 4343834, 1982.
140. N. Favre and A. C. Pierre, "Synthesis and behaviour of hybrid polymer-silica membranes made by sol gel process with adsorbed carbonic anhydrase enzyme, in the capture of CO2," Journal of Sol-Gel Science and Technology, vol. 60, pp. 177–188, 2011.
141. A. C. Pierre and G. M. Pajonk, "Chemistry of aerogels and their applications," Chemical Reviews, vol. 102, no. 11, pp. 4243–4265, 2002.
142. N. Favre, Y. Ahmad, and A. C. Pierre, "Biomaterials obtained by gelation of silica precursor with CO2 saturated water containing a carbonic anhydrase enzyme," Journal of Sol-Gel Science and Technology, vol. 58, no. 2, pp. 442–451, 2011.
143. G. W. Scherer and C. J. Brinker, Sol-Gel Science: Hydrolysis and Condensation of Silicon Alkoxides, Academic Press, 1990.
144. N. Kröger, R. Deutzmann, and M. Sumper, "Polycationic peptides from diatom biosilica that direct silica nanosphere formation," Science, vol. 286, no. 5442, pp. 1129–1132, 1999.
145. Y. Jiang, Y. Zhang, H. Wu et al., "Protamine-templated biomimetic hybrid capsules: efficient and stable carrier for enzyme encapsulation," Chemistry of Materials, vol. 20, no. 3, pp. 1041–1048, 2008.
146. J. N. Cha, K. Shimizu, Y. Zhou et al., "Silicatein filaments and subunits from a marine sponge direct the polymerization of silica and silicones in vitro," Proceedings of the National Academy of Sciences of the United States of America, vol. 96, no. 2, pp. 361–365, 1999.
147. H. R. Luckarift, M. B. Dickerson, K. H. Sandhage, and J. C. Spain, "Rapid, room-temperature synthesis of antibacterial bionanocomposites of lysozyme with amorphous silica or titania," Small, vol. 2, no. 5, pp. 640–643, 2006.
148. V. Abbate, A. R. Bassindale, K. F. Brandstadt, R. Lawson, and P. G. Taylor, "Enzyme mediated silicon-oxygen bond formation; The use of Rhizopus oryzae lipase,

lysozyme and phytase under mild conditions," Dalton Transactions, vol. 39, no. 39, pp. 9361–9368, 2010.
149. M. Frampton, A. Vawda, J. Fletcher, and P. M. Zelisko, "Enzyme-mediated sol-gel processing of alkoxysilanes," Chemical Communications, no. 43, pp. 5544–5546, 2008.
150. P. Buisson, H. El Rassy, S. Maury, and A. C. Pierre, "Biocatalytic gelation of silica in the presence of a lipase," Journal of Sol-Gel Science and Technology, vol. 27, no. 3, pp. 373–379, 2003.
151. K. Z. House, C. H. House, M. J. Aziz, and D. P. Schrag, "Carbon dioxide capture and related processes," International Patent WO 2008/018928A2, 2008.
152. R. Pérez-López, G. Montes-Hernandez, J. M. Nieto, F. Renard, and L. Charlet, "Carbonation of alkaline paper mill waste to reduce CO2 greenhouse gas emissions into the atmosphere," Applied Geochemistry, vol. 23, no. 8, pp. 2292–2300, 2008.
153. G. Montes-Hernandez, R. Pérez-López, F. Renard, J. M. Nieto, and L. Charlet, "Mineral sequestration of CO2 by aqueous carbonation of coal combustion fly-ash," Journal of Hazardous Materials, vol. 161, no. 2-3, pp. 1347–1354, 2009.
154. A. Santos, J. A. Toledo-Fernández, R. Mendoza-Serna et al., "Chemically active silica aerogel—wollastonite composites for CO2 fixation by carbonation reactions," Industrial and Engineering Chemistry Research, vol. 46, no. 1, pp. 103–107, 2007.
155. W. Li, W. Li, B. Li, and Z. Bai, "Electrolysis and heat pretreatment methods to promote CO2 sequestration by mineral carbonation," Chemical Engineering Research and Design, vol. 87, no. 2, pp. 210–215, 2009.
156. W. J. J. Huijgen, R. N. J. Comans, and G. J. Witkamp, "Cost evaluation of CO2 sequestration by aqueous mineral carbonation," Energy Conversion and Management, vol. 48, no. 7, pp. 1923–1935, 2007.
157. Y. Soong, D. L. Fauth, B. H. Howard et al., "CO2 sequestration with brine solution and fly ashes," Energy Conversion and Management, vol. 47, no. 13-14, pp. 1676–1685, 2006.
158. G. Walenta, V. Morin, A. C. Pierre, and L. Christ, "Cementitious compositions containing enzymes for trapping CO2 into carbonates and/or bicarbonates," PCT International Application. WO 2011048335 A1 20110428, 2011.
159. X. X. Wang, H. Fu, D. M. Du et al., "The comparison of pKa determination between carbonic acid and formic acid and its application to prediction of the hydration numbers," Chemical Physics Letters, vol. 460, no. 1–3, pp. 339–342, 2008.
160. W. Dreybrodt, L. Eisenlohr, B. Madry, and S. Ringer, "Precipitation kinetics of calcite in the system CaCO3-H2O-CO2: the conversion to CO2 by the slow process H++HCO3-→ CO2+H2O as a rate limiting step," Geochimica et Cosmochimica Acta, vol. 61, no. 18, pp. 3897–3904, 1997.
161. M. L. Druckenmiller and M. M. Maroto-Valer, "Carbon sequestration using brine of adjusted pH to form mineral carbonates," Fuel Processing Technology, vol. 86, no. 14-15, pp. 1599–1614, 2005.
162. W. Li, L. Liu, W. Chen, L. Yu, W. Li, and H. Yu, "Calcium carbonate precipitation and crystal morphology induced by microbial carbonic anhydrase and other biological factors," Process Biochemistry, vol. 45, no. 6, pp. 1017–1021, 2010.

163. D. H. Kim, M. Vinoba, W. S. Shin, K. S. Lim, S. K. Jeong, and S. H. Kim, "Biomimetic sequestration of carbon dioxide using an enzyme extracted from oyster shell," Korean Journal of Chemical Engineering, vol. 28, no. 10, pp. 2081–2085, 2011.
164. N. Favre, M. L. Christ, and A. C. Pierre, "Biocatalytic capture of CO2 with carbonic anhydrase and its transformation to solid carbonate," Journal of Molecular Catalysis B, vol. 60, no. 3-4, pp. 163–170, 2009.
165. S. R. Kamhi, "On the structure of vaterite CaCO3," Acta Cristallographica, vol. 16, pp. 770–772, 1963.
166. D. L. Graf, "Crystallographic tables for rhombohedral carbonates," American Mineralogist, vol. 46, pp. 1283–1316, 1961.
167. A. C. Pierre, Introduction to Sol-Gel Processing, Colloidal particles and Sols, chapter 3, Kluwer Academic, Boston, Mass, USA, 1998.

CHAPTER 6

ON THE POTENTIAL FOR CO_2 MINERAL STORAGE IN CONTINENTAL FLOOD BASALTS—PHREEQC BATCH- AND 1D DIFFUSION-REACTION SIMULATIONS

THI HAI VAN PHAM, PER AAGAARD, AND HELGE HELLEVANG

6.1 INTRODUCTION

Underground sequestration of carbon dioxide is a potentially viable greenhouse gas mitigation option as it reduces the release rate of CO_2 to the atmosphere [1]. CO_2 can be trapped subsurface by four storage mechanisms: (1) structural and stratigraphic trapping; (2) residual CO_2 trapping; (3) solubility trapping; and (4) mineral trapping [2]. Mineral trapping has been considered as the safest mechanism in long-term storage of CO_2[3].

Mineral storage of CO_2 in basaltic rocks is favored over siliciclastic reservoirs both by the higher abundance of divalent metal ions in basalt and the faster reactivity of basaltic glass or crystalline basalt [4]. Moreover, basalts such as the Columbia River flood basalts (CRB) are abundant

This chapter was originally published under the Creative Commons Attribution License. Pham THV, Aagaard P, and Hellevang H. On the Potential for CO_2 Mineral Storage in Continental Flood Basalts – PHREEQC Batch—and 1D Diffusion–Reaction Simulations. Geochemical Transactions ***13,5*** *(2012). doi:10.1186/1467-4866-13-5.*

and in many places close to CO_2 point source emissions [5]. During the last decade several flood basalts around the world have been mapped for the possibility of CO_2 storage, and possible candidates such as CRB in USA and the Deccan traps in India have been identified [4-6].

To be a candidate for CO_2 storage, the flood basalt must have a proper sealing and sufficient injectivity, the latter limited by the available connected pore space. In flood basalts, the connected pore space is typically found at zones containing abundant vesicles or in breccias between basalt flows. Because central zones of flood basalts commonly are dense and impermeable without vesicles, and flows are laterally continuous over large areas and commonly stacked vertically for hundreds of meters, flow units can act as seals [5]. The non-porous inner parts of flows may however be penetrated by networks of vertical fractures. These fractures can be open and conductive, or closed by mineralization and non-conductive.

The main objectives of this study were to performe batch- and 1D diffusion–reaction numerical simulations to determine the geochemical potential for secondary carbonate formation and to estimate the volume changes and the possibility of self-sealing following the basalt-CO_2 interactions. The CRB system was used as an example case and our results were compared to earlier reported laboratory experiments and numerical simulations of CO_2-basalt interactions. As CO_2 stored underground will distribute spatially in the reservoir to give a range of reactive conditions, such as the potential of reactions by H_2O dissolved in supercritical CO_2[7,8] or reactions in the H_2O-rich phase from residually trapped CO_2, we divided the simulations into three systems representing different parts of CO_2 storage: (1) basalt alteration in the H_2O-rich phase at constant CO_2 pressure; (2) basalt alteration in a H_2O saturated CO_2 phase, and (3) reactions at the boundary of the CO_2 plume where CO_2 diffuses into the aquifer from the boundary of the CO_2 plume (Figure 1).

6.2 METHODS

All thermodynamic and kinetic calculations were performed using the geochemical code PHREEQC-2 [9]. This code is capable of simulating complex interactions between dissolved gases, aqueous solutions, and

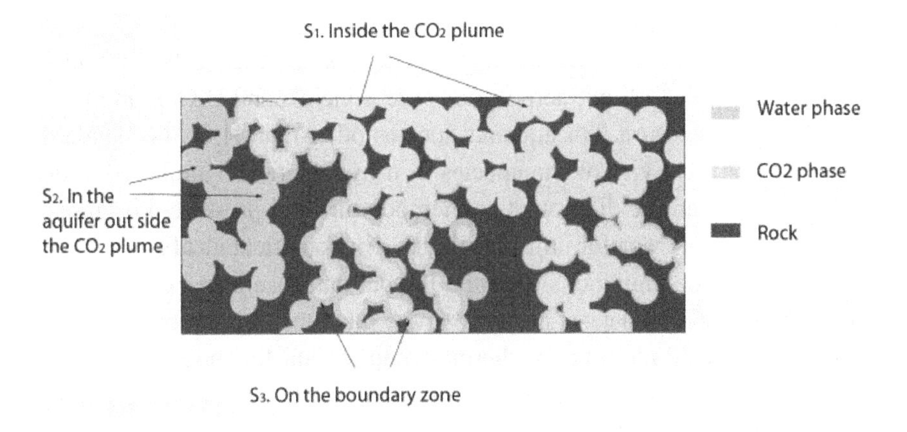

FIGURE 1: Sketch of possible reaction settings during CO_2 storage in basalt. System 1 (S1) is close to the injector and contains a wet CO_2 (0.5 mole% H_2O at 100 bar and 40 C) with no residual water; system $_2$ (S2) is fully in the H_2O rich phase with CO_2 diffusing in from the plume boundary; and system 3 (S3) is at the boundary of the CO_2 plume with both sufficient non-wetting CO_2 at a constant CO_2 partial pressure of 100 bar and with sufficient water wetting the mineral surfaces and available for reactions.

mineral assemblages in batch and 1D advection–diffusion-reaction mode. As the code can only model fully saturated systems, natural systems must be simplified to end-member situations, such as given by constant pressure boundary conditions as may be the case close to underground CO_2 plumes, or the assumption of packages (batches) of water reacting along a reaction path with a homogenous sediment or rock body. Based on these limitations we divided the simulations into three systems representing different parts of CO_2 storage: (1) basalt alteration in the H_2O-rich phase at constant CO_2 pressure; (2) basalt alteration in a H_2O saturated $scCO_2$ phase, and (3) reactions at the boundary of the CO_2 plume where CO_2 diffuses into the aquifer from the boundary of the $scCO_2$ plume (Figure 1). In the second case, we assumed that the CO_2 phase had swept through the systems and dried out residual water, giving only dissolved water in the $scCO_2$ phase. In this case an upper limit of carbonation potential was estimated as reactions were allowed to occur until (nearly) all water was consumed, passing the upper 2 mol/Kgw theoretical limit for the Truesdell-Jones activity model [9].

The standard state adopted in this study for the thermodynamic calculations was that of unit activity for pure minerals and H_2O at any temperature and pressure. For aqueous species other than H_2O, the standard state was unit activity of the species in a hypothetical 1 molal solution referenced to infinite dilution at any temperature and pressure. For gases, the standard state was for unit fugacity of a hypothetical ideal gas at 1 bar of pressure. All simulations used the llnl.dat database based on the thermo.com.V8.R6.230 dataset prepared at the Lawrence Livermore National Laboratory, with additions of thermodynamic data for those phases not present (see description below).

CO_2 fugacity coefficients were estimated according to the modified Redlich-Kwong (SRK) equation of state [10] and the solubility was adjusted for by a poynting correction term (exp(vCO2(Psat- P)/RT) where v denotes molar volume, P pressure, R the universal gas constant and T absolute temperature) [11]. The density of CO_2 at 40 C and 100 bar was approximated from Bachu and Stewart [12] to be 600 Kg/m^3 and the solubility of water in $scCO_2$ at the same conditions was approximated to 0.5 mole% [13,14]

The simulations were divided into batch simulations of the H_2O rich and CO_2 rich phases respectively, and 1D diffusion of CO_2 in the H_2O rich phase to obtain information on the CO_2-basalt interactions over a continuous range of CO_2 pressures. The latter was solved by PHREEQC using ∂tC=DL∂2xC+q, where C denotes molal (mol/Kgw) concentration, q denotes the sink term, subscripts t and x refer to derivatives in time and x-direction respectively, and an efficient diffusion coefficient DL of $0.45x10^{-9} m^2/s$ was used for CO_2[15] and all solutes.

Dissolution rates of minerals in the basaltic rock were calculated according to a kinetic equation taking into account pH and the distance from equilibrium:

$$r_+ = S\left\{k_H \exp\left(\frac{-E_{a,H}}{RT}\right) a_H^{nH} + k_N \exp\left(\frac{-E_{a,N}}{RT}\right) + k_{OH} \exp\left(\frac{-E_{a,OH}}{RT}\right) a_H^{nOH}\right\}(1-\Omega) \tag{1}$$

where S is the reactive surface area (m^2), k_i are rate constants ($moles/m^2s$), aH is the H^+ activity, n is the reaction order with respect to H^+ and OH^-, and Ω is the saturation state given by:

$$\Omega = \exp\left(\frac{\Delta G_r}{RT}\right) \tag{2}$$

Where ΔG_r is the Gibbs free energy of the reaction, R is the gas constant, and T is absolute temperature. Reaction rate constants for crystalline basalt (pyroxenes and plagioclase) were obtained from Palandri and Kharaka [16], and pH dependencies were taken from the same source. The dissolution rate of basaltic glass was calculated according to the expression suggested by Gislason and Oelkers [17]:

$$r_+ = k_+ \exp\left(\frac{-E_a}{RT}\right) S \left(\frac{a_{H+}^3}{a_{Al3+}}\right)^{0.33} (1 - \Omega) \tag{3}$$

where k_+ is the far-from-equilibrium dissolution rate coefficient. The saturation state term $1\text{-}\Omega$ was approximated to 1 (i.e., rate independent to distance from equilibrium) supported by earlier numerical estimates of glass-CO_2 reactivity suggesting an approximately linear relation between time and reaction progress for basaltic glass [18]. This expression takes into account the effect of pH as well as the effect of the concentration of solutes such as fluoride as they complex with Al^{3+} and reduce the Al^{3+} activity [19]. The specific surface area for basalt (m^2/g) was estimated by:

$$S_{sp} = \frac{\phi}{1 - \phi} \frac{A_p}{\bar{\rho}_s V_p} \tag{4}$$

where the ratio A_p/V_p denotes the ratio between pore surface and pore volume (m^{-1}), ϕ is connected porosity, and ρ_s (g/m^3) is the density of the basalt solid estimated from the fraction of the individual basalt components.

A S_{sp} value of $1.52 \times 10^{-5} m^2/g$ basalt (= $0.137 m^2/Kg$ water) was obtained for the CRB using an average basalt solid density of 2.93×106 g/m^3 with 10% connected pore space and a A_p/V_p ratio of $400 m^{-1}$[5]. The reactive surface area was calculated from the mass of the glass and minerals present according to:

$$S_i = M_i n_i S_{sp} X_r \quad (5)$$

where M and n are molar mass and moles of mineral i, and X_r is the fraction of the total mineral surface that is reactive. As X_r is highly uncertain and is suggested to vary by orders of magnitude [20,21], we used a value of 0.1 for the base case and varied X_r from 1 to 10^{-3}. The use of mass or mass fractions of the individual basalt components to estimate the release rates of elements from the basalt is supported by a recent experimental study which suggests that release rates estimated from the sum of volume fractions of the constituent minerals are within one order of magnitude from measured values [22]. A list of kinetic parameters is given in Table 1. All secondary phases were allowed to form according to the local equilibrium assumption [23].

Changes in solid-phase volumes and porosities ϕ caused by the mineral reactions were calculated according to:

$$\Delta\varphi_t = \left(1 - \frac{\sum_i n_{i,t}\bar{v}_i}{V_{total}}\right) - \varphi_{t=0} \quad (6)$$

where ϕ_t=0 is the initial porosity, n and v are moles and molar volume of mineral i respectively, and V_{total} is the total volume of the system.

The basalt was defined to consist of a mixture of glass and crystalline basalt with mineral and glass fractions chosen based on reported data from CRB [6,24,25]. To represent the crystalline basalt, plagioclase ($Ca_{0.5}Na_{0.5}Al_{1.5}Si_{2.5}O_8$) and the pyroxenes augite ($Ca_{0.7}Fe_{0.6}Mg_{0.7}Si_2O_6$) and pigeonite ($Ca_{1.14}Fe_{0.64}Mg_{0.22}Si_2O_6$) were chosen. The hydrolysis equilibrium constants of these phases were estimated using the PHREEQC program assuming ideal solid solutions of the end-members enstatite, ferrosilite and wollastonite for the pyroxenes, and albite and anorthite for the plagioclase. Equilibrium constants for the solid solutions for temperatures up to 100C were estimated with PHREEQC and from these data coefficients a to e for the PHREEQC built-in analytical expression ($\log K = a + bT + c/T + d\log_{10}(T) + e/T_2$) were estimated using non-linear regression in MATLAB.

TABLE 1: Kinetic parameters for dissolution of primary minerals based on empirical data given in Palandri and Kharaka[16]and for basaltic glass from[17]

	$k_{\_H}$ (mol/m²s)	Ea_H kJ/mol	n_H	$k_{\_N}$ (mol/m²s)	Ea_N kJ/mol	$k_{\_OH}$ (mol/m²s)	Ea_{OH} kJ/mol	n_{OH}	References
Augite	1.58e-7	78	0.7	1.07e-12	78	-	-		[16]
Pigeonite	1.58e-7	78	0.7	1.07e-12	78	-	-		[16]
Feldspar	1.58e-9	53.5	0.541	3.39e-12	57.4	4.78e-15	59	-0.57	[16]
glass	1e-10	25.5	1	-	-	-	-	-	[17]
Magnetite	2.57e-9	18.6	0.279	1.66e-11	18.6	-	-	-	[16]

The glass composition ($Ca_{0.015}Fe_{0.095}Mg_{0.065}Na_{0.025}K_{0.01}Al_{0.105}S_{0.003}Si_{0.5}O_{1.35}$) was taken from [6] and modified by adding a small fraction of sulfur which is a common minor constituent of the CR basaltic glass [26].

The secondary mineral assemblage was chosen based on reports on basalt weathering [27-30], with additional carbonates that could potentially form at elevated CO_2 pressures from the release of Fe, Mg and Ca. The ankerite composition chosen for this work was $CaFe_{0.6}Mg_{0.4}(CO_3)_2$ which corresponds to a solid solution of 0.6 ankerite ($CaFe(CO_3)_2$) and 0.4 dolomite ($CaMg(CO_3)_2$). Because ankerite ($CaFe_{0.6}Mg_{0.4}(CO_3)_2$) was not listed in the thermodynamic database, we estimated values using the same approach as in [31]. The full list of secondary minerals is given in Table 2.

To simulate the CRB-CO_2 interaction we used the average concentrations of solutes reported for the Grand Ronde Formation (Table 3). As supercritical CO_2 ($scCO_2$ at T>31.1 C; P>73.9 bar) is the preferred choice for CO_2 storage, based on higher density compared to gaseous CO_2, we simulated aqueous-phase basalt-CO_2 interaction at a depth of 800 meters at a CO_2 pressure of 100 bar and temperatures of 40 to 100C. The reactivity of basalt and a H_2O saturated $scCO_2$ phase was simulated using an

TABLE 2: Mineralogy included in the model

	Initial Weight %	Density (g/cm^3)	[2,3]Log K^0
Primary minerals			
[1]Augite (En0.35Fs0.3Wo0.35)	16	3.40	21.00
[1]Pigeonite (En0.57Fs0.32Wo0.11)	3	3.38	21.40
[1]Plagioclase (An50)	35	2.68	14.20
Glass $Ca_{0.015}Fe_{0.095}Mg_{0.065}Na_{0.025}K_{0.01}Al_{0.105}S_{0.003}Si_{0.5}O_{1.35}$	45	2.92	-99.00
Magnetite	1	5.15	10.47
Secondary minerals			
SiO_2(am)	0	2.62	-2.71
Albite	0	2.62	2.76
Goethite	0	3.80	0.53
Calcite	0	2.71	1.85
Hematite	0	5.30	0.11
Kaolinite	0	2.60	6.81
Smec high Fe-Mg	0	2.70	17.42
Saponite-Mg	0	2.40	26.25
Celadonite	0	3.00	7.46
Stilbite	0	2.15	1.05
Dawnsonite	0	2.42	4.35
Siderite	0	3.96	-0.19
[1]Ankerite (Ank0.6Do0.4)	0	3.05	-19.51
Dolomite	0	2.84	4.06
Magnesite	0	3.00	2.29

The mineralogy of the CRB has been described in [25,32] and the weight fraction of pyroxene, feldspar and glass was estimated as average values from the reported data.

[1]Solid solutions. En (enstatite), Fs (ferrosilite), Wo (wollastonite), An (anorthite), Ank (ankerite), Do (dolomite). For details on the calculations of the ankerite solid solution see [31].

[2]Superscript 0 denotes Standard state (T=298K, P=1 atm). The equilibrium constant log K value is that for the forward dissolution reaction for one mole unit of the mineral.

[3]All thermodynamic data (log K and coefficients for the PHREEQC analytical temperature expression) from the llnl.dat PHREEQC database, except for the solid-solutions first estimated in PHREEQC by ideal solid solutions and then added to the PHREEQC database as new solid solution phases.

estimated 0.5 mol% H_2O and a CO_2 density of 600 g/cc giving an initial mass of 0.003 Kg H_2O per 1 liter pore space.

TABLE 3: Composition of initial formation water

Elements (totals)	Mol/kgw
Na	1.0×10^{-3}
Ca	6.0×10^{-4}
K	1.0×10^{-4}
Mg	2.0×10^{-5}
Fe	1.2×10^{-6}
Alkalinity (HCO_3^-)	2.0×10^{-3}
Cl	3.0×10^{-4}
S (SO_4^{2-})	1.0×10^{-4}
Si	2.0×10^{-4}
Al	1.0×10^{-6}
Log(O_2)	-10.68
pH	7.5

6.3 RESULTS

6.3.1 SYSTEM 1: BASALT ALTERATION IN THE H_2O-RICH PHASE AT CONSTANT CO_2 PRESSURE

6.3.1.1 CRB MINERAL AND GLASS DISSOLUTION AND FORMATION OF SECONDARY MINERALS

Following the injection of CO_2 into the system, pH immediately decreased from 9.5 to below 4, and thereafter gradually increased to 5.8 at the end of 10000 years (Figure 2a). At the acidic pH secondary phases such as saponite ($Ca_{0.165}Mg_3Al_{0.33}Si_{3.67}O_{10}(OH)_2$), celadonite ($KMgAlSi_4O_{10}(OH)_2$) and zeolite (stilbite) were thermodynamically stable and formed (Figure 2b). Glass dissolved orders of magnitude faster than the crystalline basaltic constituents and more than half dissolved after 10000 years (Figure 2c). The dissolution rate of glass was not increased by the aqueous fluoride as the Al^{3+} activity was fixed by the kaolinite and amorphous silica equilibria. The fluoride therefore only increased the total soluble aluminium. The steady release of Fe from the basalt saturated the water with respect to siderite and a total amount of 10 moles/kgw formed after 10000 years (Figure 2d). Other carbonates, such as ankerite, dolomite, magnesite, and dawsonite did not form as elements such as Mg and Ca was consumed by the non-carbonate secondary phases.

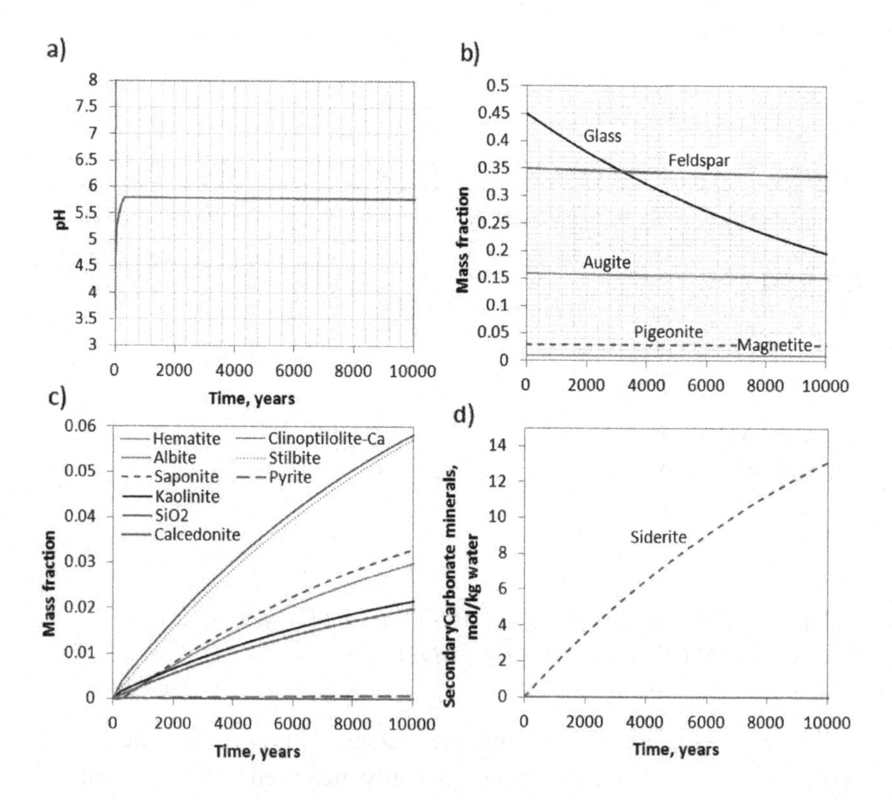

FIGURE 2: Basalt alteration at 40C and 100 bar CO_2 pressure over 10000 years. a) pH changes; b) mass fractions of basaltic glass and crystalline basalt components; c) secondary phases formed; and d) moles of secondary carbonates (siderite) formed per kgw.

The effect of temperature on the basalt hydration and carbonation was investigated by simulating the system at 60, 80 and 100 C (Figure 3). As for the 40 C simulation we see that basaltic glass dissolves orders of magnitude faster than the crystalline basalt components and the glass is the major source for the secondary phases. The dissolution rates of the basalt components scale exponentially with temperature, and the glass is almost completely dissolved after 10000 years at 60 C, whereas the time for a complete dissolution takes 4000 and 1500 years at 80 and 100 C respectively (Figure 3a, d, g). The secondary mineral assemblages were largely the same for all temperatures. Stilbite dominated together with amorphous

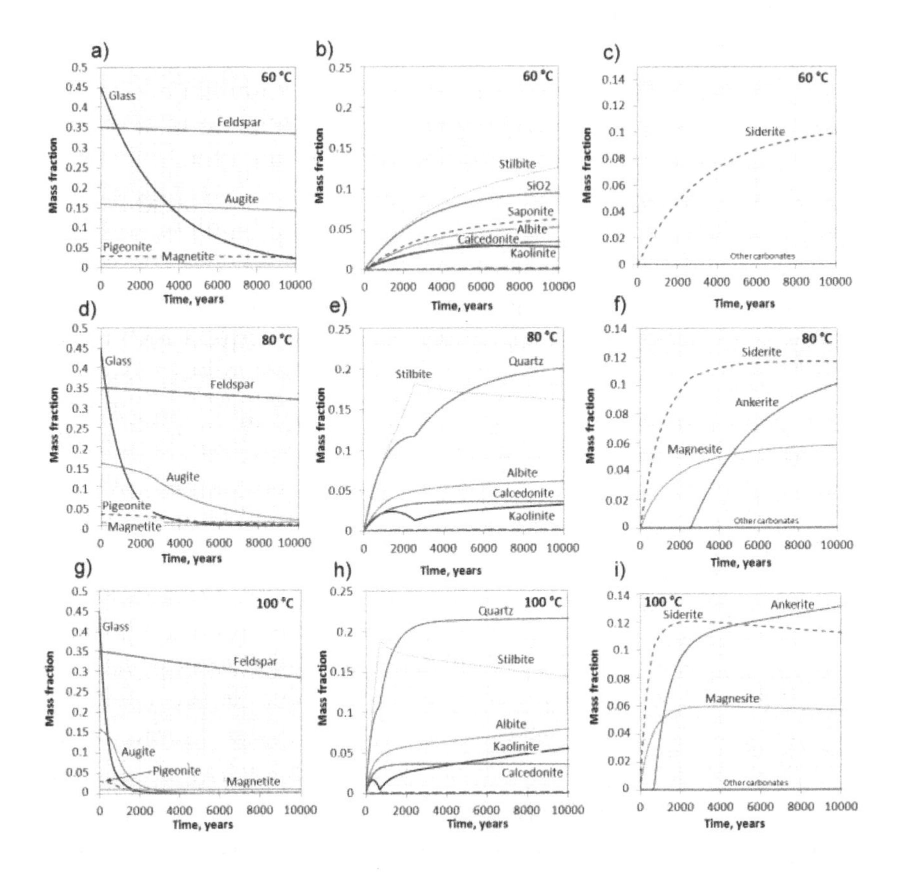

FIGURE 3: Mass fractions of minerals following basalt alteration at 60, 80, and 100 C over 10000 years: a, d, g) primary basalt minerals and glass; b,e, h) secondary phases except the carbonates; and c, f, j) carbonates.

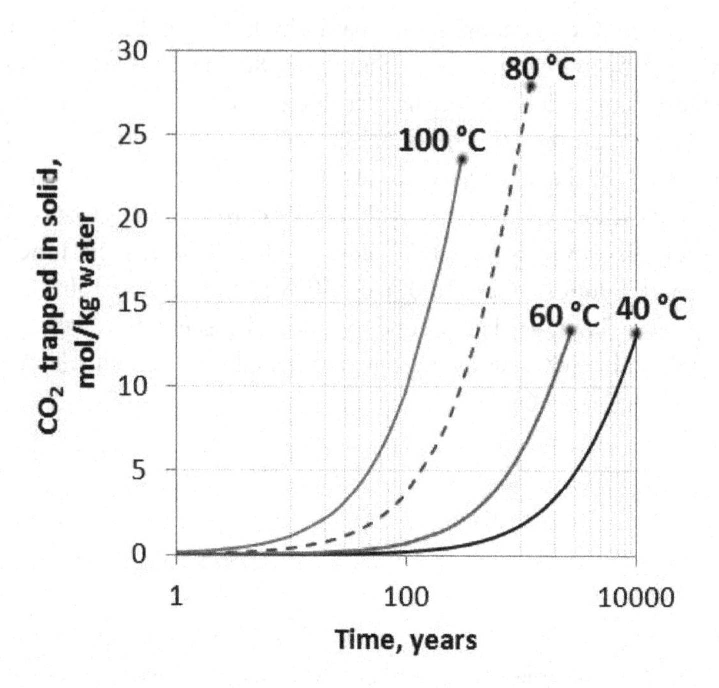

FIGURE 4: CO_2 trapped in solids for 40 to 100 C simulations at 100 bar CO_2 pressure. A cut-off value is used when all pore space is filled up with the secondary phases (see Figure 5). The simulations suggest that the total amount of secondary carbonates that form is dictated by the available pore space and the thermodynamic stability of secondary phases rather than temperature, whereas carbonate generation rates depend on the exponential increase of basalt dissolution rates with temperature.

silica (40 and 60 C) and quartz (80 and 100 C) (Figure 3b, e, h). Saponite formed at 40 and 60 C, but not at higher temperatures. Other secondary minerals such as albite, celadonite, and kaolinite formed at all conditions. At 60 C, magnesite and dolomite were still considered to be too slow to form (see [29]) and siderite was the only phase that formed. At 80 and 100 C, magnesite and later ankerite formed together with siderite. Taking zero porosity as the maximum extent of possible reactions we see that the total amount of CO_2 trapped as solid carbonates did not change much with temperature (Figure 4). The reaction rates increased however with temperature and the time needed to reach the maximum potential therefore decreased with temperature (Figure 4).

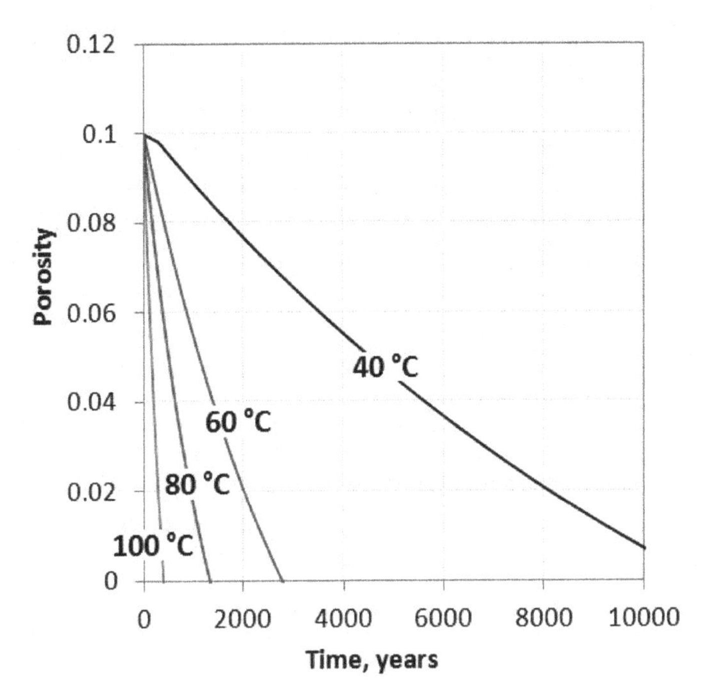

FIGURE 5: Porosity changes caused by the basalt alteration at 40 to 100C. Secondary hydrated species and carbonates incorporate the H_2O and CO_2 masses into the solids and clogs the pore space. As reaction rates increase exponentially with temperature, the pore space is filled up faster at the higher temperatures.

6.3.1.2 ON THE LIMITATION OF PORE-SPACE FOR THE BASALT CARBONATION

Secondary phases such as stilbite and amorphous silica have lower density than the basalt components and alteration therefore leads to a reduction of pore space. At the presence of CO_2, secondary carbonates further reduce the pore space. For the volume calculations we used expression (6) with the molar volumes listed in Table 2. At 40C, the starting porosity of 10% is reduced to 0.85% after 10000 years. At the higher temperatures all porosity is lost after 2700, 1200, and 300 years respectively at 60, 80, and

100 C (Figure 5). Taking the extreme of 0% porosity as the limit for the reactions we obtain a maximum carbonation potential (mol CO_2 stored/Kgw) at the different temperatures of 13.5, 29.3, and 28.5 moles for 60, 80, and 100 C (Figure 5). The simulated clogging of the pore space fits well with short term laboratory percolation experiments on open-system basalt-CO_2 alteration which shows loss of porosity and a rapid reduction of permeability during CO_2-basalt interactions (e.g., [33]).

6.3.1.3 REDUCTION OF PORE-SPACE AS A FUNCTION OF REACTIVE SURFACE AREA

As the reactive surface area is a large uncertainty we simulated the changes of porosity over a range of values from a maximum being equal to the

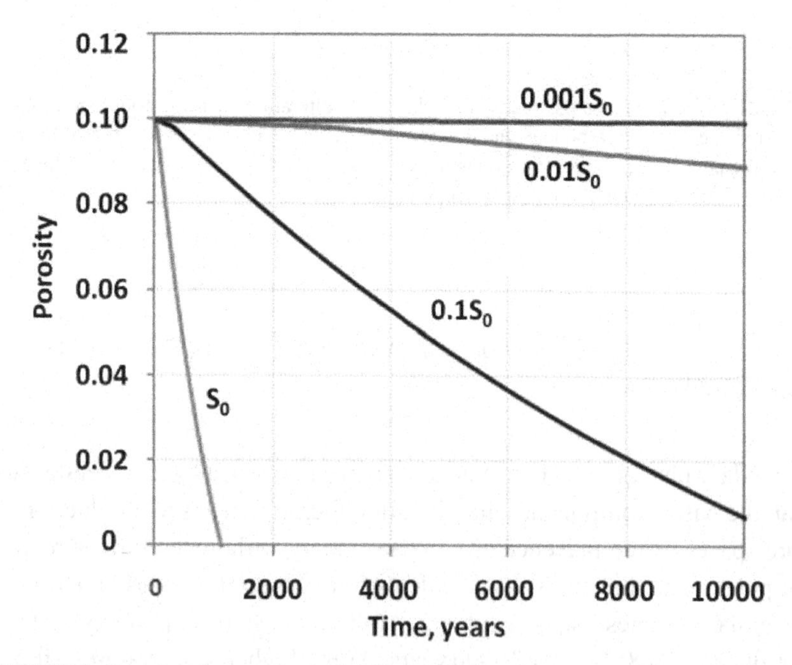

FIGURE 6: Porosity changes caused by the basalt alteration at 40 C and specific surface areas ranging from So(estimated total basalt surface area) down to a three orders of magnitude reduction. The base case specific surface used was So/10.

estimated physical surface area S0 (equation (5) with $X_r = 1$) to a three orders of magnitude reduction (Figure 6). The physical conditions of the simulated system was the same as for the base-case at 40 C and a CO_2 pressure of 100 bars. At a reactive surface area that is equal to the estimated S0 all porosity is lost after approximately 1000 years as stilbite and siderite fills the pore space. If the reactive surface area is reduced by one order of magnitude (i.e. the base case) nearly 1/10 of the original 10% porosity is preserved. Further reductions by one and two orders of magnitude lead to smaller changes and at three orders of magnitude reduction relative to S0 almost no change is observed (Figure 6).

6.3.2 SYSTEM 2: THE POTENTIAL FOR CARBONATE GROWTH IN A H_2O-SAURATED $SCCO_2$ PHASE

The reaction between H_2O dissolved in $scCO_2$ and basalt was simulated at 100 bar pressure and 40 C. The initial amount of water was 0.003 Kg and no H_2O was allowed to enter the system. This is an ideal end-member case

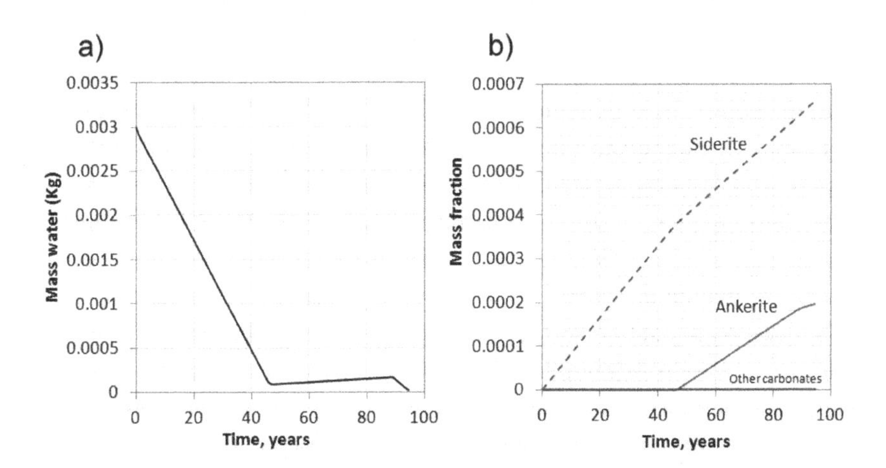

FIGURE 7: Basalt alteration in the $scCO_2$rich phase with initial 3 grams of water per liter pore space. A) as zeolites form H_2O is consumed and the water activity is reduced. After approximately 45 years most water is consumed, whereas all is gone before 100 years. B) siderite formed the secondary carbonate initially followed by ankerite.

and serves to illustrate the carbonation potential in a volume with limited hydration potential.

As secondary phases such as stilbite formed, water was rapidly consumed and most gone after 45 years (Figure 7a). At this point stilbite was unstable and supplied water until all water was consumed after approximately 100 years (Figure 7a). Following the basalt hydration, siderite and ankerite formed from the released Ca, Mg, and Fe, with a final total amount of 0.2 moles CO_2 consumed per liter pore space after 100 years (Figure 7b). If H_2O had been allowed to dissolve into the $scCO_2$ phase from residual aqueous phases trapped in the smaller pores, the carbonation potential would have been larger. This process is however, to the knowledge of the authors, not possible to simulate using the PHREEQC code, and was therefore outside the scope of this study.

6.3.3 SYSTEM 3: 1D DIFFUSION OF CO_2 INTO THE CRB AQUIFER

To see how the basalt reacted under different CO_2 pressures, we defined a 1D diffusion–reaction simulation. This provided us with basalt-CO_2 interactions over a continuous range of CO_2 pressures from the background 1 bar up to the maximum 100 bars. The system corresponds to a stagnant zone presented as a column with one end close to the boundary of the injected CO_2 plume and the other end further away from the plume (Figure 1). The distance reached for the CO_2 into the column is given by the balance between diffusive transport and removal of carbon by secondary carbonate formation. We therefore varied reaction rates from no reaction giving the maximum lengtht of diffusive transport, and up to the base-case rate given by a reaction surface area 1 order of magnitude lower than the estimated physical surface area. Figure 8 shows pH, dissolved CO_2 (mol/Kgw) and amount of secondary carbonate formed in the 1D column. As CO_2 diffuses into the column pH drops to approximately 4 at full saturation. The depth of diffusion into the 1D column is approximately 40 meters at 1000 years if no carbonate forming reactions occur (Figure 8a). The penetration depth decreased rapidly if reactions were allowed as siderite formed and pulled carbon out of aqueous system (Figure 8b, c). At the

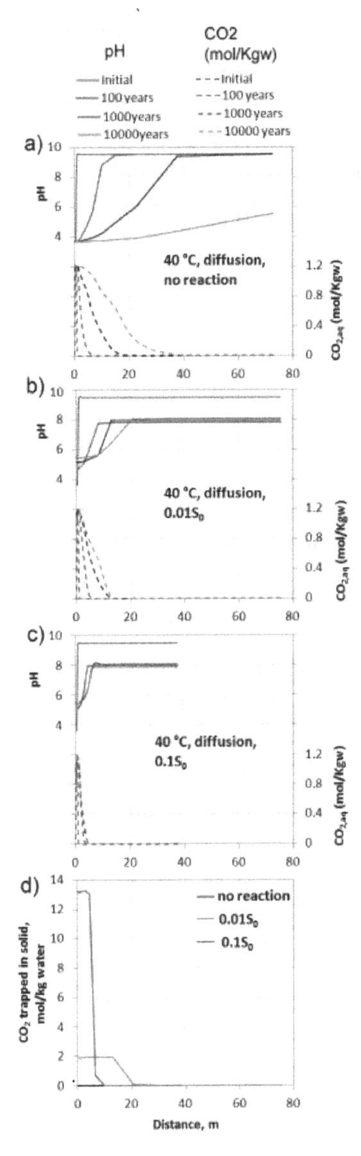

FIGURE 8: 1D reaction–diffusion of CO_2into permeable basalt. Partial pressure of the inlet boundary was fixed at 100 bar and with a column temperature of 40 C. As consumption of CO_2 by siderite growth affects the depth of CO_2 diffusion, we ran a sensitivity study on reactive surface area going from no reaction (a) up to the base case (c). Finally the amount of CO_2 trapped as solid carbonate (siderite) was compared for the base case and reduced reactive surface area (d). We see that the reaction rates strongly constraint the depth of the column affected by the CO_2 diffusion.

highest reaction rate (base-case), CO_2 diffused less than 10 meters into the column as approximately 13 moles/Kgw of siderite formed at the end of the 10000 years simulation (Figure 8d). Siderite formed at greater depth if the reactive surface area was reduced by another order of magnitude, but less formed in total (Figure 8d).

6.4 DISCUSSION

6.4.1 UNCERTAINTY ON THE REACTIVE SURFACE AREA

The reactive surface area is considered as a major source of uncertainty (e.g., [20,34]) and this leads to corresponding high uncertainties in timing and extent of reactions as dissolution rates have a first order dependence on reactive surface areas. Weathering rates in nature are commonly observed to be 1–3 orders of magnitude lower than in laboratory experiments (e.g., [20,21,34]), and this may in part be explained by differences in reactive and physical (total) surface area between experimental and natural systems. We assumed in this study a base-case reactive surface area 1 order of magnitude lower than the estimated physical surface area for the basalt. A further two orders of magnitude reduction in the reactive surface area, which is within the range of values expected for natural systems, resulted in little basalt alteration and only minor reduction of porosity (see Figure 5). A better understanding of the surface area of porous basalt and the effect of time (aging) on features such as dislocation densities and reactive surface areas are therefore required to understand the potential for CO_2 mineral storage in basaltic rocks.

6.4.2 UNCERTAINTY ON THE CHOICE OF SECONDARY PHASES USED IN THE MODEL

Growth rate experiments of carbonates such as magnesite and dolomite have shown that the activation energy is high and that growth is negligible at low temperatures (e.g., [35-37]). Dissolution rate studies of siderite suggests that the reaction rate is intermediate between calcite and magnesite [38,39], and growth rate data suggest that siderite may form down to room temperature [40]. Data on ankerite dissolution and growth is to the knowledge of the authors not known. The crystallographic and physical characteristics of ankerite do resemble those of dolomite and siderite, and the chemistry is related to dolomite with the Mg^{2+} substituted by various amounts of Fe^{2+} and Mn^{2+}. If the growth rate is close to the magnesian carbonates such as dolomite and magnesite [41,42], the amount that may form during low-temperature alteration is likely low. In this case, more iron would be available for siderite growth. If on the other hand the growth rate is closer to siderite, we would expect ankerite or other FeMg solid solution carbonates to grow during low-temperature alteration.

One uncertainty related to the local-equilibrium assumption is on the growth retention time for the secondary carbonates. The local-equilibrium assumption predicts growth of the secondary phases as soon as an infinitesimally small supersaturation is reached [23]. The time it takes to nucleate sufficient mass to initiate a significant growth may however be hundreds to thousands of years for some secondary phases [31]. There are no nucleation rate data for siderite and ankerite and the retention time is hence unknown.

Finally, the total potential for secondary carbonate growth may be affected by the amount of magnesium and iron that enters ferromagnesian calcites. As a significant fraction of the metal cations may substitute for calcium (e.g., [43]), a iron-magnesium rich calcite may potentially form rather than ankerite and thereby reduce the amount of siderite formed.

6.4.3 COMPARISONS TO EXPERIMENTS, NUMERICAL SIMULATIONS AND NATURAL ANALOGUES OF BASALT-CO_2 INTERACTIONS

Our simulations suggest that the potential for carbonate growth is limited to siderite or FeMg carbonates at low temperatures as secondary phases such as zeolites outcompeted the carbonates for calcium. We here compare our simulated results with reported data on CO_2-basalt interactions from laboratory experiments, natural analogues, and other reported numerical simulations.

The reactivity of CRB and other continental flood basalts are available from the long-term (months to years) laboratory experiments done by Schaef and co-workers [6,24]. In these experiments basalt samples from USA, India, South Africa, and Canada were reacted with CO_2 at about 100 bars and 60 to 100 C. Reacted samples from these experiments showed generation of Ca-rich carbonates interpreted as calcites with minor siderite and magnesite. In experiments on CRB using mixtures of H_2S and CO_2 at 60 C and 100 bar and run for 181 days, pyrite (FeS_2) formed together with Mg-Fe poor calcite and a Ca-poor Fe-carbonate [6]. Our simulations at the same temperatures show rapid formation of siderite (60 C) or siderite and magnesite at higher temperatures (Figure 3). Our simulations do not predict any calcite growth as the calcium activity is lowered by zeolite formation. Calcite would however form in our models if the zeolites were not allowed to form at local equilibrium, and possibly if a magnesian ferroan (solid solution) calcite was used in the model instead of the pure end-member calcite. Therefore, the apparent difference between our model and the experiment may be caused by our use of the local equilibrium assumption, whereas the zeolites in the laboratory experiments did not form at low temperatures due to slow kinetics. Recent experiments on basalt dissolution support the preferential release of Mg and Fe over Ca at acidic conditions [22], suggesting that the MgFe-carbonates will dominate as secondary carbonates during CO_2 storage in basalt.

Our numerical simulations share some similarities to other works such as by Marini [18] and Gysi [44], but our model and hence the outcome is different in several aspects. The most comprehensive work done earlier is the numerical simulations done by Marini [18] on the reactivity of crystalline and glassy CFB following CO_2 storage. The initial mineralogy was similar to our study whereas the temperature of 60 C was slightly higher than our base case 40 C. In [18] the CO_2-basalt interactions were stretched to last for more than 280000 years compared to our 10000 years perspective. The main differences between our model and [18] are on the choice of secondary mineral assemblage, and on the focus of limiting factors such as the availability of water for hydration in the present work. The lack of zeolites and hydrous phases other than kaolinite and goethite in [18] made

Ca available for secondary carbonates and the total potential for carbonate formation was higher than in our work. Marini allowed dolomite and magnesite to form at 60 C, whereas our simulations only produced siderite at the similar conditions. Moreover, the formation of dawsonite in [18] is still uncertain and possibly limited at high silica activities and with an assemblage of stable NaAl-silicates defined to form [45]. Based on two different approaches, the reactive surface area for basalt was estimated to quite similar values. We estimated a specific surface area of approximately 1.5×10^{-5} m^2/gbasalt (= 0.14 m^2/Kg water at 10% porosity) based on the Ap/Vp values estimated by [46] and reported in [5], and reduced this value by one order of magnitude to get the reactive surface area. Marini used a geometric model giving a reactive surface area of 0.41 m^2/Kg water. The higher reactive surface area and higher temperature of [18] resulted in faster reactions and more rapid clogging of the pore space (within a few years). Studies of natural basalt systems at similar or higher temperatures may give some insight into how fast pore space is clogged by basalt hydration or carbonation, and this should be used to improve the estimates of reactive surface areas of basalt for future studies.

Another numerical study on low-temperature (25 C, 30 bar CO_2) basaltic glass alteration was presented by Gysi et al. [44]. Again a main difference is on the choice of secondary minerals. Gysi et al. [44] allowed dolomite, magnesite, and Fe-Mg carbonate to form together with calcite and siderite, whereas we did not allow other Mg-Fe carbonates to form than ankerite. As previously stated, the low-temperature formation of dolomite and magnesite is not likely because of the high apparent activation energy and small kinetic coefficients for the growth of Mg-carbonates [35-37]. Other carbonates such as siderite and potentially FeMg-calcites are more likely to form at these low temperatures. The high reactive surface area used in [44] is based on a geometric model for glass fragments, and is hence not directly comparable with the surface area estimated for a vesicle pore space of a solid basalt. Although no inverse modeling was done to estimate the reactive surface area of the basalt in [44], fragmented basaltic rocks such as hyaloclastite breccias are expected to have significantly higher reactive surface areas than porous solid basalts, and they are therefore correspondingly more reactive.

One example of a natural analogue that shed light on CO_2 basalt interactions is the CO_2 charged basalt hosted groundwaters at Hekla, Iceland. Solution aqueous species sampled from natural cold springs and rivers here showed a drop in total inorganic carbon (TIC) that was interpreted to result from considerable formation of secondary carbonate phases such as calcite [47]. Reaction path modeling of the system suggests however that the carbonate formation is associated with high pH in accordance with the low TIC in the sampled waters. This system is therefore different from basalt CO_2 storage projects where higher CO_2 pressures may be maintained over time and the pH is lower. In addition to calcite, dolomite was also suggested as a potential storage host for the low temperature reactions in Hekla [47]. This may however be questionable as long-term laboratory experiments at room temperature have failed to form dolomite even at significant super saturations [48], explained by the high activation energy for dolomite growth [32,41]. Another natural analogue that more closely corresponds to industrial CO_2 storage is the basalt-hosted petroleum reservoir on Nuussuaq, West Greenland. In this system the bulk carbonate formation appears to have occurred as secondary weathering products. Other alteration products such as zeolites and oxides were replaced by dolomite, magnesite, siderite, and calcite at temperatures of 70–120 C [49]. Therefore, taking into account the basalt weathering products and not only primary basalt minerals appears to be vital in estimating the total potential for secondary carbonate formation and the long-term potential for CO_2 storage in basalt systems.

6.5 SUMMARY AND CONCLUSIONS

Simulations of closed-system (PCO_2 = 100 bar, 40 C) and 1D reaction–diffusion (PCO_2 = 0–100 bar, 40 C) alteration of basalt suggest that the potential of secondary carbonate formation is limited to siderite at low temperatures as divalent metal cations are preferentially consumed by zeolites and oxides. Higher temperatures 60 – 100 C appear to be in favor of secondary carbonate formation, allowing the precipitation of carbonates such as magnesite, siderite and possibly dolomite and other FeMg carbonates (ankerite). Given an unlimited source of CO_2 (fixed CO_2 pressure), the total amount of CO_2 stored as solid carbonates is orders of magnitude higher than the 1–2 mol/Kg water solubility in the formation water (Figure 4). The total amount trapped might however be reduced if CO_2, H_2O or pore space are limiting factors. The formation of secondary hydrous and carbonate phases increases the volume of solids and the porosity is correspondingly reduced (Figure 5). This together with the immobilization of CO_2 by solid carbonate formation is in favor of safe long-term storage of CO_2 in basaltic aquifers.

REFERENCES

1. Holloway S: Underground sequestration of carbon dioxide–a viable greenhouse gas mitigation option. Energy 2004, 30:2318-2333.
2. Bachu S, Bonijoly D, Bradshaw J, Burruss R, Holloway S, Christensen NP, Mathiassen OM: CO2 storage capacity estimation: Methodology and gaps. Int J Greenhouse Gas Control 2007, 1:430-443.
3. Metz B, Davidson O, Coninck H, Loos M, Meyer L (Eds): IPCC Special Report on Carbon Dioxide Capture and Storage In In Prepared by Working Group III of the Intergovernmental Panel on Climate Change. Cambridge University Press, Cambridge, United Kingdom and New York, NY, USA; 2005::442 pp.
4. Oelkers EH, Gislason SR, Matter J: Mineral Carbonation of CO2. Elements 2008, 4:333-337.
5. McGrail BP, Schaef HT, Ho AM, Chien Y-J, Dooley JJ, Davidson CL: Potential for carbon dioxide sequestration in flood basalts. J Geophysical Res-Solid Earth 2006, 111:1-13.
6. Schaef HT, McGrail BP, Owen AT: Carbonate mineralization of volcanic province basalts. Int J Greenhouse Gas Control 2010, 4:249-261.
7. Schaef HT, Windish CF, McGrail BP, Martin PF, Rosso KM: Brucite [Mg(OH)2] carbonation in wet supercritical CO2: An in situ high pressure x-ray diffration study. Geochim Cosmochim Acta 2011, 75:7458-7471.
8. White MD, McGrail BP, Schaef HT, Hu JZ, Hoyt DW, Felmy AR, Rosso KM, Wurstner SK: Multiphase sequestration geochemistry: Model for mineral carbonation. Energy Procedia 2011, 4:5009-5016.
9. Parkhurst DL, Appelo CAJ: User's guide to PHREEQC (version 2) - a computer program for speciation, reaction-path, 1D-transport, and inverse geochemical calculations. U.S. Geological Survey, Water-Resources Investigation Report; 1999::312.
10. Soave G: Equilibrium constants from a modified Redlich-Kwrong equation of state. Chem Eng Sci 1972, 27:1197-1203.
11. Hellevang H, Kvamme B: ACCRETE - Geochemistry solver for CO2-water-rock interactions. Proceedings GHGT 8 conference 2006, :8p.
12. Bachu S, Stewart S: Geological sequestration of anthropogenic carbon dioxide in the western Canada sedimentary basin: Suitability analysis. J Can Pet Technol 2002, 41:32-40.

13. Coan CR, King AD: Solubility of Water in Compressed Carbon Dioxide, Nitrous Oxide, and Ethane - Evidence for Hydration of Carbon Dioxide and Nitrous Oxide in Gas Phase. J Am Chem Soc 1971, 93:1857-1862.
14. Sabirzyanov AN, Ilìn AP, Akhunov AR, Gumerov FM: Solubility of water in Supercritical carbon dioxide. High Temp 2002, 40:203-206.
15. Bahar M, Lui K: Measurement of the diffusion coefficient of CO2 in formation water under reservoir conditions: Implication for CO2 storage. Perth, Australia; 2008.
16. Palandri JL, Kharaka YK: A Complilation of Rate Parameters of Water-Mineral Interaction Kinetics for Applicatin to Geochemical Modeling. U.S. Geological survey, Open report. 2004, :1068.
17. Gislason SR, Oelkers EH: Mechanism, rates, and consequences of basaltic glass dissolution: II. An experimental study of the dissolution rates of basaltic glass as a function of pH and temperature. Geochim Cosmochim Acta 2003, 67:3817-3832.
18. Marini L: Chapter 7: Reaction path Modelling of geological CO2 sequatration. Dev Geochem; 2006::319-394.
19. Wolff-Boenisch D, Gislason SR, Oelkers EH: The effect of fluoride on the dissolution rates of natural glasses at pH 4 and 25 degrees C. Geochim Cosmochim Acta 2004, 68:4571-4582.
20. White AF, Peterson ML, et al.: Role of reactive-surface-area characterization in geochemical kinetic models. In In. Chemical Modeling of Aqueous Systems II. Edited by Melchior D. ACS Symposium Series, ACS, Washington, DC; 1990.
21. White AF, Brantley SL: The effect of time on the weathering of silicate minerals: why do weathering rates differ in the laboratory and field? Chem Geol 2003, 202:479-506.
22. Gudbrandsson S, Wolff-Boenisch D, Gislason SR, Oelkers EH: An experimental study of crystalline basalt dissolution from $2 \leq pH \leq 11$ and temperatures from 5 to 75 C. Geochim Cosmochim Acta 2011, 75:5496-5509.
23. Helgeson HC: Evaluation of Irreversible Reactions in Geochemical Processes Involving Minerals and Aqueous Solutions .I. Thermodynamic Relations. Geochim Cosmochim Acta 1968, 32:853-877.
24. Schaef HT, McGrail BP, Owen AT: Basalt-CO2-H2O Interactions and Variability in Carbonate Mineralization Rates. Energy Procedia 2009, 1:4899-4906.
25. Schaef HT, McGrail BP: Dissolution of Columbia River Basalt under mildly acidic conditions as a function of temperature: Experimental results relevant to the geological sequestration of carbon dioxide. Appl Geochem 2009, 24:980-987.
26. Blake S, Self S, Sharma K, Sephton S: Sulfur release from the Columbia River Basalts and other flood lava eruptions constrained by a model of sulfide saturation. Earth Planet Sci Lett 2010, 299:328-338.
27. Neuhoff PS, Fridriksson T, Arnorsson S, Bird DK: Porosity evolution and mineral paragenesis during low-grade metamorphism of basaltic lavas at Teigarhorn, eastern Iceland. Am J Sci 1999, 299:467-501.
28. Neuhoff PS, Rogers KL, Stannius LS, Bird DK, Pedersen AK: Regional very low-grade metamorphism of basaltic lavas, Disko-Nuussuaq region, West Greenland. Lithos 2006, 92:33-54.
29. Stefansson A, Gislason SR, Arnorsson S: Dissolution of primary minerals in natural waters - II. Mineral saturation state. Chem Geol 2001, 172:251-276.

30. Reidel SP, Johnson VG, Spane FA: Natural gas storage in basalt aquifers of the Colombia Basin, Pacific Northwest USA: A guide to site characterization. Report Pacific Northwest National Laboratory; PNNL-13962, H; 2002.
31. Pham VTH, Lu P, Aagaard P, Zhu C, Hellevang H: On the potential of CO2-water-rock interactions for CO2 storage using a modified kinetic model. International Journal of Greenhouse Gas Control 2011, 5:1002-1015.
32. Reidel SP: A lava flow without a source: the cohassett flow and its compositional components, Sentinel Bluffs Member, Columbia River Basalt Group. J Geol 2005, 113:1-21.
33. Peuble S, Godard M, Gouze P, Luquot L: CO2 sequestration in olivine rich basaltic aquifers: a reactive percolation experimental study. Geochim Cosmochim Acta 2009, 73:A1635.
34. Velbel MA: Constancy of silicate-mineral weathering-rate ratios between natural and experimental weathering: implications for hydrologic control of differences in absolute rates. Chem Geol 1993, 105:89-99.
35. Saldi GD, Jordan G, Schott J, Oelkers EH: Magnesite growth rates as a function of temperature and saturation state. Geochim Cosmochim Acta 2009, 73:5646-5657.
36. Arvidson RS, Mackenzie FT: The dolomite problem; control of precipitation kinetics by temperature and saturation state. Am J Sci 1999, 299:257-288.
37. Arvidson RS, Mackenzie FT: Tentative kinetic model for dolomite precipitation rate and its application to dolomite distribution. Aquat Geochem 1996, 2:273-298.
38. Golubev SV, Bébnézeth P, Schott J, Dandurand JL, Castillo A: Siderite dissolution kinetics in acidic aqueous solutions from 25 to 100 C and 0 to 50 atm pCO2. Chem Geol 2009, 265:13-19.
39. Pokrovsky OS, Golubev SV, Schott J, Castillo A: Calcite, dolomite and magnesite dissolution kinetics in aqueous solutions at acid to circumneutral pH, 25 to 150 C and 1 to 55 atm pCO2: New constraints on CO2 sequestration in sedimentary basins. Chem Geol 2009, 265:20-32.
40. Greenberg J, Tomson M: Precipitation and dissolution kinetics and equilibria of aqueous ferrous carbonate vs temperature. Appl Geochem 1992, 7:185-190.
41. Arvidson RS, Mackenzie FT: Tentative Kinetic Model for Dolomite Precipitation Rate and Its Application to Dolomite Distribution. Aquat Geochem 1997, 2:273-298.
42. Saldi GD, Jordan G, Schott J, Oelkers EH: Magnesite growth rates as function of temperature and saturation state: An HAFM study. Geochim Cosmochim Acta 2009, 73:A1149.
43. Busenberg E, Plummer LN: Thermodynamics of magnesian calcite solid-solutions at 25 C and 1 atm total pressure. Geochim Cosmochim Acta 1989, 53:1189-1208.
44. Gysi AP, Stefansson A: Numerical modelling of CO2-water-basalt interaction. Mineral Mag 2008, 72:55-59.
45. Hellevang H, Declercq J, Aagaard P: Why is dawsonite absent in CO2 charged reservoirs? Oil & Gas Science and Technology - Re. IFP Energies nouvelles 2011, 66:119-135.
46. Saar MO, Manga M: The relationship between permeability, porosity, and microstructure in vesicular basalts. Master Thesis, Univ. of Oregon, Eugene; 1998::91.
47. Flaathen TK, Gislason SR, Oelkers EH, Sveinbjornsdottir AE: Chemical evolution of the Mt. Hekla, Iceland, groundwaters: a natural analogue for CO2sequestration in basaltic rocks. Appl Geochem 2009, 24:463-474.

48. Land LS: Failure to precipitate dolomite at 25 degrees C from dilute solution despite 1000-fold oversaturation after 32 years. Aquat Geochem 1998, 4:361-368.
49. Rogers KL, Neuhoff PS, Pedersen AK, Bird DK: CO2metasomatism in a basalt-hosted petroleum reservoir, Nuussuaq, West Greenland. Lithos 2006, 92:55-82.

CHAPTER 7

EXPERIMENTAL STUDY OF CEMENT-SANDSTONE/SHALE-BRINE-CO_2 INTERACTIONS

SUSAN A. CARROLL, WALT W. MCNAB, AND SHARON C. TORRES

7.1 BACKGROUND

Carbon dioxide is actively being stored at depth in a sandstone saline reservoir as part of the In Salah Gas Project in Krechba, Algeria [1]. It is one of few commercial scale CO_2 storage projects and serves as an important platform to study the scientific and technical issues for safe and effective long-term CO_2 storage in deep saline reservoirs [2-11].

Wellbores are a potential risk pathway for leakage of CO_2 from the storage reservoir to overlying drinking water aquifers and back into the atmosphere. Carbonation of cements, used in wellbores to seal off fluid flow from the reservoir, can bring about changes in permeability and alter the movement of fluids within the wellbore environment. Field, experimental and modeling studies suggest that carbonation of hydrated cements lowers porosity and has the potential to heal fractures within the cement [12-18].

Risk of leakage from a CO_2 storage reservoir would be significantly reduced if the CO_2 could be stored as a solid carbonate mineral and if these reactions improved the seal within the cap rock above the reservoir.

This chapter was originally published under the Creative Commons Attribution License. Carroll SA, McNab WW, and Torres SC. Experimental Study of Cement - Sandstone/Shale - Brine - CO_2 Interactions. Geochemical Transactions ***12****,9 (2011). doi:10.1186/1467-4866-12-9.*

Field and laboratory experiments have shown mineral dissolution in CO_2-rich brines leads to increased concentrations of Ca, Fe, and Mg and, in some cases, to the formation of carbonate minerals [19-24]. The amount of CO_2 stored as carbonate minerals over geologic times estimated from reactive transport simulations varies substantially and depends on the reaction rates and the amount of CO_2 injected into the subsurface [25-30]. The possibility of even small amounts of carbonate mineralization in shale cap rock may significantly improve seal integrity by reducing porosity. Simulation results suggest seal integrity is enhanced due to carbonate mineral precipitation after 100 years of reaction with CO_2-rich fluids [31]. Another modeling study predicts that redistribution of calcite within 0.1 m of the cap rock - reservoir interface effectively seals reservoir from the overlying strata [32].

The focus of this work was to determine the key geochemical reactions involving common cements used in wellbore construction, formation mineralogy, and supercritical CO_2 stored at the Krechba site. We reacted the end member components of the heterolithic sandstone and shale unit that forms the upper section of the carbon storage reservoir with supercritical CO_2 and representative brine with and without cement at 95°C and 10 MPa in gold bag autoclaves. Separate cement experiments without CO_2 were conducted to measure cement hydration at temperature prior to the injection of CO_2. The experimental results can be used to develop geochemical models for estimating long-term trapping of CO_2, and wellbore and cap rock integrity at the Krechba site.

7.2 METHODS

7.2.1 MATERIALS

The heterolithic sandstone and shale in units C10.2 and C10.3 form the upper section of the carbon storage reservoir at the Krechba Field, In

Salah, Algeria. Reported mineralogy among 17 core samples collected from Krechba reservoir ranged from a quartz-dominated sandstone to a shale-like material containing abundant illite clay [33]. Iron-rich chlorite appears as coatings on quartz grains in the sandstone and in abundances as high as 30 percent by volume in the shale. Other aluminosilicates include minor quantities (e.g., typically less than 5 percent) of kaolinite and feldspar. Carbonate phases have been described as siderite [33], ankerite plus dolomite (unpublished mineralogical analyses conducted by Statoil), or calcite (unpublished XRD analyses of Krechba core samples). The shale end member, Sample 14, consisted of 44% illite, 30% chlorite, 20% quartz, 4% kaolinite, 2% feldspar and trace amounts of pyrite by weight. The sandstone end member, Sample 7, consisted of 88% quartz, 6% chlorite, 4% kaolinite and 2% siderite by weight. Limited availability of the heterolithic sandstone and shale necessitated the use of rock fragments in the experiments rather than a well-defined powdered size fraction.

The powdered class G oil well cement used in the experiments was provided by Mountain Cement Company and consists of 56% Ca_3SiO_5, 39% Ca_2SiO_4, 5% $Ca_3AlO_{4.5}$, + 0.5% Na_2O and K_2O by weight as determined by standard ASTM C 150. In some experiments small amounts of bentonite were added to the cement to reflect mixtures identified in well logs from the Krechba site (bentonite to cement ratio = 1:39 by weight). Any curing of the cement occurred in the reaction vessels at the experimental conditions. Combination of powdered cement and rock fragments did not compromise the results because the primary objective of the experiments was to determine the dominant geochemical reactions controlling the solution composition.

Initial solutions were distilled and deionized water, 0.13 m $CaCl_2$, and synthetic Krecha brine consisting of 1.8 molal NaCl, 0.55 molal $CalCl_2$, and 0.1 molal $MgCl_2$. All salts used to synthesize the brines were reagent grade. The experiments were conducted in a synthetic brine to capture the major ion chemistry measured at the site. A more complex reservoir brine was not used to avoid masking relevant geochemical reactions. High purity liquid CO_2 was pressurized at temperature and pressure to generate supercritical CO_2 for the experiments.

7.2.2 CEMENT HYDRATION EXPERIMENTS

Distilled and deionized, 0.13 m $CaCl_2$, and synthetic Krechba solutions were used to determine ion activity products for cement hydration at different solid:solution ratios at 115 and 95°C (Table 1). Solutions and solids were reacted in teflon-lined Parr reaction vessels, sealed, and placed into an oven to maintain temperature. Sealed reaction vessels were quenched in cold water prior to taking filtered aqueous samples for chemical analyses. Solids were washed with distilled and deionized water and dried at 60°C prior to analysis by an environmental scanning electron microscopy with energy dispersive x-ray spectroscopy (ESEM/EDX) and powder x-ray diffraction (XRD).

TABLE 1: Cement Hydration Experiments.

ID	Solid	Solution	Solid:Soln (g/g)	T °C	Days	pH(c,25)
G3	A	0.13 m $CaCl_2$	1:10	115	43	11.9
G6	A	Brine	1:10	115	43	not measured
G7	B	MQ water	1:10	115	58	12.2
G8	B	MQ water	1:10	115	87	12.3
G9	B	MQ water	2:10	115	58	12.3
G10	B	MQ water	2:10	115	87	12.1
G11	B	MQ water	1:10	95	74	12.3
G12	B	MQ water	2:10	95	43	12.3
G13	B	MQ water	2:10	95	74	12.1
G14	B	Brine	1:10	95	88	11.4
G15	B	Brine	2:10	95	88	12.3

A indicates Class G cement, B indicates Class G cement plus bentonite (39 g cement to 1 g bentonite), and Brine indicates 1.8 molal NaCl, 0.55 molal $CaCl_2$ and 0.1 molal $MgCl_2$ solution. Note that brine pH values are conditional (c) because of the high ionic strength.

7.2.3 CEMENT - ROCK - BRINE - CO_2 EXPERIMENTS

Static Dickson-type Au reactors housed in water-filled pressure vessels were used to react cement, sandstone, shale, synthetic brine and supercritical

CO_2 at 95°C and 10 MPa. Specific weights of cement, sandstone, shale, and brine are listed in Table 2. Coherent sandstone or shale rock fragments were used in the experiments due to limited availability from core. We monitored reaction kinetics and the approach to equilibrium by sampling the solution as a function of time. The reactor setup allows sequential sampling of the aqueous phase while the experiment is at pressure and temperature. All metals measured in solution were from the rock-fluid interactions, because the supercritical CO_2 and the brine contact only gold or passivated titanium. After one month of reaction, supercritical CO_2 was injected into the gold bag and reacted for an additional month. About 20 grams of supercritical CO_2 were added to the reaction vessel to ensure excess CO_2 during the reaction. To add the CO_2, liquid CO_2 was pressurized above the run pressure and injected into the reaction vessel through the sample tube. The liquid CO_2 transitions to supercritical CO_2 at the run pressure and temperature. The amount of CO_2 injected was estimated from change in volume of the liquid CO_2. Several brine samples were taken and analyzed for solution chemistry over the duration of the experiment. At the end of the experiment, the reaction vessel was cooled to room temperature, excess CO_2 was removed, and solid reactants were rinsed with distilled and deionized water several times to remove brine. The solids were dried at 60°C prior to XRD and ESEM/EDX analysis. Samples for dissolved Al, Ca, Fe, Mg, and Si analyses were filtered, and directly diluted with acidified distilled and deionized water (using high purity HNO_3).

TABLE 2: Cement-Rock-Brine-CO_2 Experiments.

ID	Cement (g)	Shale (g)	Sandstone (g)	Brine (g)	Days reacted before CO_2	Days reacted after CO_2
GBCO2_1	4.0	280.5	11	30		
GBCO2_2	20.0	204.5	21	22		
7CO2	5.6	301.2	33	28		
14CO2	6.4	301.2	31	31		
GB7CO2	4.8	4.9	252.5	26	44	
GB14CO2	8.6	8.7	246.5	40	35	

All experiments were conducted in 1.8 molal NaCl, 0.55 molal $CaCl_2$ and 0.1 molal $MgCl_2$ brine with 20 grams of supercritical CO_2 at 95°C and 10 MPa.

Samples for total dissolved inorganic carbon were injected directly into 1 N NaOH to trap the CO_2, filtered to remove any solids that precipitated, and analyzed for dissolved inorganic carbon, calcium, and magnesium. Total dissolved carbon should be equal to the measured inorganic carbon in the filtered sample plus the amount of carbon trapped as calcite minerals in the NaOH extraction. Comparison of results from the NaOH extraction with estimates from Duan and Sun (2003) caused us to question the viability of using the extraction technique to quantify dissolved carbon in the experiments. Although the median value of the dissolved carbon concentrations estimated from the extraction technique (0.78 molal) agrees with the theoretical prediction (0.69 molal), there is a significant amount of scatter in the extracted values over time (ranging from 1.12 to 0.53 molal). The scatter largely correlates to changes in measured dissolved calcium, because calcium is predicted to be trapped as CaCO3 solid rather than some mixture of $CaCO_3 + Ca(OH)_2$ solids (any Mg is predicted to be trapped as $Mg(OH)_2$ and was not considered in the NaOH extraction method). We have chosen to use the theoretical dissolved CO_2 concentrations in the development of the geochemical model because of the uncertainty associated with the NaOH extraction chemistry. However, we report both the extraction and theoretical values, because caustic extractions are commonly used to quantify total dissolved CO_2.

7.2.4 ANALYSIS

Major and trace metals in the aqueous samples and the stock solution were analyzed using inductively coupled plasma mass spectroscopy (ICP-MS, Make/Model: Thermo Electron Corp/X Series Q-ICPMS). Samples were prepared volumetrically using an internal standard solution in 2% nitric acid. A fully quantitative analysis using a linear calibration curve based on known standards was performed. The internal standard was corrected for instrument drift and suppression from the sodium chloride matrix. Silica was run in collision cell technology (CCT) mode to avoid polyatomic interferences. Detection levels were established from duplicate blanks and serial dilution preparations. Matrix spike samples were analyzed for quality control. Detection limits were about 3, 0.2, 0.5, 4, and 0.35 ng/g for Ca, Mg, Al, Si, and Fe, respectively.

Total inorganic carbon (TIC) concentrations are determined using an automated OI Analytical Aurora 1030W Carbon Analyzer. The Aurora 1030W uses a syringe pump to transfer samples and reagents to a temperature-controlled reaction chamber. TIC samples are reacted with 5% phosphoric acid to evolve CO_2 gas purged by a stream of N_2 gas and quantified using a NDIR detector.

Solid mineralogy was determined from data collected from random orientation powder samples with a Scintag PAD V instrument using a Cu-Kα source at 45 kV and 35 mA from 5° to 70° 2Θ in 0.02° steps. XRD cannot detect amorphous solids or minerals that are present at less than 2 wt%.

Unreacted and reacted samples were analyzed using a Quanta 200 Environmental Scanning Electron Microscope in low vacuum mode with EDX. Images were collected with secondary and backscatter detectors from pressures ranging from 0.23 to 0.90 torr and 20-25 kV. EDX was use to determine the local chemical composition of materials using at 20kV and 11 mm working distance. All analyses are semi-quantitative.

TABLE 3: Mineral weight percents used in geochemical simulations.

Phase	Sandstone	Shale
Albite	1.5%	2%
Chlorite	6%	30%
Dolomite	0.5%	0.75%
Illite	1.5%	44%
Kaolinite	4%	4%
Quartz	86%	20%
Siderite	1.5%	0.75%

(Ankerite was modeled as a 25 - 75 siderite-dolomite mixture for sandstone and a 50 - 50 siderite-dolomite mixture for the shale.)

7.2.5 GEOCHEMICAL MODELING

Solution compositions from the batch experiments were modeled using the PHREEQC 2.15.0 geochemical code [34], and the SUPCRT92 thermodynamic database [35] augmented by CEMDATA07v2 [36]. The CEMDATA provides Gibbs free energies, heat capacity, and volume data for cement phases as a function of temperature and pressure [37,38]. The standard state for minerals and pure water is unit activity, and for all aqueous species other than dissolved CO_2 is unit activity in a hypothetical 1 molal solution referenced to infinite dilution at any pressure and temperature. The dissolved CO_2 concentrations were calculated assuming equilibrium with fCO_2 estimated from CO_2 equation of state at 10 MPa [39]. No other mass balance reactions were corrected for pressure. This introduces an error of about ± 0.1 log K. pH was estimated from the forward model simulations. The B-dot ion interaction model was used to approximate the non-ideal behavior of solutions at elevated ionic strength and temperature. The B-dot equation is an extended form of the Debye-Huckel equation and was used in this study because it can be applied to NaCl based solutions with high ionic strengths (3 molal) over a wide range of temperature. However, it is generally recognized that the B-dot equation becomes increasing less accurate at I > 0.5 molal. Despite these limitations, we chose to use B-dot equation to correction for species activity because the Pitzer equations are lacking for many elements at temperatures above 25°C. The use of the B-dot equation typically yields brines with slightly higher solution pH (≈ 0.1 pH units). The thermodynamic and kinetic inputs to the geochemical model for reaction of cement, sandstone and shale with CO_2-rich Na-Ca-Mg chloride brines are shown in Tables 3, 4, 5, 6, and 7.

TABLE 4: Surface areas used in the modeling calculations.

Phase	Surface Area (cm^2/g)	
	Shale	Sandstone
Boehmite	0.02	0
Smectite	9505	317.7
Ripidolite	224.2	87.9
Dolomite	0.5	1.0
Illite	9505	317.7
Kaolinite	18.3	17.9
Low-albite	45.3	33.4
Quartz	9.1	39.1
Siderite	0.4	2.2

TABLE 5: Geochemical model for reaction of cement, sandstone and shale with CO_2 and Na, Ca, Mg chloride brines.

Cement Hydration			
Phase	Mass Balance	Log K 95°C	Log SI 95°C
Portlandite	$Ca(OH)_2 + 2H^+ \; Ca^{2+} + 2H_2O$	18.30	0.04 ± 0.04
Psuedowollas-tonite	$CaSiO_3 + 2H^+ \leftrightarrows SiO_2 + Ca^{2+} + H_2O$	10.97	0.3 ± 1.3
Brucite	$Mg(OH)_2 + 2H^+ \leftrightarrows Mg^{2+} + 2H_2O$	12.65	0.7 ± 0.1
1Hydrotalcite	$Mg_4Al_2 O_7(OH)_2{:}10H_2O + 14H^+ \leftrightarrows 2Al^{3+} + 4Mg^{2+} + 17H_2O$	53.67	2.8 ± 0.2
1Fe-Hydrog-arnet	$Ca_3Fe_2(OH)_{12} + 12H^+ \leftrightarrows 3Ca^{2+} + 2Fe^{3+} + 12H_2O$	68.50	-3.4 ± 2.7
Anhydrite	$CaSO_4 \leftrightarrows Ca^{2+} + SO_4^{2-}$	-5.08	-0.3 ± 0.1
Cement, Sandstone, and Shale Carbonation			
Phase	Mass Balance		Log K 95°C
Albite	$NaAlSi_3O_8 + 4H^+ \leftrightarrows Al^{3+} + Na^+ + 2H_2O + 3SiO_2$		0.46
1Amorphous Al(OH)3	$Al(OH)_3 + 3H^+ \leftrightarrows Al^{3+} + 3H_2O$		5.42
1Amorphous Fe(OH) 3	$Fe(OH)_3 + 3H^+ \leftrightarrows Fe^{3+} + 3H_2O$		2.86
Boehmite	$AlO(OH) + 3H^+ \leftrightarrows Al^{3+} + 2H_2O$		3.75
Calcite	$CaCO_3 + H^+ \leftrightarrows Ca^{2+} + HCO_3^-$		0.85
Chalcedony	$SiO_2 \leftrightarrows SiO_2{,}aq$		-2.88
Dolomite	$CaMg(CO_3)_2 + 2H^+ \leftrightarrows Ca^{2+} + Mg^{2+} + 2HCO_3^-$		1.41
Ripedolite 14Å	$Mg_3Fe_2Al_2Si_3O_{10}(OH)_8 + 16 H^+ \leftrightarrows 2Al^{3+} + 3SiO_2{,}aq + 3Mg^{2+} + 2Fe^{2+} + 12H_2O$		41.45
Illite	$K_{0.6}Mg_{0.25}Al_{2.3}Si_{3.5}O_{10}(OH)_2 + 8H^+ \leftrightarrows 0.25Mg^{2+} + 0.6K^+ + 2.3Al^{3+} + 3.5SiO_2 + 5H_2O$		2.56
Kaolinite	$Al_2Si_2O_5(OH)_4 + 6 H^+ \leftrightarrows 2 Al^{3+} + 2SiO_2 + 5H_2O$		1.35
Magnesite	$MgCO_3 + H^+ \leftrightarrows Mg^{2+} + HCO_3^-$		0.71
Quartz	$SiO_2 \leftrightarrows SiO_2{,}aq$		-3.10
Siderite	$FeCO_3 + H^+ \leftrightarrows Fe^{2+} + HCO_3^-$		-1.40
Smectite (Ca-Beidellite)	$Ca_{0.165}Al_{2.33}Si_{3.67}O_{10}(OH)_2 + 7.32H^+ \leftrightarrows + 0.165Ca^{2+} + 2.33Al^{3+} + 4.66H_2O + 3.67SiO_2$		-0.62

Unless otherwise noted the values are estimated from SUPCRT92 [35]; 1 values from CEMDATA [36].

TABLE 6: Rate parameters from Palandri and Kharaka [40].

Phase	Neutral		Acid			Base		
	k25°C (mol/m²/s)	Ea (mol/KJ)	k25°C (mol/m²/s)	Ea (mol/KJ)	N	k25°C (mol/m²/s)	Ea (mol/KJ)	n
[a]Boehmite	-11.5	61.2	-7.7	47.5	0.99	-16.7	80.1	-0.78
Dolomite	-7.5	52.2	-3.2	36.1	0.50			
[b]$Fe(OH)_3$								
Ripedolite	-12.5	88	-11.1	88.0	0.50			
Illite	-12.8	35	-11.0	23.6	0.34	-16.5	58.9	-0.40
Kaolinite	-13.2	22.2	-11.3	65.9	0.78	-17.0	17.9	-0.47
Low-albite	-12.6	69.8	-10.2	65.0	0.46	-15.6	71.0	-0.57
Quartz	-14.0	87.7						
Siderite	-8.9	62.76	-3.8	45.0	0.90			
Smectite	-12.8	35	-11.0	23.6	0.34	-16.5	58.9	-0.40

[a]Gibbsite dissolution rates were applied for boehmite. [b]Fe(OH)3 rate was estimated from match to the solution chemistry normalized to mineral moles and is listed in Table 7.

All sandstone and shale reactions are assumed to be kinetically controlled and are modeled using

$$r = \pm kA\left(1 - \frac{LAP}{K_{sp}}\right) \tag{1}$$

where r is the dissolution or precipitation rate per unit time per unit area, k is the kinetic rate constant, A is the mineral surface area, IAP is the ion activity product, and K_{sp} the solubility constant. Surface area is adjusted to fit the solution composition (Table 4). All of reactive surface area was assumed to be available for reaction for the sandstone experiments, whereas only 10 percent of the shale reactive surface area was assumed to participate in reactions. The rate constant for $Fe(OH)_3$ precipitation in the sandstone experiments was fitted to the solution composition and is normalized the mineral moles (Table 7). All other rate constants are from Palandri and Kharaka [40] (Table 6):

TABLE 7: Conditional rate constants for the cement phases estimated from fits of the solution composition and anhydrous cement composition for each experiment.

Phase	Experiment	log k (mol/s/mol-mineral)
Anhydrite	[a]Cement:Brine	-6.2
	[b]Cement:Brine	-7.3
	Sandstone ± cement	-6.3
	Shale ± cement	-6.9
Hydrogarnet-Fe	[a]Cement:Brine	-6.2
	[b]Cement:Brine	-7.3
	Sandstone ± cement	-6.3
	Shale ± cement	-6.9
Hydrotalcite	[a]Cement:Brine	-5.5
	[b]Cement:Brine	-6.6
	Sandstone ± cement	-5.6
	Shale ± cement	-6.2
Brucite	[a]Cement:Brine	-5.5
	[b]Cement:Brine	-6.6
	Sandstone ± cement	-5.6
	Shale ± cement	-6.2
Portlandite	[a]Cement:Brine	-5.5
	[b]Cement:Brine	-6.6
	Sandstone ± cement	-5.6
	Shale ± cement	-6.2
Pseudowollastonite	[a]Cement:Brine	-6.2
	[b]Cement:Brine	-7.3
	Sandstone ± cement	-6.3
	Shale ± cement	-6.9
Amorphous Fe(OH)3	Sandstone	-8.6
Calcite	[b]Cement:Brine	-8.0
Calcite	Shale ± cement	-7.0
$FeCO_3$	Shale ± cement	-10.0

The range in values reflects incomplete hydration and diffusion-controlled transport. [a]Cement:Brine refers to experiment GBCO2_1 and [b]Cement:Brine refers to experiment GBCO2_2

$$k = k_{25}^{nu} \exp\left(\frac{-E_a^{nu}}{R}\left(\frac{1}{T} - \frac{1}{298.15}\right)\right) + k_{25}^{H} \exp\left(\frac{-E_a^{H}}{R}\left(\frac{1}{T} - \frac{1}{298.15}\right)\right)\{H+\}^n \quad (2)$$

where R is the universal gas constant, T the absolute temperature, {H+} the activity of the hydrogen ion, n an exponential factor, and k_{25}^{nu} and k_{25}^{H} the neutral and acid rate constants at 25°C and E_a^{nu} and E_a^{H} the neutral and acid activation energies, respectively.

A few additional constraints to the geochemical model were needed to match Al, Fe, and Si solution chemistry and involve the precipitation of secondary phases. We added dissolved O_2 to account for the attenuation of the initial spike in Fe in the sandstone experiment as secondary precipitation of $Fe(OH)_3$. Quartz precipitation is suppressed and dissolved Si concentrations are limited by chalcedony equilibrium with no kinetic controls. Smectite precipitation is tied to illite surface area and is modeled as Ca-beidellite, $Ca_{0.165}Al_{2.33}Si_{3.67}O_{10}(OH)_2$. Boehmite, AlO(OH), precipitation is suppressed until the solution is slightly supersaturated (log SI = 1). Uptake of Fe(II) during cement carbonation is modeled as an ideal continuous $FeCO_3$ - $CaCO_3$ solid solution. No other solid solutions are included in this model. We model cement carbonation from an assemblage of portlandite, psuedowollastonite, brucite, hydrotalcite, Fe-hydrogarnet, and anhydrite estimated from the hydration of initial anhydrous cement composition using the experimental cement to brine ratios (Table 2). We use pseudowollastonite to represent the amorphous hydrated calcium silicate ("CSH") because its solubility is consistent with our experimental measurements. Brucite and hydrotalcite were needed to account for the observed removal of Mg from the brine during cement hydration and its enhanced solubility during carbonation, even though these phases were not identified in the XRD pattern. Attempts to model the poorly crystalline phase as Mg-Ca silica hydrate introduced excessive amounts of dissolved silica. Anhydrite accounts for the sulfate noted in the anhydrous components.

Carbonation of the hydrated mineral assemblage was modeled with calcite, amorphous SiO_2 (as chalcedony), $Fe(OH)_3$, and boehmite for cement: brine < 1:50 or amorphous $Al(OH)_3$ for cement: brine > 1:25 (Table 2).

Fitted rate constants for the cement phases reflect the varied extent of cement hydration and carbonation in each experiment and are normalized to mineral moles (Table 7). It is important to note that the fitted reaction rates are conditional to these experiments and likely represent diffusion of the reactants into the cement, which had a tendency to solidify at the bottom of the reaction vessel even though the experiments were rocked.

7.2.6 MODEL UNCERTAINTY

It is important to note that the lithology and cement geochemical models represent possible realizations of the dominant geochemical reactions. The non-uniqueness of the lithology model reflects the wide range of minerals that can be used to describe major element chemistry and was made apparent in our efforts to fit the dissolved Si prior to the injection of CO_2 in the sandstone experiment. The best matches required either using unrealistically high illite in the sandstone (higher than in the shale) or alternatively excessive high quartz surface areas (which were also higher than in the shale). Presumably, some combination of aluminosilicate dissolution and enhanced quartz dissolution (in the sandstone) is responsible for the Si accumulation prior to CO_2 injection. Compositional variations among carbonates, chlorite, and illite could not be considered owing to a lack of thermodynamic data. Additionally, phases such as boehmite or $Fe(OH)_3$ represent idealizations of aluminum- or ferric oxyhydroxides that may, in reality, may be characterized by different stoichiometries or crystal structure than those phases represented in the thermodynamic database.

A significant number of factors contribute to the uncertainty in the cement carbonation geochemical model. Mineralogy, including the pertinent stoichiometry and equilibrium constants, of the hydrated cement is poorly constrained by the experimental data. Moreover, the mineralogical sink for magnesium is unconstrained and may or may not be associated with the CSH phase. Solid solutions may be important for a number of phases in the hydrated cement and carbonated mineral datasets as well as in the reservoir lithology. We employ either pure phase or ideal solid solution because data to constrain non-ideal solid solutions are lacking.

7.3 EXPERIMENTAL RESULTS

7.3.1 CEMENT HYDRATION AT 95 AND 115°C

Cement hydration with reservoir brines altered both the brine chemistry and the hydrated cement phases in experiments at 95 and 115°C (Table 8). Our results show that when Class G cement reacts with distilled water or 0.13 molal $CaCl_2$ brine, the water composition is largely controlled by the solubility of portlandite and an amorphous calcium silica hydrate (CSH). Portlandite was identified as the only crystalline phase by XRD. ESEM analysis also showed crystalline portlandite, as well as an amorphous calcium silicate, which we assume to be CSH (Figure 1).

TABLE 8: Measured solution composition for cement hydration and the experiment end.

ID	Al molal	Ca molal	Fe molal	Mg molal	Si molal	F^- molal	Cl^- molal	NO^{3-} molal	SO_4^{2-} molal
G3	1.39E-05	0.109	3.68E-07	ND	1.73E-05				
G6	1.91E-05	4.68E-01	9.64E-07	1.34E-05	4.87E-06				
G7	1.70E-05	7.83E-03	1.07E-07	ND	8.53E-06	5.86E-05	1.73E-05	2.38E-05	1.81E-03
G8	1.89E-05	6.62E-03	2.21E-07	ND	3.99E-05	6.01E-05	1.66E-04	2.21E-05	1.77E-03
G9	3.30E-05	4.42E-03	6.30E-08	ND	9.89E-06	8.06E-05	1.54E-05	2.91E-05	3.64E-03
G10	2.76E-05	6.01E-03	7.99E-08	ND	3.44E-05	1.21E-04	2.30E-05	2.60E-05	3.85E-03
G11	2.05E-05	7.82E-03	7.30E-08	ND	3.30E-05	5.51E-05	2.42E-05	2.43E-05	2.33E-03
G12	4.22E-05	5.74E-03	6.89E-08	ND	8.56E-06	7.06E-05	1.93E-05	2.59E-05	4.10E-03
G13	3.12E-05	5.90E-03	7.43E-08	ND	5.31E-05	6.96E-05	3.89E-05	2.77E-05	4.85E-03
G14	3.05E-06	5.88E-01	1.46E-06	5.41E-06	1.89E-07		2.97		
G15	3.35E-06	5.82E-01	1.38E-06	5.95E-06	2.05E-07		3.00		

Blank values indicate that the ions were not measured. ND indicates concentrations were below detection.

The storage reservoir brine is likely to be distinct from the dilute waters used to mix the cement prior to injection in the well. The experiments in synthetic brine show that mineral assemblage at the cement - brine interface will be different than in the interior of the cement. The primary difference is that Mg has a very low solubility when the brine reacts with the anhydrous cement. Cement hydration produced a less alkaline and

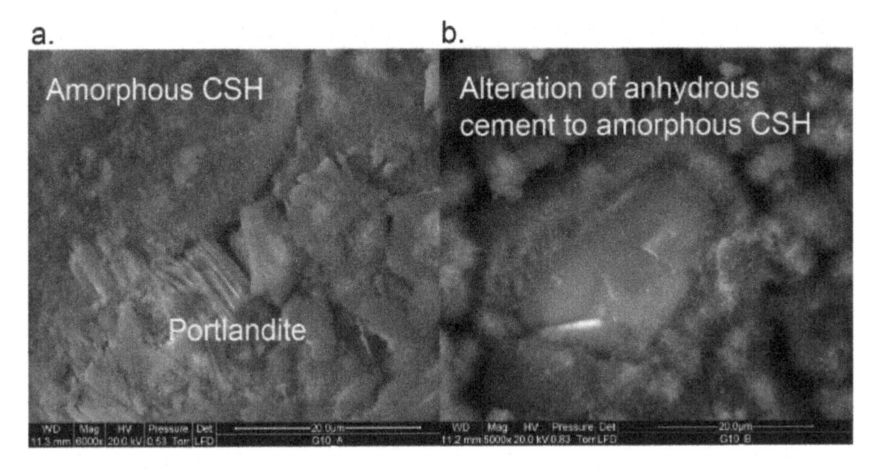

FIGURE 1: ESEM images of Class G cement reacted in 0.13 molal $CaCl_2$ solution showing crystalline portlandite, residual anhydrous Ca_2SiO_4 and amorphous CSH.

Mg-poor brine (Mg = 10^{-5} molal). Although the solution was supersaturated with respect to brucite, $Mg(OH)_2$, the resulting Mg-bearing phase was poorly crystalline and could not be identified by XRD.

There is a large amount of uncertainty associated with the identification of stable or metastable cement minerals in the wellbore environment, because cements can range in composition and are often amorphous. Possible cement phases include anhydrous belite (Ca_2SiO_4) which was present as a residual reactant in most of the experiments, and CSH phases such as hillebrandite ($Ca_2SiO_3(OH)_2*0.17H_2O$), jennite ($Ca_{1.67}SiO_2(OH)_{3.33}$: $0.43H_2O$), and tobermorite-CSH ($Ca_{0.83}SiO_2(OH)_{1.7}*0.5H_2O$). Average ion activity products calculated from the solution chemistry at 95 and 115°C are listed in Table 9. It is likely that cements will alter to crystalline phases with time, because the transformation from amorphous to stable phases has been observed at 150°C for tobermorite [41]. Given the amorphous nature of CSH in our experiments, we have chosen to model it as psuedowollastonite ($CaSiO_3$) because measured solution compositions are near psuedowollastonite equilibrium (Table 5).

TABLE 9: Calculated ion activity products (IAP) for select CSH phases

Cement Hydration			
Phase	Mass Balance	Log IAP 95°C	Log IAP 115°C
Belite	$Ca_2SiO_4 + 4H^+ = 2Ca^{2+} + SiO_2(aq) + 2H_2O$	29.6 ± 1.2	28.9 ± 0.4
Hillibrandite	$Ca_2SiO_3(OH)_2{*}0.2H_2O + 4H^+ = 2Ca^{2+} + SiO_2(aq) + 3.2H_2O$	29.6 ± 1.2	28.9 ± 0.4
Jennite - CSH	$Ca_{1.67}SiO_2(OH)_{3.33}$: $0.43H_2O + 3.33H^+ = 1.67Ca^{2+} + SiO_2(aq) + 3.76H_2O$	27.8 ± 1.4	27.3 ± 0.5
Tobermorite	$Ca_{0.83}Si_2(OH)_{1.7}{*}0.5H_2O + 1.66H^+ = 0.83Ca^{2+} + SiO_2(aq) + 2.2H_2O$	11.7 ± 0.7	11.1 ± 0.5
Pseudowollastonite	$CaSiO_3 + 2H^+ \leftrightarrows SiO_2 + Ca^{2+} + H_2O$	11.3 ± 1.3	11.5 ± 0.5

7.3.2 CEMENT - ROCK - BRINE - CO_2 EXPERIMENTS

In this section we describe the experimental results from the reaction of supercritical CO_2 and synthetic brine with reservoir rock, cap rock, and wellbore cement. The sandstone and shale used in the experiments are meant to represent storage reservoir and the cap rock respectively (although the materials themselves are from the heterolithic sandstone and shale unit that forms the upper section of the carbon storage reservoir at the Krechba Field, In Salah, Algeria). Interpretation of the rate and solubility controlling reactions will be discussed in Geochemical Model.

7.3.2.1 SANDSTONE - BRINE - CO_2

The sandstone consists of tightly carbonate-cemented quartz grains with about 7 wt % chlorite lining the pores [33]. Figures 2 and 3 show the solution profiles with time and images of the unreacted and reacted sandstone. Reaction of sandstone with brine produced a fairly neutral solution with low dissolved CO_2, dissolved Ca and Mg near the initial brine concentrations, dissolved Si and Fe that increased slowly with time, and very low dissolved Al. Injection of supercritical CO_2 into the reaction vessel resulted in marked increases in dissolved CO_2, Si, and Fe. Dissolved CO_2 and Si increase rapidly to a constant concentration. Dissolved Fe peaked upon

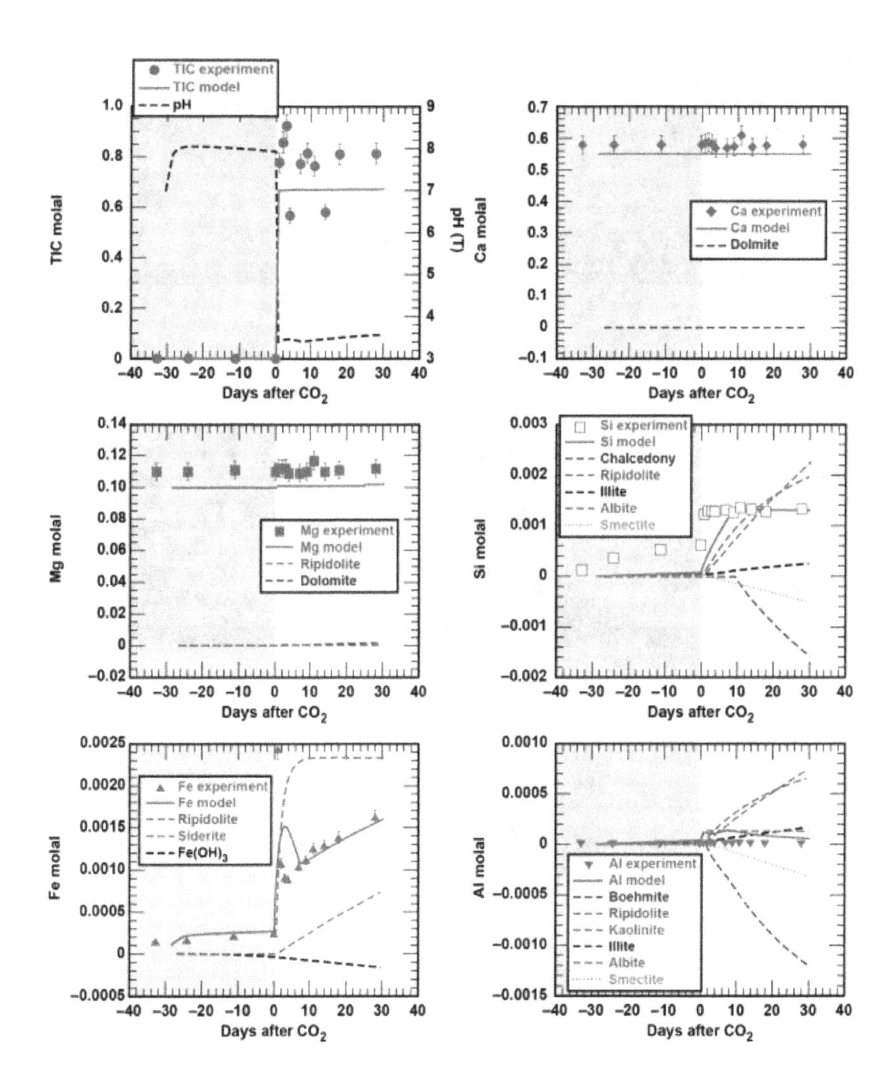

FIGURE 2: Carbonation of sandstone plotted as solution composition versus reaction time. Lines are the modeled results.

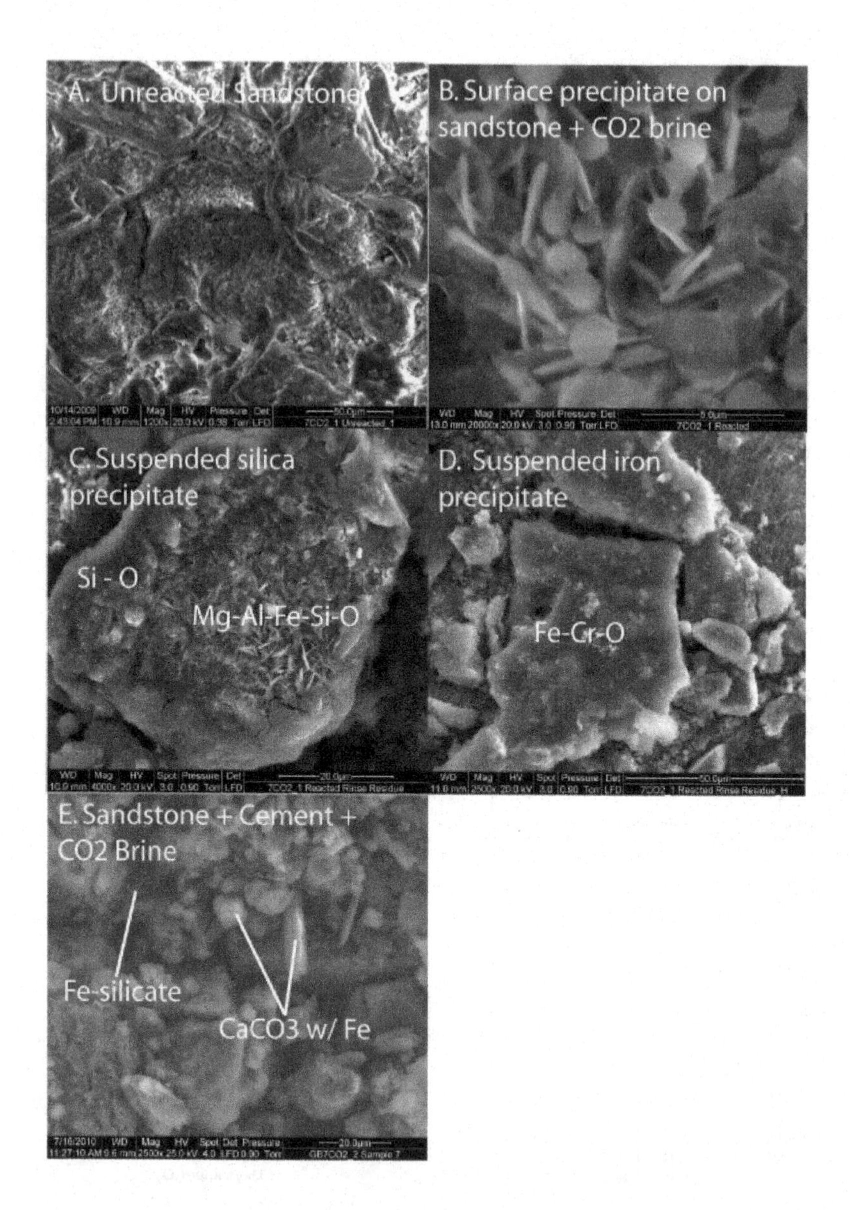

FIGURE 3: ESEM images of reacted sandstone. (A) Unreacted sandstone consisted of quartz, carbonate cement, and chlorite which lines the pore spaces. Reaction of sandstone with CO_2-rich brine produced (B) aluminum hydroxide or aluminosilicate reaction products deposited on the sandstone surface and (C-D) coagulated Si-rich and Fe-rich precipitates in the brine. Reaction of the sandstone with cement and CO_2-rich brine produced (E) Fe - bearing $CaCO_3$ precipitates.

injection of CO_2, dropped to a minimum value and then increased linearly with time. No abrupt changes were observed in dissolved Ca, Mg, or Al.

Secondary precipitates were either amorphous to XRD or present in amounts below the XRD detection limit for crystalline phases to be identified. ESEM images show relatively large amounts of silica and iron precipitates as well as thin hexagonal sheet silicates (Figure 3). It is not possible to identify the composition with EDX because the beam samples an area larger than the surface precipitates. It is also possible that the precipitates formed as the solution was cooled prior to taking apart the reaction vessel and recovering the solids for analysis.

7.3.2.2 SHALE - BRINE - CO_2

Figures 4 and 5 show the solution profiles with time and images of the unreacted and reacted shale. Similar to the sandstone experiment, reaction of shale with brine produced a fairly neutral solution with low dissolved CO_2, dissolved Ca and Mg near the initial brine composition, dissolved Si that increased slowly, and very low dissolved Fe and Al. Injection of supercritical CO_2 into the reaction vessel produced a marked increase in dissolved CO_2, Si, Fe, and Al, with no change in the dissolved Ca and Mg.

We see very little indication of alteration of the shale by CO_2-rich brines at the experiment end by ESEM/EDX analysis (Figure 5). We detect only small precipitates on the shale surface and in the suspension, which may have formed when the sample was quenched from 95°C to room temperature. Particle size was too small to confirm the chemical composition with EDX.

7.3.2.3 CEMENT - BRINE - CO_2

Cement altered to aragonite, calcite, and amorphous silica by the CO_2-rich brines. XRD analyses show aragonite, calcite, and residual anhydrous Ca_2SiO_3. We assume that Si from the CSH phase was altered to amorphous silica. Figures 6 and 7 show the change in solution composition for the reaction of cement, brine, and supercritical CO_2 at 95°C and 10 MPa as

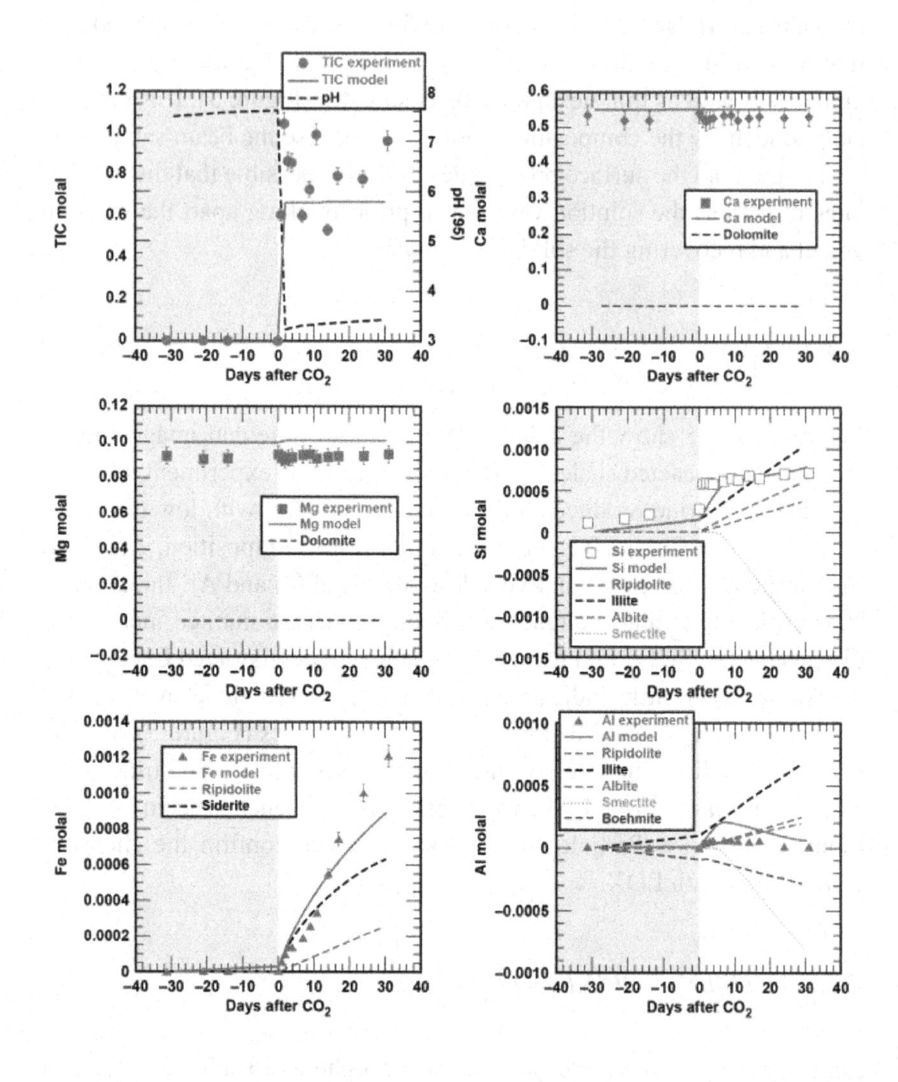

FIGURE 4: Carbonation of shale plotted as solution composition versus reaction time. Lines are the modeled results.

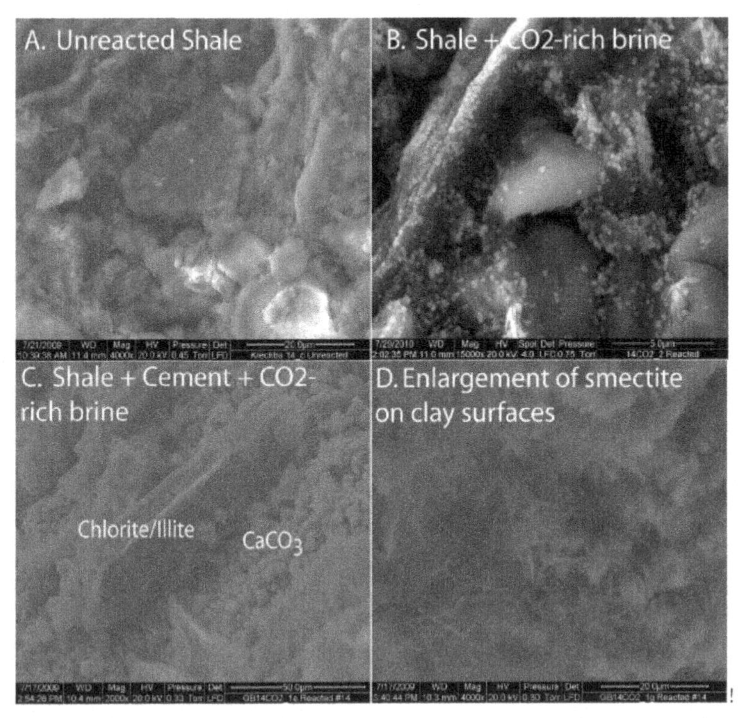

FIGURE 5: ESEM images of reacted shale. (A) Unreacted shale consisted of fine-grained quartz, illite, and carbonate. When the shale was reacted with CO_2-rich brine (B) submicron reaction products deposited on the shale surface and in solution. When the shale reacted with cement and the CO_2-rich brine (C-D) there was extensive clay dissolution and precipitation of smectite and calcium carbonate.

solution pH (95°C), total dissolved CO_2, Ca, Mg, Si, Fe, and Al. These experiments had solid to brine ratios of 1:68 and 1:10 on g/g basis. Trends in dissolved Ca and Mg suggest that starting materials may not have been fully hydrated prior to the injection of CO_2. Extrapolation of the linear decrease in Mg to values measured in the solubility experiments, suggest that the cements would fully equilibrate with the brine within 20 days of reaction at 95°C. The lack of complete hydration is of little consequence because the cement system is very reactive in brines with supercritical CO_2. Upon injection of supercritical CO_2, there was a marked increase in dissolved CO_2, dissolved Ca decreased, dissolved Mg increased toward their initial brine concentration, dissolved Si increased to a constant value, and dissolved Fe and Al were quite low.

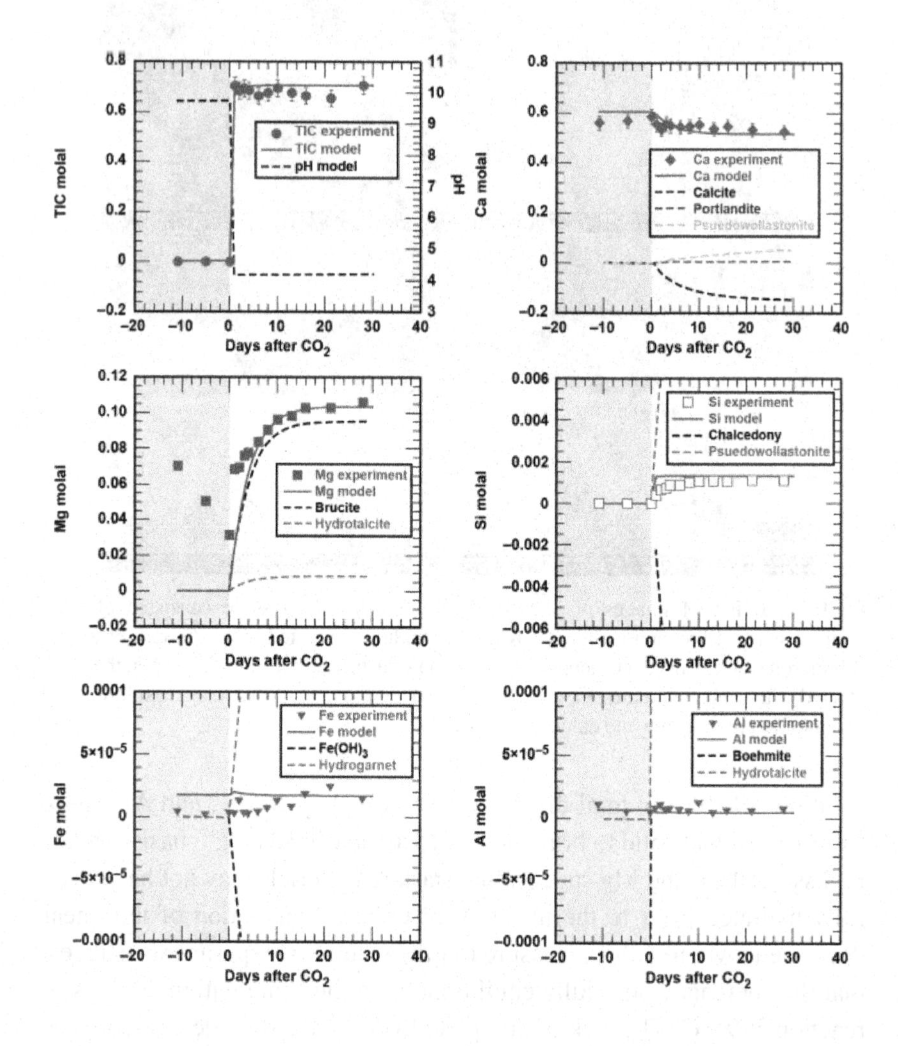

FIGURE 6: Carbonation of class G cement as solution composition versus reaction time with a solid(g): brine(g) ≈ 1:68. Lines are the modeled results.

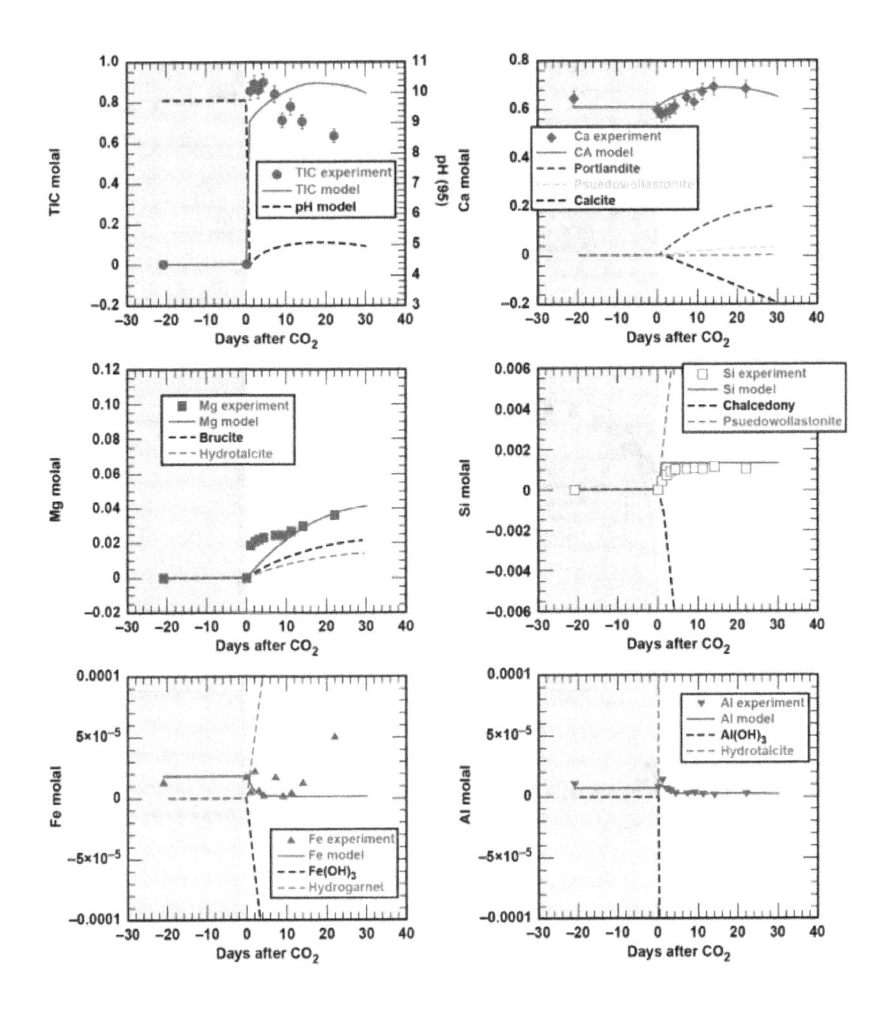

FIGURE 7: Carbonation of class G cement as solution composition versus reaction time with a solid(g): brine(g) ≈ 1:10. Lines are the modeled results.

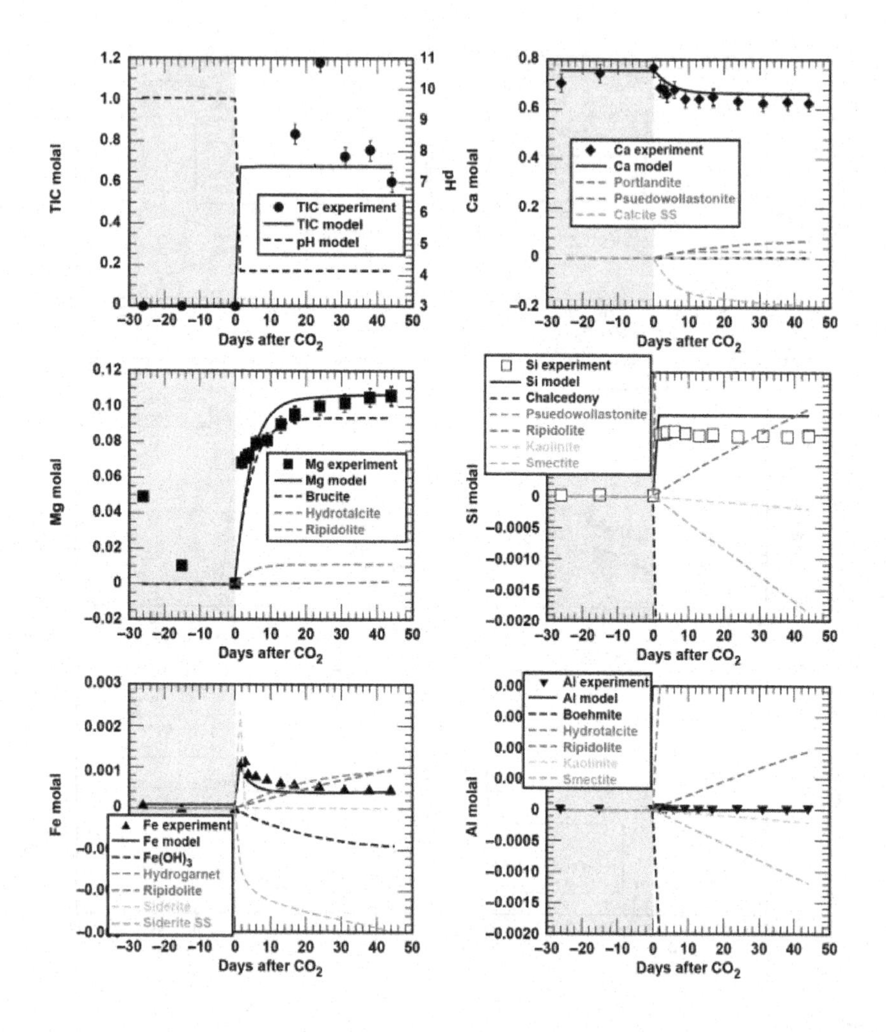

FIGURE 8: Carbonation of class G cement and sandstone plotted as solution composition versus reaction time. Lines are the modeled results.

Experiments with higher amounts of cement resulted in solidification of cement at the bottom of the reactor. We suspect that dissolution was ultimately limited by diffusion at the cement-solution interface. Less than 40% of the cement reacted with CO_2-rich brine based on the dissolved Mg. We estimate the extent of the cement carbonation reaction from the recovery of Mg in solution, because the brines are undersaturated with respect to magnesite ($MgCO_3$) and there is no indication of Mg in the carbonate precipitates. Another difference between this experiment and the one at lower cement: brine was that the dissolved Ca increased with time.

7.3.2.4 CEMENT - SANDSTONE - BRINE - CO_2

Comparison of the solution chemistry profiles from the cement and cement - sandstone experiments suggest that cement carbonation will drive reaction chemistry in the wellbore environment where the cement contacts sandstone geology (Figures 2 and 8). In the first phase of the experiment, cement hydration produced alkaline solutions with elevated Ca and depleted Mg. The cement mineral assemblage underwent rapid carbonation when supercritical CO_2 was injected into the brine. Dissolved CO_2 increased by several orders of magnitude to values between 0.6 and 0.8 molal, dissolved Ca decreased, dissolved Mg increased to the initial brine concentration, dissolved Si also increased to a constant value, and dissolved Al was quite low.

Despite the dominance for the cement carbonation reactions, there is a chemical signature from the sandstone. Upon injection of the CO_2, the dissolved Fe increased by 3 orders of magnitude to a peak concentration, and then decreased over time to a constant value. This is in sharp contrast to the continued increase in dissolved Fe when sandstone was reacted with CO_2-rich brines in the absence of cement. Qualitative EDX analyses show some Fe in rhombahedral and bladed shaped calcium carbonate alteration products (Figure 3). Fe was also detected in the thin bladed micron-sized silicates that are either residual chlorite or a secondary smectite or iron hydroxide. Cr is also detected in these micron-sized crystals.

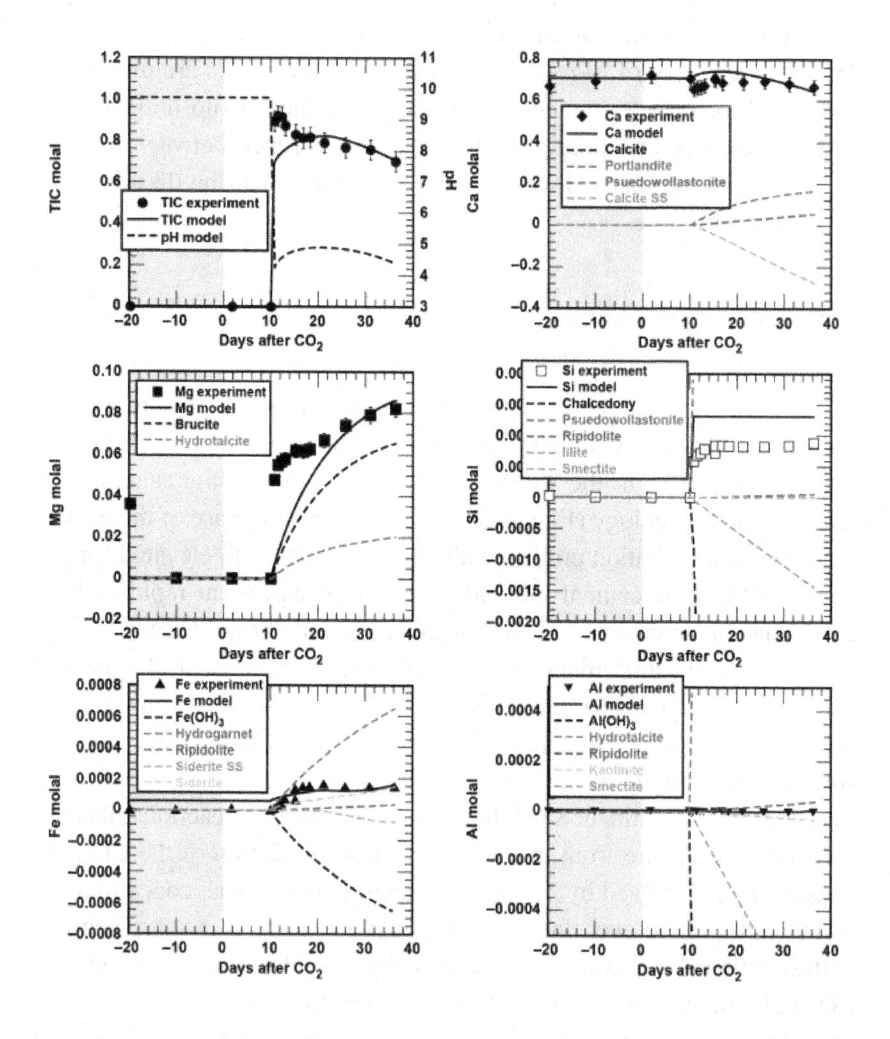

FIGURE 9: Carbonation of class G cement and shale plotted as solution composition versus reaction time. Lines are the modeled results.

7.3.2.5 CEMENT - SHALE - BRINE - CO_2

Similar to cement - sandstone - CO_2 experiment, the cement carbonation chemistry drives the dominant alteration products when shale is reacted with cement and CO_2-rich brines (Figure 9). Cement hydration in this experiment was analogous to the other experiments, producing an alkaline solution with elevated Ca and depleted Mg. Reaction of supercritical CO_2, brine, shale and cement yielded Mg, Si, and Fe profiles that are different from their respective profiles in the cement and cement - sandstone experiments. About 80% of the Mg removed from the brine during the cement hydration phase of the experiment was recovered in the solution at the experiment end suggesting that 80% of the bulk cement was carbonated. Dissolved Si increased to a level below that observed for experiments with cement and cement - sandstone. Dissolved Fe approached a value similar to the final concentrations measured in the cement - sandstone experiments, and dissolved Al was quite low.

SEM images show that sheet silicates were altered when the shale reacts with the cement and CO_2-rich brines (Figure 5). Extensive dissolution groves formed along the edges of the sheet silicates and fibrous precipitates formed on the planar surfaces of the sheet silicates, in addition to calcium carbonate precipitation from cement carbonation.

7.4 GEOCHEMICAL MODEL

Geochemical modeling was used to identify a plausible set of reactions consistent from (1) the reported mineralogy from the Krechba reservoir, (2) the changes in brine chemistry observed during each of the experiments, and (3) the alteration products identified at the end of the experiments. Our objective in creating the geochemical model was to preserve key attributes, such as mineral composition and dissolution rates, across all experiments to better constrain conceptual models for the assessment of long-term CO_2 trapping mechanisms and wellbore and cap rock integrity using reactive-transport simulations. It is important to note that the geochemical model represents one realization that describes six experiments.

Details of the modeling approach and related uncertainties can be found in Modeling Uncertainty.

7.4.1 SANDSTONE RESERVOIR AND SHALE CAP ROCK GEOCHEMICAL MODEL

The geochemical model for the reaction of sandstone and shale with CO_2 and brine is a simple one, in which chlorite, illite, albite, quartz and carbonate minerals partially dissolve and boehmite, smectite, $Fe(OH)_3$ and amorphous silica precipitate (Table 5). The same geochemical model is used to describe both the storage reservoir and the cap rock, because the mineralogy is the same for both rock types, although the relative proportion of the minerals differs.

Comparison of the measured and simulated data shows that this simple model adequately describes the experiments (Figures 2 and 4). In both experiments dolomite dissolution predicts dissolved Ca and Mg to within 5% of the experimental values. Upon injection of CO_2 into the sandstone experiment dissolved Si from albite, chlorite, and lesser amounts of illite dissolution is offset by silica precipitation once chalcedony saturation is exceeded and some smectite precipitation. In the sandstone experiment small amounts of illite limit the amount of smectite precipitation, which is tied to the illite surface area. However, in the shale experiment, where there is significantly more illite (44%) than in the sandstone experiment (1.5%), smectite precipitation effectively limits the dissolved Si to concentrations below chalcedony saturation. Pre-CO_2 and the abrupt changes in dissolved Si, possibly from the dissolution of fines that occurred when CO_2 was first injected into the sandstone experiment, are not captured in the model fits. Both siderite and chlorite dissolution contribute to the dissolved Fe concentrations in both experiments. The sharp decrease in dissolved Fe in the sandstone experiment can be modeled as $Fe(OH)_3$ precipitation and the depletion of dissolved oxygen present in the stock solutions that were prepared at atmospheric conditions. Any dissolved oxygen appears to have been quickly consumed in the shale experiment as no concentration peaks were observed.

The low dissolved aluminum concentrations are a product of secondary precipitation of boehmite, kaolinite, and smectite.

7.4.2 *WELLBORE GEOCHEMICAL MODEL*

We derive the wellbore geochemical model by combining the lithologic model with cement hydration and carbonation models described below. Carbonation of the hydrated cement assemblage was modeled with a set of carbonate minerals, amorphous SiO_2 (as chalcedony), $Fe(OH)_3$, and boehmite or amorphous $Al(OH)_3$ (dependent on the cement: brine; Table 5, Figures 6 and 7). Comparison of the measured and simulated data shows that this simple model adequately describes the data and captures the effects of reacting varying amounts of cement with the CO_2-rich brine. At low solid to brine ratios (1:68 g/g), calcite precipitation results in a decrease in dissolved Ca, brucite and hydrotalcite dissolution result in the recovery of dissolved Mg to initial values, and SiO_2, $Fe(OH)_3$ and boehmite precipitation limit the amount of dissolved Si, Fe, and Al as CSH, Fe-hydrogarnet and hydrotalcite dissolved during the carbonation process. At higher solid to brine ratios (1:10 g/g), where roughly 60% more cement reacted with brine (based on percent recovery of dissolved Mg and the initial amount of cement), the model captures the increase in dissolved Ca with cement carbonation and higher dissolved Al concentrations when amorphous $Al(OH)_3$ is used to control Al solubility. Recall, that rates used here are conditional to the experiments and scale with dissolved Mg recovery.

The combination of the lithology - brine - CO_2 and cement carbonation models reproduces brine chemistry evolution observed during the carbonation phases of the composite experiments (Figures 8 and 9). As might be expected, cement carbonation dominates the geochemical reactions in the wellbore environment, largely because cement reactivity masks contributions from the much less reactive sandstone and shale minerals. Dissolved Ca can be accounted for by the carbonation of portlandite and CSH. Dissolved Mg can be accounted for by dissolution of brucite and hydrotalcite (where the extent of cement carbonation is fit to the proportion of Mg recovered). Although chalcedony precipitation accounts the bulk of Si during carbonation, the higher inputs of Si and Ca result in smectite precipitation in both the cement - sandstone and cement - shale experiments. This model result agrees with the appreciable amount of smectite observed in the cement - shale experiment. One added parameter specific to the lithology - cement - brine CO_2 experiments was the introduction of a ferroan calcite solid solution, which limited the dissolved Fe from chlorite dissolution in the sandstone and shale experiments.

7.5 CONCLUSIONS

Our research shows that relatively simple geochemical models can describe the dominant reactions that will occur when CO_2 is stored in deep saline aquifers sealed with overlying shale cap rocks, and when CO_2 reacts at the interface between cement and reservoir and shale cap rock. Although the experiments and modeling reported here are specific to the CO_2 storage at the Krechba site, the model may be applicable to other storage sites with similar geology. Development of these relatively simple geochemical models is needed to assess long-term CO_2 trapping mechanisms, cap rock and wellbore integrity in more computationally intensive reactive-transport simulations that couple chemistry, flow, and possibly geomechanics. As is expected, Al/Fe silicate dissolution drives the geochemical alterations within the reservoir and cap rock pore space. Addition of CO_2 lowers the pH and promotes silicate dissolution and amorphous silica, smectite and boehmite precipitation. The dissolved Fe may be a source of long-term mineral trapping of CO_2 and the precipitation of secondary Fe-carbonates, clays and hydroxides could alter reservoir and seal permeability by clogging pores and fracture networks. In agreement with other studies we find that alkaline cements are highly reactive in the presence of CO_2-rich brines and are quickly transformed to carbonate minerals and amorphous silica. These reactions can be easily modeled as the transformation of portlandite, and Ca- and Mg-silicates to aragonite or calcite and amorphous silica. Finally, we find that dissolved Mg common in deep saline brines will react with the wellbore cement to form poorly-crystalline solids. Additional research is required to assess mineral structure of the Mg-rich cement phase, as it could not be identified in this study and to assess what the impact of the Mg - induced alteration may have on wellbore integrity.

REFERENCES

1. In Salah Gas Stockage de CO2 [http://www.insalahco2.com]
2. Oldenburg CM, Jordan PD, Nicot J-P, Mazzoldi A, Gupta AK, Bryant SL: Leakage risk assessment of the In Salah CO2 storage project: Applying the certification framework in a dynamic context. Energy Procedia 2011, 4:4154-4161.
3. Dodds K, Watson M, Wright I: Evaluation of risk assessment methodologies using the In Salah CO2 stroage project as a case history. Energy Procedia 2011, 4:4161-4169.
4. Iding M, Ringrose P: Evaluating the impact of fractures on the performance of the In Salah CO2 storage site. I J Greenhouse Gas Control 2010, 4:242-248.
5. Rutqvist J, Vasco DW, Myer L: Coupled reservoir-geomechanical analysis of CO2 injection and ground deformations at In Salah, Algeria. I J Greenhouse Gas Control 2010, 4:225-230.
6. Michael K, Globab A, Shulakova V, Ennis-King J, Allinson G, Sharma S, Aiken T: Geological storage of CO2 in saline aquifers - A review of the experience from existing storage operations. I J Greenhouse Gas Control 2010, 4:659-667.
7. Vasco DW, Ferretti A, Novali F: Reservoir monitoring and characterization using satellite geodetic data: interferometric synthetic aperture radar observations from the Krechba field Algeria. Geophysics 2008, 73:WA113-WA122.
8. Vasco DW, Ferretti A, Novali F, Bissel F, Ringrose P, Mathieson A, Wright I: Satellite-based measurements of surface deformation reveal fluid flow associated with the geological storage of carbon dioxide. Geophysical Research Letters 2010, 37:L03303.
9. Morse JP, Hao Y, Foxall W, McNab W: A study of injection-induced mechanical deformation at the In Salah CO2 storage project. I J Greenhouse Gas Control 2011, 5:270-280.
10. Mathieson A, Wright IW, Roberts D, Ringrose P: Satellite imaging to monitor CO2 movement at Krechba, Algeria. Energy Procedia 2009, 1:2201-2209.
11. Onuma T, Ohkaws S: Detection of surface deformation related with CO2 injection by DInSAR at In Salah, Algeria. Energy Procedia 2009, 1:2177-2184.
12. Kutchko BG, Strazisar BR, Dzombak DA, Lowry GV, Thaulow N: Degradation of well cement by CO2 under geologic sequestration conditions. Environmental Science and Technology 2007, 41:4787-4792.
13. Barlet-Gouedard V, Rimmele G, Porcherie O, Quisel N, Desroches J: A solution against well cement degradation under CO2 geological storage environment. I J Greenhouse Gas Control 2009, 3:206-216.
14. Carey JW, Wigand M, Chipera SJ, WoldeGabriel G, Pawar R, Lichtner PC, Wehner SC, Raines MA, Guthrie GD: Analysis and performance of oil well cement with 30

years of CO2 exposure from the SACROC Unit, West Texas, USA. I J Greenhouse Gas Control 2007, 1:75-85.

15. Crow W, Williams B, Carey JW, Celia M, Gasda S: Wellbore integrity of a natural CO2 producer. Energy Procedia 2009, 1:3561-3569.
16. Duguid A: An estimate of the time to degrade the cement sheath in a well exposed to carbonated brine. Energy Procedia 2009, 1:3181-3188.
17. Wigand M, Kazuba JP, Carey JW, Hollis WK: Geochemical effects of CO2 sequestration on fractured wellbore cement at the cement cap rock interface. Chemical Geology 2009, 265:122-133.
18. Huet BM, Prevost JH, Scherer GW: Quantitative reactive transport modeling of Portland cement in CO2-saturated water. I J Greenhouse Gas Control 2010, 4:561-574.
19. Emberley S, Hutcheon I, Shevalier M, Durocher K, Mayer B, Gunter WD, Perkins EH: Monitoring of fluid-rock interaction and CO2 storage through produced fluid sampling at the Weyburn CO2-injection enhanced oil recovery site, Saskatchewan, Canada. Appl Geochem 2005, 20:1131-1157.
20. Kaszuba JP, Janecky DR, Snow MG: Carbon dioxide reaction processes in a model brine aquifer at 200°C and 200 bars: implications for geologic sequestration of carbon. Appl Geochem 2003, 18:1065-1080.
21. Janecky DR, Snow MG: Experimental evaluation of mixed fluid reactions between supercritical carbon dioxide and NaCl brine: relevance to the integrity of a geologic carbon repository. Chemical Geology 2005, 217:277-293.
22. Palandri JL, Rosenbauer RJ, Kharaka YK: Ferric iron in sediments as a novel CO2 mineral trap: CO2-SO2 reaction with hematite. Appl Geochem 2005, 20:2038-2048.
23. Kharaka YK, Hovorka SD, Gunter WD, Knauss KG, Freifeld BM: Gas-water-rock interactions in Frio Formation following CO2 injection: Implications for the storage of greenhouse gases in sedimentary basin. Geology 2006, 34:577-580.
24. Lu P, Fu Q, Seyfried WE Jr, Hereford A, Zhu C: Navajo Sandstone-brine-CO2 interaction: implications for geologic carbon sequestration. Environ Earth Sci 2011, 62:101-118.
25. Xu T, Apps JA, Pruess K: Reactive geochemical transport simulation to study mineral trapping for CO2 disposal in deep arenaceous formations. J Geophysical Research-Solid Earth 2003, 108:2071.
26. Xu T, Kharaka YK, Daughty C, Freifeld BM, Daley TM: Reactive transport modeling to study changes in water chemistry induced by CO2 injection at the Frio-I Brine Pilot. Chemical Geology 2010, 271:153-164.
27. Johnson JW, Nitao JJ, Knauss KG: Reactive transport modeling of CO2 storage in saline aquifers to elucidate fundamental processes, trapping mechanisms and sequestration partitioning. In Geological Storage of Carbon Dioxide. Volume 233. Edited by Baines SJ, Worden RH. Geological Society, London, Special Publications; 2004::107-128.
28. White SP, Allis RG, Moore J, Chidsey T, Morgan C, Gwynn W, Adams M: Simulation of reactive transport of injected CO2 on the Colorado Plateau, Utah, USA. Chemical Geology 2005, 217:387-405.
29. Zerai B, Saylor BZ, Matisoff G: Computer simulation of CO2 trapped through mineral precipitation in the Rose Run Sandstone, Ohio. Appl Geochem 2006, 21:223-240.

30. Lui F, Lu P, Zhu C, Xiao Y: Coupled reactive flow and transport modeling of CO2 sequestration in the Mt. Simon sandstone formation, Midwest U.S.A. Intern J Greenhouse Gas Control 5:294-307.
31. Gherardi F, Xu T, Pruess P: Numerical modeling of self-limiting and self-enhancing cap rock alteration induced by CO2 storage in a depleted gas reservoir. Chemcial Geology 2007, 244:103-129.
32. Guas I, Azaroual M, Czernichowski-Lauriol I: Reactive transport modeling of the impact of CO2 injection on the clayey rock at Sleipner (North Sea). Chemical Geology 2005, 217:319-337.
33. Armitage PJ, Worden RH, Faulkner DR, Aplin AC, Butcher AR, Iliffe J: Diagenetic and sedimentary controls on porosity in Lower Carboniferous fine-grained lithologies, Krechba field, Algeria: A petrological study of a cap rock to a carbon capture site. Marine and Petroleum Geology 2010, 27:1395-1410.
34. Parkhurst DL, Appelo CAJ: User's Guide to PHREEQC (Version 2) - A Computer Program for Speciation, Batch-reaction, One-dimensional Transport, and Inverse Geochemical Calculations. U.S. Geological Survey Water-Resources Investigations Report 99-4259 1999, :312.
35. Johnson JW, Oelkers EH, Helgeson HC: SUPCRT92: A software package for calculating the standard molal thermodynamic properties of minerals, gases, aqueous species, and reactions from 1 to 5000 bar and 0 to 1000°C. Computers and Geoscience 1992, 18:899-947.
36. CEMDATA Thermodynamic data for hydrated solids in Portland cement system (CaO-Al2O3-SiO2-CaSO4-CaCO3-Fe2O3-MgO-H2O) [http://www.empa.ch/cemdata]
37. Matschei T, Lothenbach B, Glasser FP: The AFM phase in Portland cement. Cement and Concrete Research 2007, 37:118-130.
38. Lothenbach B, Matschei T, Möschner G, Glasser FP: Thermodynamic modeling of the effect of temperature on the hydration and porosity of Portland cement. Cement and Concrete Research 2008, 38:1-81.
39. Duan ZH, Sun R: An improved model calculating CO2 solubility in pure water and aqueous NaCl solutions from 273 to 533 K and from 0 to 2000 bar. Chemical Geology 2003, 193:257-271.
40. Palanrdi JL, Kahraka YK: A complication of rate parameters of water-mineral interactions kinetics for application to geochemical modeling. USGS Open File Report 2004-1068;
41. Houston J, Maxwell RS, Carroll S: Transformation of meta-stable calcium silicate hydrates to tobermorite: reaction kinetics and molecular structure from XRD and NMR spectroscopy. Geochemical Transactions 2009, 10:1.

PART III

BIOLOGICAL SEQUESTRATION OF CO_2

CHAPTER 8

IDENTIFICATION OF A CO_2 RESPONSIVE REGULON IN *BORDETELLA*

SARA E. HESTER, MINGHSUN LUI, TRACY NICHOLSON, DARYL NOWACKI, AND ERIC T. HARVILL

8.1 INTRODUCTION

Many cues, such as temperature, oxygen (O_2), iron, pH, osmolarity and bicarbonate, allow bacteria to distinguish between environments within a host and outside of a host, as well as various microenvironments within a host [1]. In sensing multiple cues, bacteria are able to synchronize gene expression to adapt and ultimately thrive [2]. One cue, carbon dioxide (CO_2), has been shown to affect regulation of virulence factor expression in many bacterial pathogens. *Bacillus anthracis* responds to elevated levels of CO_2 by increasing expression of the genes encoding edema toxin, lethal factor and protective antigen [3]–[5]. In response to 10% CO_2, *Streptococcus pyogenes* increases transcription of M protein, an important virulence factor that prevents the deposition of complement onto the bacterial surface [6]. In increased CO_2, M protein has been shown to be regulated by a trans-acting positive regulatory protein that binds to the promoter of the emm gene [6], [7]. CO_2 regulation in *B. anthracis* appears to be more complicated since the transcriptional regulator of the toxins is not increased transcriptionally in response to growth in CO_2 [3]. Additionally, *Staphylococcus aureus*, *Salmonella enterocolitica* and *Borrelia burgdorferri* are responsive to increased CO_2 concentrations, suggesting this ability is useful to a variety of pathogens [8]–[11].

Bordetella bronchiseptica is a Gram-negative bacterium that infects a wide range of hosts causing respiratory disease varying from asymptomatic persistence in the nasal cavity for the life of the host to lethal pneumonia [12]–[14]. *B. bronchiseptica* is very closely related to the other two classical *Bordetella*e, *Bordetella pertussis* and *Bordetella parapertussis*, the causative agents of whooping cough in humans [13], [15], [16]. Several virulence factors are produced by *B. bronchiseptica* such as, pertactin (PRN), filamentious hemaglutinin (FHA), two serotypes of fimbriae, and the two cytotoxic mechanisms, adenylate cyclase toxin (ACT) and the Type III Secretion System (TTSS) [13], [16]. ACT, a member of the repeats-in-toxin (RTX) family, is a bi-functional adenylate cyclase/hemolysin that converts ATP to cAMP, disrupting oxidative burst, phagocytosis, chemotaxis and eventually leads to apoptosis in macrophages and neutrophils [17]–[19]. ACT has also been shown to contribute to pathology, efficient colonization and persistence of *B. bronchiseptica* and *B. pertussis* species [20]–[22].

Regulation of virulence factors in *Bordetella*e occurs via the BvgAS two-component system [23]. BvgS, the sensor in the cytoplasmic membrane, is thought to directly sense changes in the environment and, through a phosphorylation-transfer mechanism, activates BvgA, the response regulator [24]–[26]. Once BvgA is activated (Bvg^+ phase), it binds to high and low affinity motifs in the genome, resulting in increased expression of the genes encoding toxins and adhesins, while expression of Bvg^- phase genes involved in motility and uptake of certain nutrients are repressed; the opposite occurs in the Bvg^- phase [27]–[32]. An intermediate phase has been described in which a subset of virulence factors are expressed, along with a unique set of factors [33]–[35]; however the Bvg^+ phase has been shown to be necessary and sufficient for host colonization [36]. Although BvgAS appears to be sufficient for regulation of virulence factors the ability to respond to multiple signal inputs to differentially regulate transcriptional networks likely allows for adaptation to different microenvironments within the host. *Bordetella* species have multiple putative transcription factors within their genomes, indicating that gene regulation is likely to be a more complex regulatory system than is currently appreciated [16].

Here we identify, through screening of a collection of *B. bronchiseptica* isolates, strains that only produce ACT in response to growth in elevated CO_2 conditions. Both strain 761 and the sequenced laboratory reference strain RB50 increased transcription of cyaA and production of ACT when grown in 5% CO_2 conditions, although only strain 761 was dependent on 5% CO_2 for efficient expression. Several other virulence factor genes were increased in transcription in response to growth in elevated CO_2. BvgAS was required for ACT production, but cyaA and fhaB were transcriptionally increased in response to 5% CO_2 conditions in the absence of BvgS. Together this indicates that an additional regulatory system increases production of ACT and other virulence factors in various *Bordetella* species.

8.2 MATERIALS AND METHODS

8.2.1 ETHICS STATEMENT

This study was carried out in strict accordance with the recommendations in the Guide for the Care and Use of Laboratory Animals of the National Institutes of Health. The protocol was approved by the Institutional Animal Care and Use Committee at The Pennsylvania State University at University Park, PA (#31297 *Bordetella*-Host Interactions). All animals were anesthetized using isoflourane or euthanized using carbon dioxide inhalation to minimize animal suffering.

8.2.2 BACTERIAL STRAINS AND GROWTH

B. bronchiseptica strain RB50 is an isolate from a rabbit [36]. RB54 and RB50Δ*bsc*NΔ*cya*A are previously described derivatives of strain RB50

[33], [37]. *B. bronchiseptica* strain 761, 448, and 308 were obtained from the CDC in Atlanta, Georgia and have been previously described [15], [38], [39]. *B. bronchiseptica* strain JC100 has been previously described [15]. *B. parapertussis* strain 12822 was isolated from German clinical trials and has been previously described [40], [41]. *B. pertussis* strain 536 is a streptomycin resistant derivative of Tohama I [42] and strain 18323 has been previously described [43]. *B. pertussis* strain CHOC 0012 was isolated on Regan-Lowe media by the Eunice Kennedy National Insitute of Child Health and Human Development (NICHD) Collaborative Pediatric Critical Care Research Network from a child displaying severe Pertussis. Bacteria were maintained on Bordet-Gengou agar (Difco, Sparks, MD) containing 10% sheep blood (Hema Resources, Aurora OR) and 20 μg/mL

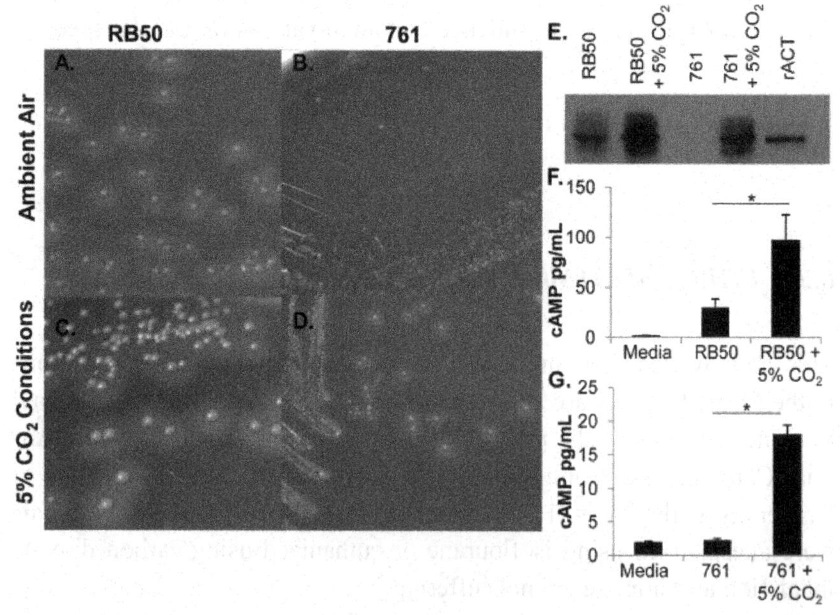

FIGURE 1: Differential production of ACT in *B. bronchiseptica* strains grown in 5% CO_2. Strains RB50 (A, C) and 761 (B, D), grown in normal atmospheric oxygen conditions (A,B) or in elevated 5% CO_2 conditions (C, D). (E) RB50 and 761 grown in atmospheric conditions or grown in 5% CO_2 conditions, or recombinant ACT (2.5 ng) were probed with a monoclonal antibody to CyaA protein at a dilution of 1:1000. J774 murine macrophage cells were stimulated with media or media containing RB50 (F) or 761 (G) at an MOI of 1 for 30 minutes, and cAMP levels were assessed. * indicates a p-value less than 0.05..

streptomycin (Sigma Aldrich, St. Louis, MO). Liquid cultures were grown at 37°C overnight in a shaker to mid-log phase (O.D. 0.7–1.0) in Stainer-Scholte (SS) broth. Bacteria were grown overnight with constant shaking (250 rpm) in standard glass test tubes in either atmospheric concentrations of oxygen and carbon dioxide (atmospheric conditions) or in atmospheric levels of oxygen with the constant controlled addition of 5% carbon dioxide into a sealed incubator 37°C (5% CO_2 conditions).

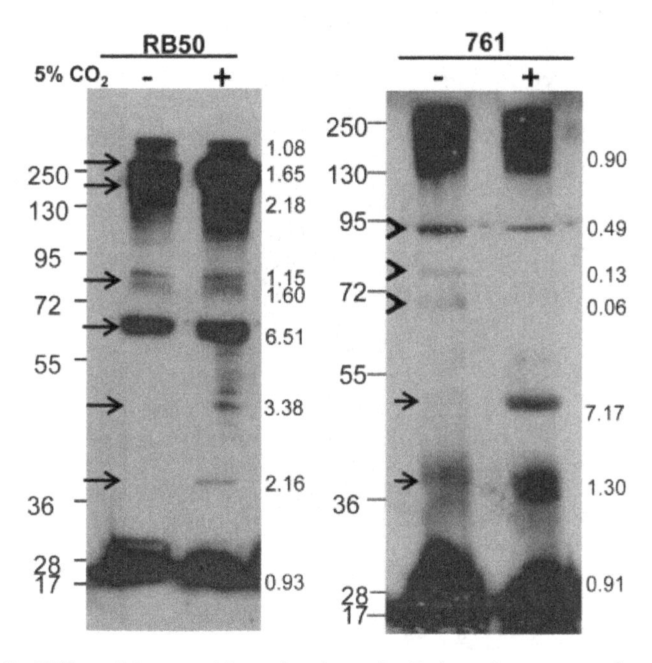

FIGURE 2: Differential recognition of antigens in *B. bronchiseptica* strains grown in atmospheric or 5% CO_2 conditions. C57BL/6 mice were inoculated with 5×105 CFU *B. bronchiseptica* strain RB50, and serum was collected 28 days later. Strains RB50 and 761 were grown in atmospheric or in elevated CO_2 concentrations, and were probed with serum against RB50. Increased (arrows) or decreased (arrowheads) production of antigens in response to 5% CO_2 conditions is denoted. The ratio of band intensity between antigens produced in ambient air and 5% CO_2 conditions is indicated in the margins: 1 = equal amounts produced in either conditions, <1 = more produced in ambient air, >1 = more produced in 5% CO_2 conditions.

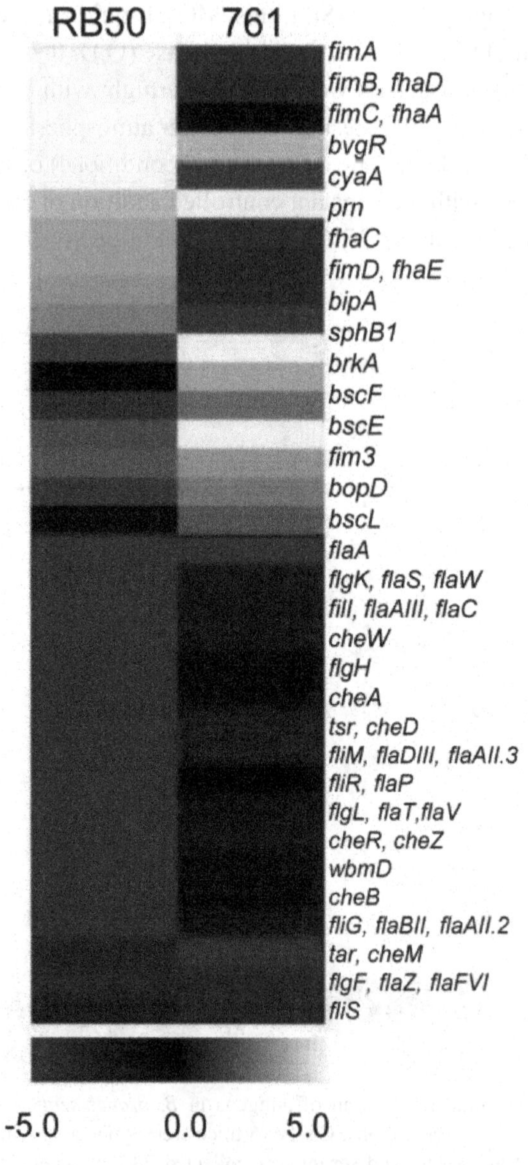

FIGURE 3: Defining the CO_2 responsive regulon in *B. bronchiseptica*. Changes in gene expression of *B. bronchiseptica* in response to 5% CO_2 are analyzed by MeV analysis [48]. Several known virulence factor genes reported to be regulated by BvgAS in prototypical *B. bronchiseptica* strain RB50 are shown for strain RB50 (left) and strain 761 (right), with yellow representing increased transcription and blue indicative of decreased transcription in growth in 5% CO_2 conditions, compared to growth in normal atmospheric conditions.

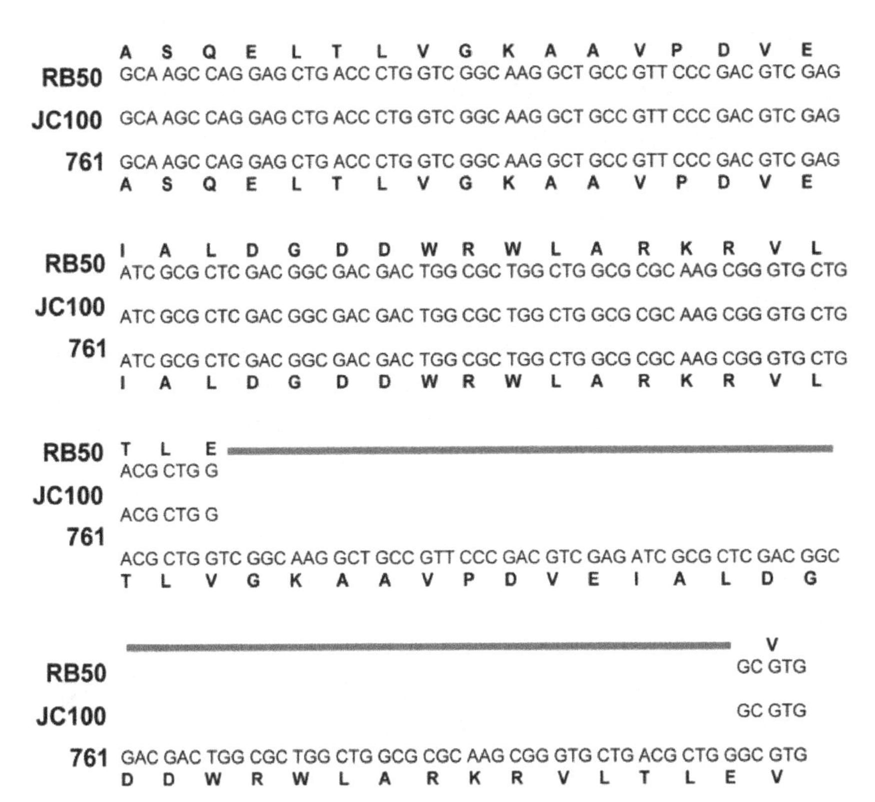

FIGURE 4: Duplication in the bvgS gene in strain JC100. The DNA and protein sequences of bvgS gene from JC100 are aligned against those of bvgS gene from RB50 and 761. The red line indicates the $_{2}$9 amino acid duplication in the bvgS locus in JC100 compared to RB50 and 761.

8.2.3 CAMP ASSAY

Murine macrophage-like cell line, J774, was cultured in Dulbecco modified Eagle medium (DMEM) with 10% fetal bovine serum (FBS) (Hyclone Laboratories, Inc., Logan, UT). Cells were grown to approximately 80% confluency, and bacteria were added at a multiplicity of infection (MOI) of 1. After a 5 minute centrifugation at 250×g, the mixture was incubated for 30 minutes at 37°C. cAMP was measured with a cyclic AMP ELISA system (Tropix, Bedford, MA) according to the manufacturer's instructions.

Results were analyzed using analysis of variance with a Tukey simultaneous test, and a P value of <0.05 was considered significant.

8.2.4 ANIMAL EXPERIMENTS

C57BL/6 mice were obtained from Jackson Laboratories (Bar Harbor, ME). Mice were bred in our *Bordetella*-free, specific pathogen-free breeding rooms at The Pennsylvania State University. All animal experiments were performed in accordance with institutional animal care and use committee (IACUC) guidelines. 4 to 6 week old mice were lightly sedated

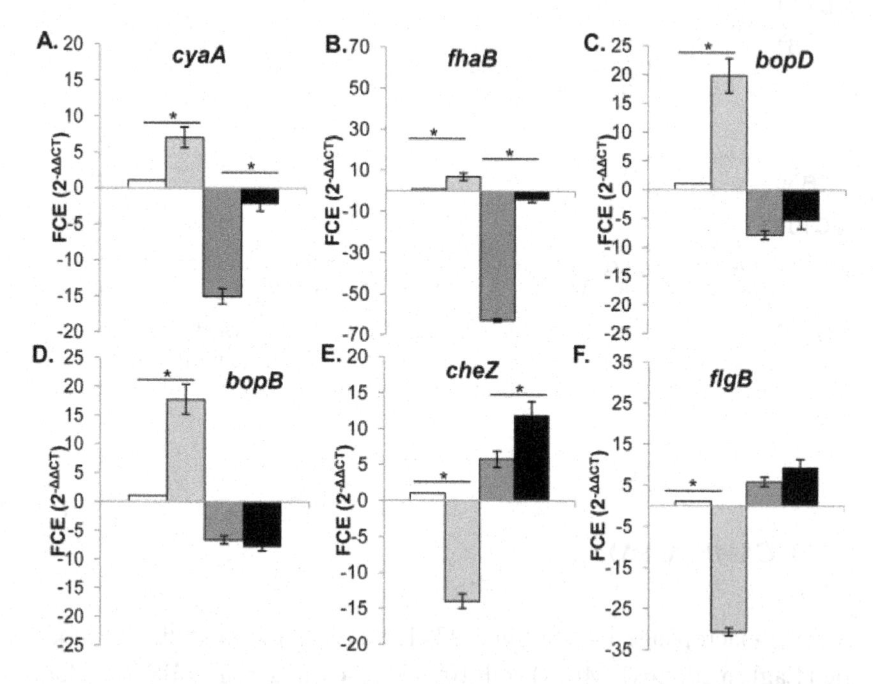

FIGURE 5: Differential transcription of 5% CO_2 responsive genes independent of bvgS expression. qRT-PCR analysis was performed on RB50 grown in 5% CO_2 (light grey bars), RB54 in ambient air (dark gray bars), and RB54 in 5% CO_2 (black bars) compared to RB50 grown in ambient air (white bars). Fold-change expression (FCE) in all strains was expressed as mean ± standard deviation for cyaA (A), fhaB (B), bopD (C), bopB (D), cheZ (E), and flgB (F). Data shown are averages obtained from quadruplicate cultures. * indicates a p-value less than 0.05.

with 5% isoflurane (IsoFlo, Abbott Laboratories) in oxygen and 5×105 CFU were pipetted in 50 ul of phosphate-buffered saline (PBS) (Omnipur, Gibbstown, NJ) onto the external nares. This method reliably distributes the bacteria throughout the respiratory tract [43]. To obtain serum, blood from inoculated or vaccinated mice was obtained 28 days post-inoculation and serum was separated from the blood by centrifugation at 500×g for 5 minutes.

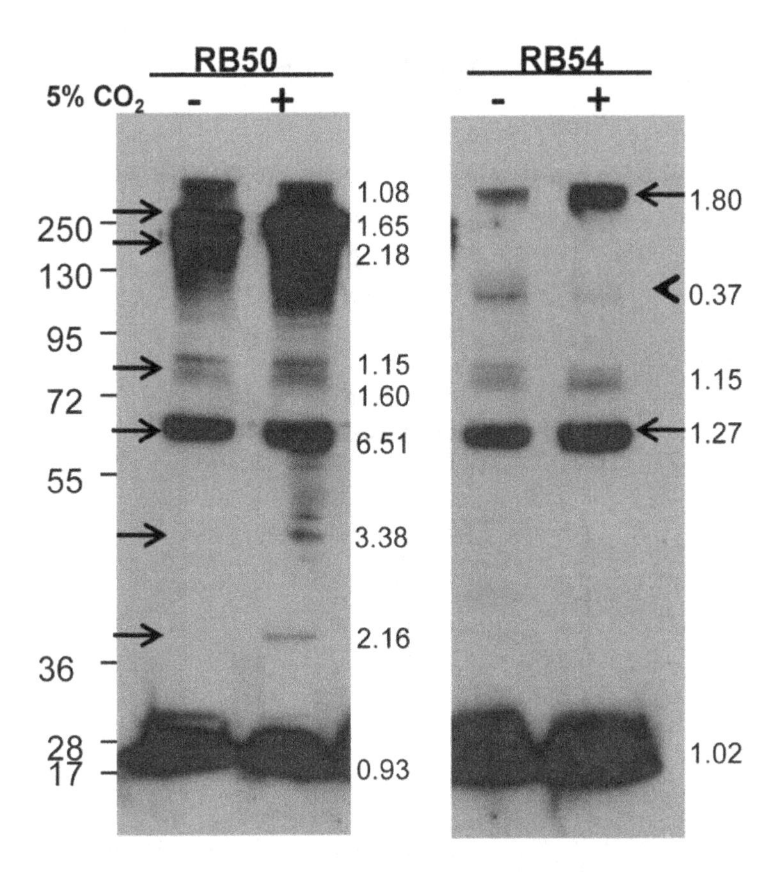

FIGURE 6: Differential recognition of antigens of a Bvg− mutant grown in atmospheric or 5% CO_2 conditions. C57BL/6 mice were inoculated with 5×105 CFU *B. bronchiseptica* strain RB50 and sera were collected $_2$8 days post-inoculation. Strains RB50 and RB54 were grown in atmospheric or in 5% CO_2 concentrations, and were probed with serum against RB50. The ratio of band intensity between antigens produced in ambient air and 5% CO_2 conditions is indicated in the margins: 1 = equal amounts produced in either conditions, <1 = more produced in ambient air, >1 = more produced in 5% CO_2 conditions.

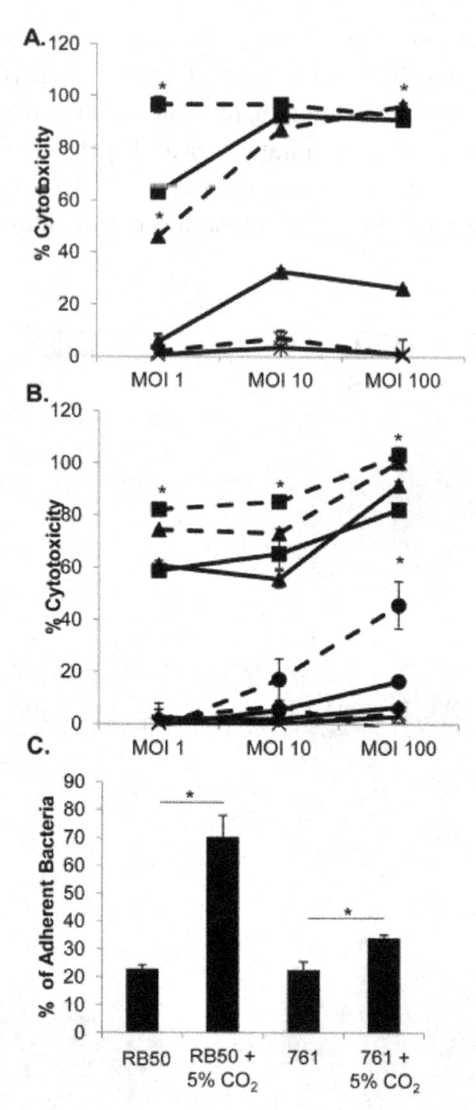

FIGURE 7: Cytotoxicity and adherence of strains grown in 5% CO_2 conditions. J774 murine macrophage cells were stimulated with media alone or media (X) containing RB50 (▪) or 761 (triangle) (A), RB50 (▪), RB50Δ*cya*A (triangle), RB50Δ*bsc*N (•), or RB50Δ*cya*AΔ*bsc*N (♦) (B) in atmospheric (solid lines) or 5% CO_2 conditions (dashed lines) at MOIs of 1, 10 or 100 for 4 hours, and LDH release was assayed. (C) Rat epithelial cells were incubated with RB50 or 761 at a MOI of 100 for 30 minutes. Adherence is expressed as the proportion of adherent bacteria to the amount in the original inoculum. The error bars represent standard deviations. * p-values less than 0.05 as compared to the same strain grown in atmospheric conditions.

8.2.5 WESTERN IMMUNOBLOTS

Western blots were performed on whole cell extracts of *B. bronchiseptica*, *B. pertussis* and *B. parapertussis* grown to mid-log phase in SS broth as described previously [38], [39]. Lysates were prepared by resuspending 1×109 CFU in 100 μl of Laemmli sample buffer; total cellular protein content were quantitated using the BCA assay to equalize protein content between samples. 1×108 CFU (10 μl) were run on an 8% sodium dodecyl sulfate-polyacrylamide electrophoresis gels in denaturing conditions and transferred to a polyvinylidene difluoride membrane (Millipore, Bedford, MA). Membranes were probed with pooled serum from mice inoculated with *B. bronchiseptica*, *B. pertussis*, *B. parapertussis*, or a monoclonal antibody against ACT (anti-ACT) at the following dilutions, 1:1000, 1:500, 1:1000 and 1:1000, respectively. A 1:10,000 dilution of goat anti-mouse Ig HRP conjugated antibody (Southern Biotech, Birmingham, AL) was used as the detector antibody. Membranes were visualized with ECL Western blotting detection reagents (Amersham Biosciences, Piscataway, NJ) and quantified using Image J software [44].

8.2.6 RNA ISOLATION

RNA was isolated from three independent biological replicates of *B. bronchiseptica* strains RB50, 761, RB54, 536 and 12822 grown in SS broth overnight. Bacteria were subcultured at a starting OD_{600} of 0.1 into 5 ml of SS broth and grown at 37°C while shaking in either atmospheric or 5% CO_2 conditions until the OD_{600} reached 0.75. Bacteria were harvested and total RNA was extracted using a RNAeasy Kit (Qiagen, Valencia, CA) and treated with RNase-free DNase I (Invitrogen, Carlsbad, CA) according to the manufacturer's instructions.

8.2.7 PREPARATION OF LABELED CDNA AND MICROARRAY ANALYSIS

RNA isolated from strains RB50 and 761 were used in microarray experiments. A 2-color hybridization format was used for the microarray

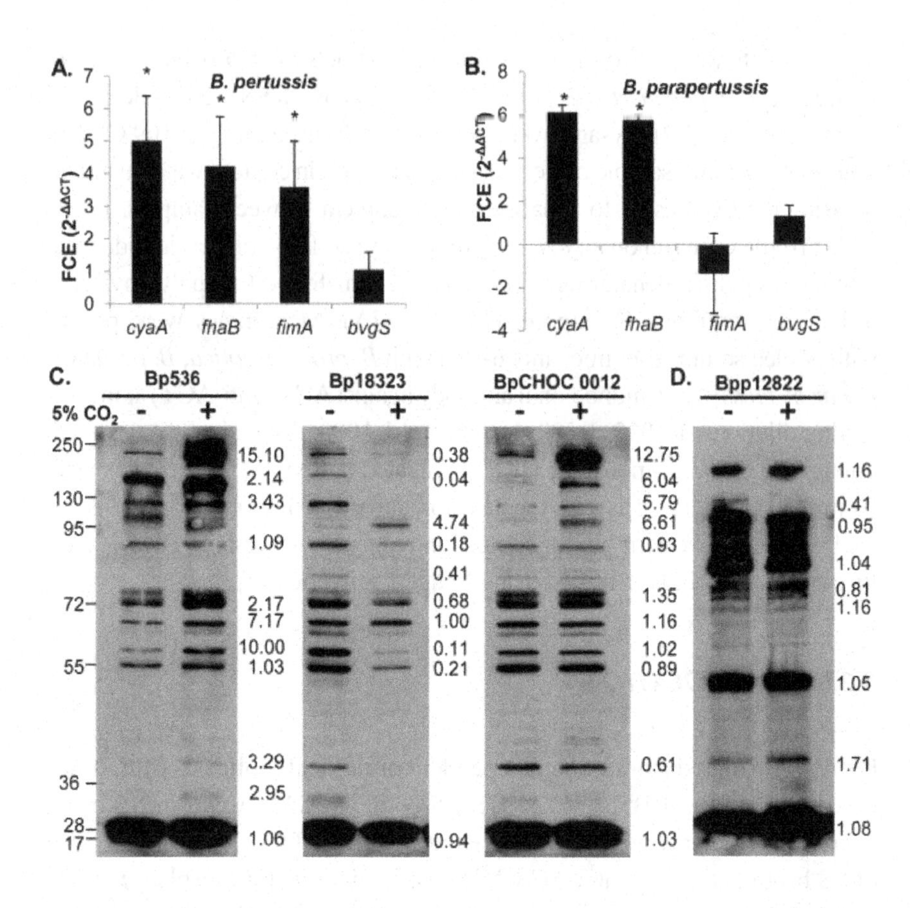

FIGURE 8: Differential expression of virulence factor genes in *B. pertussis* and *B. parapertussis* in response to 5% CO_2 conditions. qRT-PCR analysis was performed on *B. pertussis* (A) and *B. parapertussis* (B) grown in atmospheric or elevated CO_2 conditions. Fold-change expression (FCE) in *B. pertussis* and *B. parapertussis* grown in 5% CO_2 was compared to expression in atmospheric levels of CO_2 and expressed as the mean ± standard deviation. Lysates from bacteria grown in either atmospheric conditions or 5% CO_2 conditions were probed with serum from mice inoculated with either a *B. pertussis* strain 536 (C) or *B. parapertussis* strain 12822 (D). The ratio of band intensity between antigens produced in ambient air and 5% CO_2 conditions is indicated in the margins: 1 = equal amounts produced in either conditions, <1 = more produced in ambient air, >1 = more produced in 5% CO_2 conditions.

analysis. For each biological replicate, RNA extracted from cells grown in 5% CO_2 conditions was used to generate Cy5-labeled cDNA and RNA extracted from cells grown in atmospheric conditions was used to generate Cy3-labeled cDNA. Additionally, dye-swap experiments were performed analogously, in which the fluorescent labels were exchanged to ensure that uneven incorporation did not confound our results. Fluorescently-labeled cDNA copies of the total RNA pool were prepared by direct incorporation of fluorescent nucleotide analogs during a first-strand reverse transcription (RT) reaction [39], [45]–[47]. The two differentially labeled reactions were then combined and directly hybridized to a *B. bronchiseptica* strain RB50-specific long-oligonucleotide microarray [46]. Slides were then scanned using a GenePix 4000B microarray scanner and analyzed with GenePix Pro software (Axon Instruments, Union City, CA). Spots were assessed visually to identify those of low quality and arrays were normalized so that the median of ratio across each array was equal to 1.0. Spots of low quality were identified and were filtered out prior to analysis. Ratio data from the two biological replicates were compiled and normalized based on the total Cy3% intensity and Cy5% intensity to eliminate slide to slide variation. Gene expression data were then normalized to 16S rRNA. The statistical significance of the gene expression changes observed was assessed by using the significant analysis of microarrays (SAM) program [48]. A one-class unpaired SAM analysis using a false discovery rate of 0.001% was performed. Hierarchical clustering of microarray data using Euclidean Distance metrics and Average Linkage clustering was performed using MeV software from TIGR [49]. All microarray data are available in Tables S1 and S2 and have been deposited in ArrayExpress or ArrayExpress Archive under accession number E-MEXP-2875.

8.2.8 REAL-TIME QPCR (QPCR)

qPCR was performed using a modified protocol previously described [46], [47]. RNA was extracted as described, and 1 μg of RNA from each biological replicate was reverse transcribed using ImProm-II Reverse transcriptase and 0.5 μg of random oligonucleotide hexamers (Promega, Madison, WI). cDNA was diluted 1:1,000 and 1 μl was used in RT-qPCRs containing

300 nM primers designed with Primer Express software (Applied Biosystems, Foster City, CA, and Integrated DNA Technologies software, www.idtdna.com) (Primer sequences are listed in Table S3) and SYBR Green PCR master mix (Invitrogen, Carlsbad, CA). Samples without reverse transcriptase were included to confirm lack of DNA contamination and dissociation curve analysis was performed to determine cycle threshold (C_T) for each reaction. Amplification of the recA RNA amplicon was used as an internal control and for data normalization. Change in transcript level was determined using the relative quantitative method ($\Delta\Delta C_T$) [50]. Results were analyzed using analysis of variance with a Tukey simultaneous test, and a P value of <0.05 was considered significant.

8.2.9 CYTOTOXICITY ASSAY

Cytotoxicity assays were carried out as previously described [45]. J774A.1 cells, a murine macrophage cell line, obtained from the ATCC were cultured in DMEM with 10% FBS. Cells were grown to approximately 80% confluency, and bacteria were added at a MOI of 100, 10 or 1. After a 5 minute centrifugation at 250×g, the mixture was incubated at 37°C for the indicated times. Cytotoxicity was determined by measuring lactate dehydrogenase (LDH) release using the Cytotox96 (Promega, Madison, WI) kit according to the manufacturer's protocol. Results were analyzed using analysis of variance with a Tukey simultaneous test, and a P value of <0.05 was considered significant.

8.2.10 ADHERENCE ASSAY

Adherence assays were modified from a previously described protocol [47]. Rat epithelial cell line L2, obtained from the ATCC, was cultured in DMEM/Ham's F12 50–50 mixture with 10% FBS. Cells were grown to approximately 80% confluency, and bacteria were added at an MOI of 100. After a 5 minute centrifugation at 250×g, the mixture was incubated for 30 minutes. Cell culture supernatant was removed and cells were washed 4 times with PBS to remove unbound bacteria. Epithelial cells were then trypsinized and resuspended in 1 mL of tissue culture media. The mixture of cells and bacteria were diluted in PBS and plated on BG agar to determine CFU. Results were analyzed using analysis of variance with a Tukey simultaneous test, and a P value of <0.05 was considered significant.

8.3 RESULTS

8.3.1 A B. BRONCHISEPTICA *ISOLATE REGULATES ACT EXPRESSION IN RESPONSE TO 5% CO_2 CONDITIONS*

B. bronchiseptica isolates are generally β-hemolytic when grown in Bvg^+ conditions due to the production of ACT, which causes lysis of red blood cells. It was recently discovered that some *B. bronchiseptica* isolates do not have the genes required to produce a functional ACT and therefore are not hemolytic on blood agar plates [39]. However, through screening of 73 isolates based on hemolysis on blood agar plates and PCR amplification of the genes encoding ACT, 4 *B. bronchiseptica* isolates were found to be non-hemolytic, but still retained the genes for production of ACT (Fig. 1, data not shown). When one isolate displaying this phenotype, *B. bronchiseptica* strain 761, was grown in a tissue culture incubator where the CO_2 concentration is increased to 5%, it was hemolytic (Fig. 1, compare panels B and D). RB50 was hemolytic even in ambient air (~0.03% CO_2) (Fig. 1, compare panels A and C), but there appeared to be more hemolysis when it was grown in 5% CO_2 conditions, suggesting both strains produce more ACT in response to growth in 5% CO_2 conditions. Growth rate and pH were not significantly affected by additiona of 5% CO_2 (Figure S1.), although other indirect effects are possible.

To more directly assess the production of ACT, lysates of strains RB50 and 761 grown in liquid cultures (mid-log phase, O.D. 0.7–0.9) in normal atmospheric conditions or 5% CO_2 conditions were probed with a monoclonal antibody to the cyaA protein product, ACT. RB50 produced more ACT when grown in 5% CO_2 conditions than in normal atmospheric conditions (Fig. 1E). Strain761 grown in normal atmospheric conditions produced no detectable ACT, while 761 in 5% CO_2 conditions did produce ACT (Fig. 1E). To determine if strain 761 produces a functional ACT, cyclic-AMP (cAMP) was measured in murine macrophages stimulated with bacteria grown in either normal or 5% CO_2 conditions. Cells were stimulated for 30 minutes with strains RB50 or 761 to assess their effects on cAMP levels. Both strains grown in 5% CO_2 induced significantly more cAMP than the same strains grown in normal atmospheric conditions (Fig. 1F, G). Together, these data demonstrate by different measures that the prototypical *B. bronchiseptica* strain, RB50, and strain 761 increase production of functional ACT when grown in 5% CO_2 conditions, but only strain 761 appears to be dependent on 5% CO_2 conditions for production of ACT.

Since ACT production was increased in 5% CO_2, we hypothesized that other antigens might also be differentially regulated in response to these conditions. To test this, Western blot analysis was performed by probing RB50 and 761 lysates, grown in ambient air or in 5% CO_2 conditions, with serum antibodies from animals convalescent from RB50 infection. No antigens appeared to be produced in greater amounts in RB50 grown in ambient air, while strain 761 produced antigens of between 72 and 95 kDa in greater amounts in 5% CO_2 growth conditions (Fig. 2, arrowheads). Antigens greater than 130 kDa were produced in greater amounts when RB50 was grown in 5% CO_2 conditions compared to ambient air (Fig. 2, arrows). Additionally, both 761 and RB50 produced antigens between 36 and 55 kDa in greater amounts when grown in elevated CO_2 concentrations than in ambient air (Fig. 2). These data indicate that additional antigens besides ACT are differentially regulated in response to 5% CO_2.

8.3.2 DEFINING A CO_2 RESPONSIVE REGULON IN B. BRONCHISEPTICA

To determine which genes are differentially regulated in response to different CO_2 concentrations, microarray analyses were performed comparing RB50 grown in atmospheric concentrations of CO_2 to growth in elevated CO_2 conditions. Transcript abundance of 35 genes increased in RB50 in response to 5% CO_2, based on SAM analysis (Table S1, Fig. 3), including genes encoding ACT, as well as genes encoding other known virulence factors such as members of the TTSS locus (*bsc*E, *bsc*F, *bop*D), FHA (*fha*C, *fha*B, *fha*D, *fha*A), fimbriae (*fim*A, *fim*3), and Prn, (Fig. 3). Expression of 452 genes were decreased when RB50 was grown in 5% CO_2, many of which are known to be expressed in the Bvg$^-$ phase including, *fla*A, *che*W, *che*B, *wbm*D, *flg*H, *fli*S, and *che*D (Fig. 3). A similar trend was observed for strain 761; genes encoding known virulence factors were increased in 5% CO_2 growth conditions while genes for flagellar assembly and chemotaxis were decreased in these conditions (Table S2, Fig. 3). Expression of 6 genes, including *cya*A, was increased in both strains and expression of 41 genes decreased in both strains when grown in 5% CO_2 conditions (Table S4). qPCR of 12 genes confirmed the microarray results (Table S3). Overall, these data indicate there is a CO_2 responsive regulon in *B. bronchiseptica* that includes several virulence factors, suggesting a role during infection.

Among the 35 genes in strain RB50 increased in expression in 5% CO_2 conditions, 19 genes were reported to be positively regulated under Bvg$^+$ conditions, 4 genes negatively regulated by BvgAS, and 13 genes not previously known to be regulated by BvgAS, based on previous analysis [46], [51]. Similarly, of the 452 genes negatively regulated by 5% CO_2, 252 were known to be negatively regulated under Bvg$^+$ conditions, 19 were positively regulated and 181 were not previously known to be regulated by BvgAS. The CO_2-responsive regulon appears to contain genes that are Bvg-regulated, as well as genes that are not, suggesting a regulatory mechanism that functions independently or cooperatively with BvgAS rather than subordinate to it.

8.3.3 CO_2 RESPONSIVENESS IS NOT CONFERRED BY DIFFERENCES IN THE BVGAS LOCI BETWEEN STRAINS

Three additional *B. bronchiseptica* strains, JC100, 308 and 448, were also observed to be hemolytic only when grown in 5% CO_2, but not in normal atmospheric conditions. Changes in virulence factor expression have previously been attributed to variation in bvgAS and since some CO_2-responsive genes are Bvg-regulated, the bvg locus of these strains was analyzed revealing that strain JC100 carries a $_2$9 amino acid duplication in the region of the *bvg*S gene encoding the periplasmic domain (Fig. 4). To determine if this duplication is involved in the CO_2/ACT dependent phenotype in JC100, the *bvg*AS locus from JC100 was expressed in a RB50 knockout of bvgAS (RB55::pBvgAS$_{JC100}$). This strain was hemolytic in the absence of 5% CO_2, indicating that transfer of the *bvg*AS locus does not confer the CO_2-dependence for ACT production (Table S5). The reverse was also true; when a plasmid carrying the bvgAS locus from RB50 was introduced into JC100 (MLJC114::pEG100), ACT production remained dependent on growth in 5% CO_2 (Table S5). Furthermore, the bvgS gene of 761 did not have this duplication (Fig. 4). These data indicate that the duplication in *bvg*S in JC100 is neither necessary nor sufficient for the CO_2 requirement for hemolysis.

8.3.4 CO_2 RESPONSIVENESS IN THE B. BRONCHISEPTICA *BVG- STATE*

Since some virulence genes known to be Bvg-regulated were responsive to 5% CO_2 conditions, we sought to determine if they are differentially regulated in response to 5% CO_2 in the absence of BvgAS. Transcription of six genes responsive to CO_2, *cya*A, *fha*B, *bop*D, *bop*B, *che*Z and *flg*B, were analyzed in *B. bronchiseptica* RB50 and RB54, a Bvg–phase locked derivative of strain RB50, grown in normal atmospheric or 5% CO_2 conditions (Fig. 5). For RB50, addition of 5% CO_2 increased transcription of *cya*A, *fha*B, *bop*D and *bop*B (Fig. 5A–D), but decreased transcription of *che*Z and *flg*B (Fig. 5 E,F). In RB54, transcription of *bop*D and *bop*B was not increased in response to addition of 5% CO_2 (Fig. 5C,D), and transcription of *che*Z and *flg*B was not decreased (Fig. 5 E,F). Therefore, the differential transcription of *bop*D, *bop*B, *che*Z and *flg*B in response to 5% CO_2 is dependent on BvgS. However, in the *bvg*S mutant RB54, the transcription of genes *cya*A and *fha*B was increased in response to addition of 5% CO_2 (Fig. 5 A,B), indicating that some gene regulation in response to 5% CO_2 is independent of BvgS.

To determine whether differential transcription results in differential accumulation of antigens in the absence of BvgS, Western blots were performed. Lysates from RB54 grown in normal atmospheric or 5% CO_2 conditions were probed with serum antibodies from mice convalescent from RB50 infection (Fig. 6). Strain RB54 grown in 5% CO_2 also produced antigens >250 kDa and ~60 kDa in greater amounts (Fig. 6). RB54 grown in ambient air produced an antigen between 95 and 130 kDa in greater amounts (Fig. 6, arrowhead). These data demonstrate that antigen production is differentially regulated in response to 5% CO_2 even when a functional BvgS is absent.

8.3.5 GROWTH IN 5% CO_2 AFFECTS CYTOTOXICITY AND ADHERENCE OF B. BRONCHISEPTICA *STRAINS*

Since genes, *cya*A and the TTSS genes, associated with the cytotoxicity of *B. bronchiseptica* were increased when strains RB50 and 761 were grown in 5% CO_2 conditions (Fig. 3), we assessed the relative cytotoxicity to J774 murine macrophages of strains grown in atmospheric or 5% CO_2 conditions. Similar to previous findings [52], RB50 killed >90% of cells at an MOI of 10 or 100; however, RB50 only killed ~65% at an MOI of 1 (Fig. 7A). RB50 grown in 5% CO_2 killed >90% at all MOIs, indicating that RB50 grown in 5% CO_2 killed more macrophages at a lower MOI than RB50 grown in atmospheric conditions (Fig. 7A). Strain 761 had detectable (~30%) killing only at high MOIs (10 and 100), while 761 grown in 5% CO_2 was cytotoxic at an MOI of 1 (~45%) and comparable to RB50 at higher MOIs (Fig. 7A). These data show that growth in 5% CO_2 increased killing of murine macrophages by both strains.

ACT and TTSS have been previously shown to account for all cytotoxicity of macrophages when RB50 is grown in ambient air [37], [52]. Since 5% CO_2 increased expression of several genes, we examined whether the increased cytotoxicity is due to increases in these known factors or a new cytotoxic mechanism. Cells were exposed for 4 hours at MOIs of 1, 10 or 100 with wild-type RB50, RB50Δ*cya*A, RB50Δ*bsc*N (encoding the ATPase of the TTSS) or a mutant lacking bscN and cyaA (RB50Δ*cya*AΔ*bsc*N). Growth in 5% CO_2 increased cytotoxicity of RB50Δ*cya* to macrophages at MOIs of 1, 10 and 100, while growth 5% CO_2 caused increased cytotoxicity of RB50Δ*bsc*N only at an MOI of 100, likely indicating the differential roles of the TTSS and ACT in cytotoxicity (Figure 7B). RB50Δ*cya*AΔ*bsc*N caused very low levels of cytotoxicity as observed previously [37], and growth in 5% CO_2 did not increase cytotoxicity(Fig. 7B). These data suggest that there are no other cytotoxic mechanisms and that increased ACT and TTSS function accounts for the increased cytotoxicity when strains are grown in 5% CO_2 conditions.

Since many genes encoding adhesins were increased in transcription in response to growth in 5% CO_2 conditions, we hypothesized that strains grown under these conditions, in comparison to growth in ambient air, would be more adherent to epithelial cells. L2 cells were incubated with RB50 or 761 pre-grown in either atmospheric or 5% CO_2 conditions. Both strains pre-grown in 5% CO_2 conditions adhered to lung epithelial cells more efficiently than bacteria grown in normal atmospheric conditions (Fig. 7C).

8.3.6 B. PERTUSSIS *AND* B. PARAPERTUSSIS *MODULATE VIRULENCE FACTOR EXPRESSION IN RESPONSE TO 5% CO_2*

Since up-regulation of virulence factors in response to growth in 5% CO_2 conditions is common to multiple *B. bronchiseptica* strains, we hypothesized that *B. pertussis* and *B. parapertussis* may also regulate virulence factor expression in response to growth in 5% CO_2 conditions. Genes shown to be CO_2 responsive (cyaA, fhaB, fimA) or non-responsive to CO_2 (bvgS) in *B. bronchiseptica* (Fig. 3), were chosen to be analyzed by qPCR in *B. pertussis* and *B. parapertussis*. *B. pertussis* and *B. parapertussis* had increased expression of fhaB and cyaA, but not bvgS in response to CO_2 (Fig. 8A, B). *B. pertussis*, unlike *B. parapertussis*, also had increased expression of fimA, indicating that the 5% CO_2 responsive regulon may be different among the three classical *Bordetella* species. To further investigate this effect, Western blots with lysates of *B. pertussis* and *B. parapertussis* grown in ambient air or 5% CO_2 were probed with either *B. pertussis*-induced sera or *B. parapertussis*-induced sera (Fig. 8C, D). *B. pertussis* strains 536, 18323 and a recent clinical isolate CHOC 001_2 grown in ambient air showed a different antigenic profile from the lysates prepared from strains grown in 5% CO_2, with bands from roughly 55 to

250 kDa which were more numerous and intense in 5% CO_2 for strains 536 and CHOC 001_2 (Fig. 8C). Notably, *B. pertussis* strains 536 and CHOC 001_2 appeared to increase similar antigens in response to growth in 5% CO_2 conditions, while strain 18323 grown in 5% CO_2 decreased production of several antigens (Figure 8C). *B. parapertussis* grown in ambient air produced a more intense band between 130 and 250 kDa, while growth in 5% CO_2 produced a more intense band between 36 and 55 kDa (Fig. 8D). Overall, *B. parapertussis* did not appear to differentially regulate many antigens in response to growth in 5% CO_2 conditions (Figure 8D). These data indicate that several antigens in *B. pertussis*, but few in *B. parapertussis* isolate 12822, are differentially regulated in response to growth in 5% CO_2 compared to growth in ambient air and that there is strain variation in CO_2 responsiveness in the *Bordetella*e.

8.4 DISCUSSION

BvgAS was originally considered an ON/OFF switch, modulating *Bordetella* species between two distinct states, avirulent (Bvg^-) and virulent (Bvg^+). The discovery of an intermediate phase has led to a view of BvgAS gene regulation as a rheostat visualized as varying along a one dimensional gradient [23], [32], [34]. In this view of the two-component system few signals, temperature and some chemical cues, are known to affect virulence factor regulation through the BvgAS system. However, the respiratory tract contains many microenvironments, and within each environment there is likely to be great variation. For example, CO_2 levels are thought to vary between air and epithelial cells of the respiratory tract, although these are separated by a fraction of a millimeter of mucous. These sites also change dramatically in the course of the various stages of an infection, and there is likely to be a selective advantage to any strain that can sense these differences and modulate virulence factor expression in response.

Here we show that the classical *Bordetella*e share the ability to sense and respond to physiological changes in CO_2 concentrations likely to be encountered in the host. In mammalian tissues and blood, CO_2 concentrations are higher than inhaled ambient air concentrations of CO_2, which are approximately 0.03%. The observed changes in expression of various virulence factors (Fig. 3), and altered phenotypes (adherence and cytotoxicity, Fig. 7) provide additional evidence that the ability to respond to changes in CO_2 concentrations allow *Bordetella* species to adjust to different microenvironments within the host respiratory tract.

Recently, it has been shown that there is a zone of oxygenation between the anaerobic luminal environment and the host epithelium in the gastrointestinal tract, which can be sensed by *Shigella flexneri* [2]. The presence of oxygen alters the expression of TTSS effectors that are important for invasion of host cells, and this 'aerobic zone' may enhance secretion of these effectors thereby increasing invasion of host cells [2]. Similarly, the respiratory tract of mammals contains multiple sites where gradients of CO_2 or oxygen likely influence virulence factor expression and how respiratory pathogens interact with host cells. *B. bronchiseptica* has been isolated from multiple sites within the respiratory tract (e.g. nasopharnyx, trachea, lungs) and the ability to detect these differences could allow this pathogen to respond by expressing the array of factors optimal for success under each condition [2]. As bacteria disseminate from the nasal cavity to the trachea, lung and potentially even invade tissues, CO_2 concentrations may increase, serving as a signal for increased transcription of factors such as adhesinsand toxins that subvert the immune response, which is more robust in these regions 17,52–54.

B. bronchiseptica strains sense and differentially regulate virulence factor gene expression in response to 5% CO_2 (Fig. 3), and differential regulation was observed in multiple strains and species of *Bordetella* demonstrating that sensing and responding to carbon dioxide levels is an ability shared among the classical *Bordetella*e. Additionally, the transcription of several Bvg^+-phase genes increased in response to 5% CO_2, suggestive of BvgAS involvement in the response. Interestingly, regulation of some virulence factor genes (*bop*D, *bop*B) by BvgAS was epistatic to 5% CO_2 regulation. However, not all virulence gene expression (*cya*A, *fha*B) was dependent on *bvg*S (Fig. 5A, B), demonstrating an independent mechanism for virulence factor gene regulation in response to 5% CO_2.

Standard Bvg^+ conditions, without additional CO_2, are sufficient for production of ACT (Fig. 1) in RB50, suggesting that additional mechanisms may contribute to increases in production, but are not required. Of

73 *B. bronchiseptica* strains screened, 4 strains were identified here, 761, 308, 448 and JC100, in which Bvg^+ phase conditions are not sufficient for measurable production of ACT. In these strains both Bvg^+ phase conditions and 5% CO_2 are required for detectable production of ACT. The requirement for 5% CO_2 for the production of virulence factors may reflect evolutionary adaption of *B. bronchiseptica* strains, and suggests that the mechanism of CO_2 sensing may confer a selective advantage. Intriguingly, there also appears to be variation in responsiveness of both *B. bronchiseptica* and *B. pertussis* strains suggesting that although the ability to respond appears to be conserved the regulon may vary between species and strains.

Collectively these data demonstrate that a CO_2 response mechanism contributes to regulation of virulence factors in the classical *Bordetella*e (Fig. 5A, B). This is the first description of a CO_2 sensing mechanism that regulates virulence factor expression cooperatively with, or independently of, BvgAS. Our data support the idea that virulence factor gene expression can be fine-tuned in response to signals specific to different microenvironments within the respiratory tract or deeper tissues within the host.

REFERENCES

1. Mekalanos JJ (1992) Environmental signals controlling expression of virulence determinants in bacteria. Bacteriol. 174: 1–7.
2. Marteyn B, West NP, Browning DF, Cole JA, Shaw JG, et al. (2010) Modulation of Shigella virulence in response to available oxygen in vivo. Nature 465: 355–358. doi: 10.1038/nature08970
3. Fouet A, Mock M (1996) Differential influence of the two *Bacillus anthracis* plasmids on regulation of virulence gene expression. Infect. Immun.. 64: 4928–4932.
4. Sirard JC, Mock M, Fouet A (1994) The three *Bacillus anthracis* toxin genes are coordinately regulated by bicarbonate and temperature. J. Bacteriol. 176: 5188–5192.
5. Sterne M (1937) Variation in *Bacillus anthracis*. Onderstepoort J. Vet. Sci. Anim. Ind. 8: 271–349.
6. Caparon MG, Geist RT, Perez-Casal J, Scott JR (1992) Environmental regulation of virulence in group A streptococci: transcription of the gene encoding M protein is stimulated by carbon dioxide. J. Bacteriol. 174: 5693–5701.
7. Okada N, Geist RT, Caparon MG (1993) Positive transcriptional control of mry regulates virulence in the group A streptococcus. Mol. Microbiol. 7: 893–903.
8. Hyde JA, Trzeciakowski JP, Skare JT (2007) Borrelia burgdorferi alters its gene expression and antigenic profile in response to CO2 levels. J Bacteriol. 189: 437–445.

9. Miller SI, Mekalanos JJ (1990) Constitutive expression of the phoP regulon attenuates Salmonella virulence and survival within macrophages. J. Bacteriol. 172: 2485–2490.
10. Skaar EP, Humayun M, Bae T, Debord KL, Schneewind O (2004) Iron-Source Preference of *Staphylococcus aureus* Infections. Science 305: 1626–1628. doi: 10.1126/science.1099930
11. Torres VJ, Stauff DL, Pishchany G, Bezbradica JS, Gordy LE, et al. (2007) A *Staphylococcus aureus* Regulatory System that Responds to Host Heme and Modulates Virulence. Cell Host & Microbe 1: 109–119. doi: 10.1016/j.chom.2007.03.001
12. Goodnow RA (1980) Biology of *Bordetella bronchiseptica.* Microbiol. Rev. 44: 722–738.
13. Mattoo S, Cherry JD (2005) Molecular Pathogenesis, Epidemiology, and Clinical Manifestations of Respiratory Infections Due to *Bordetella pertussis* and Other *Bordetella* Subspecies. Clin. Microbiol. Rev. 18: 326–382.
14. Stavely CM, Register KB, Miller MA, Brockmeier SL, Jessup DA, et al. (2003) Molecular and antigenic characterization of *Bordetella bronchiseptica* isolated from a wild southern sea otter (Enhydra lutris nereis) with severe suppurative bronchopneumonia. J. Vet. Diagn. Invest. 15: 570–574.
15. Diavatopoulos DA, Cummings CA, Schouls LM, Brinig MM, Relman DA, et al. (2005) *Bordetella pertussis*, the Causative Agent of Whooping Cough, Evolved from a Distinct, Human-Associated Lineage of *B. bronchiseptica.* PLoS Pathog 1(4): e45. doi: 10.1371/journal.ppat.0010045
16. Parkhill J, Sebaihia M, Preston A, Murph LD, Thomson N, et al. (2003) Comparative analysis of the genome sequences of *Bordetella pertussis*, *Bordetella parapertussis* and *Bordetella bronchiseptica.* Nat. Genet. 35: 32–40. doi: 10.1038/ng1227
17. Hewlett EL, Donato GM, Gray MC (2006) Macrophage cytotoxicity produced by adenylate cyclase toxin from *Bordetella pertussis*: more than just making cyclic AMP! Mol. Microbiol. 59: 447–459. doi: 10.1038/ng1227
18. Khelef N, Zychlinsky A, Guiso N (1993) *Bordetella pertussis* induces apoptosis in macrophages: role of adenylate cyclase-hemolysin. Infect. Immun. 61: 4064–4071. doi: 10.1038/ng1227
19. Weingart CL, Weiss AA (2000) *Bordetella pertussis* virulence factors affect phagocytosis by human neutrophils. Infect. Immun. 68: 1735–1739. doi: 10.1038/ng1227
20. Harvill ET, Cotter PA, Yuk MH, Miller JF (1999) Probing the function of *Bordetella bronchiseptica* adenylate cyclase toxin by manipulating host immunity. Infect. Immun. 67: 1493–1500. doi: 10.1038/ng1227
21. Khelef N, Sakamoto H, Guiso N (1992) Both adenylate cyclase and hemolytic activities are required by *Bordetella pertussis* to initiate infection. Microb. Pathog. 12: 227–235. doi: 10.1016/0882-4010(92)90057-u
22. Weiss AA, Goodwin MS (1989) Lethal infection by *Bordetella pertussis* mutants in the infant mouse model. Infect. Immun. 57: 3757–3764. doi: 10.1016/0882-4010(92)90057-u
23. Cotter PA, Jones AM (2003) Phosphorelay control of virulence gene expression in *Bordetella.* Trends Microbiol. 11: 367–373. doi: 10.1016/0882-4010(92)90057-u
24. Steffen P, Goyard S, Ullmann A (1996) Phosphorylated BvgA is sufficient for transcriptional activation of virulence-regulated genes in *Bordetella pertussis.* EMBO J 15(1): 102–109. doi: 10.1016/0882-4010(92)90057-u

25. Stibitz S, Yang MS (1991) Subcellular localization and immunological detection of proteins encoded by the vir locus of *Bordetella pertussis*. J. Bacteriol. 173: 4288–4296. doi: 10.1016/0882-4010(92)90057-u
26. Uhl MA, Miller JF (1994) Autophosphorylation and phosphotransfer in the *Bordetella pertussis* BvgAS signal transduction cascade. Proc. Natl. Acad. Sci. 91: 1163–1167. doi: 10.1073/pnas.91.3.1163
27. Boucher PE, Maris AE, Yang MS, Stibitz S (2003) The response regulator BvgA and RNA polymerase a subunit C-terminal domain bind simultaneously to different faces of the same segment of promoter DNA. Mol. Cell 11: 163–173. doi: 10.1073/pnas.91.3.1163
28. Boucher PE, Stibitz S (1995) Synergistic binding of RNA polymerase and BvgA phosphate to the pertussis toxin promoter of *Bordetella pertussis*. J. Bacteriol. 177(22): 6486–6491. doi: 10.1073/pnas.91.3.1163
29. Karimova G, Bellalou J, Ullmann A (1996) Phosphorylation-dependent binding of BvgA to the upstream region of the cyaA gene of *Bordetella pertussis*. Mol. Microbiol. 20: 489–496. doi: 10.1073/pnas.91.3.1163
30. Merkel TJ, Barros C, Stibitz S (1998) Characterization of the bvgR locus of *Bordetella pertussis*. J Bacteriol. 180: 1682–1690. doi: 10.1073/pnas.91.3.1163
31. Merkel TJ, Boucher PE, Stibitz S, Grippe VK (2003) Analysis of bvgR expression in *Bordetella pertussis*. J Bacteriol. 185: 6902–6912. doi: 10.1128/jb.185.23.6902-6912.2003
32. Williams CL, Boucher PE, Stibitz S, Cotter PA (2005) BvgA functions as both an activator and a repressor to control Bvg phase expression of bipA in *Bordetella pertussis*. Mol Microbiol. 56: 175–188. doi: 10.1128/jb.185.23.6902-6912.2003
33. Deora R, Bootsma HJ, Miller JF, Cotter PA (2001) Diversity in the *Bordetella* virulence regulon: transcriptional control of a Bvg-intermediate phase gene. Mol. Microbiol. 40: 669–683. doi: 10.1128/jb.185.23.6902-6912.2003
34. Stockbauer KE, Fuchslocher B, Miller JF, Cotter PA (2001) Identification and characterization of BipA, a *Bordetella* Bvg-intermediate phase protein. Mol. Microbiol. 39: 65–78. doi: 10.1128/jb.185.23.6902-6912.2003
35. Vergara-Irigaray N, Chávarri-Martínez A, Rodríguez-Cuesta J, Miller JF, Cotter PA, et al. (2005) Evaluation of the role of the Bvg intermediate phase in *Bordetella pertussis* during experimental respiratory infection. Infect. Immun. 73: 748–760. doi: 10.1128/jb.185.23.6902-6912.2003
36. Cotter PA, Miller JF (1994) BvgAS-mediated signal transduction: analysis of phase-locked regulatory mutants of *Bordetella bronchiseptica* in a rabbit model. Infect. Immun. 62: 3381–3390.
37. Mann P, Goebel E, Barbarich J, Pilione M, Kennett M, et al. (2007) Use of a genetically defined double mutant strain of *Bordetella bronchiseptica* lacking adenylate cyclase and type III secretion as a live vaccine. Infect. Immun. 75(7): 3665–3672.
38. Buboltz AM, Nicholson TL, Karanikas AT, Preston A, et al. (2009) Evidence for horizontal gene transfer of two antigenically distinct O antigens in *Bordetella bronchiseptica*. Infect. Immun. 77(8): 3249–3257.
39. Buboltz AM, Nicholson TL, Parette MR, Hester SE, Parkhill J, et al. (2008) Replacement of adenylate cyclase toxin in a lineage of *Bordetella bronchiseptica*. J Bacteriol. 190: 5502–5511.

40. Heininger U, Cotter PA, Fescemyer HW, Martinez de Tejada G, Yuk MH, et al. (2002) Comparative phenotypic analysis of the *Bordetella parapertussis* isolate chosen for genomic sequencing. Infect. Immun. 70: 3777–3784.
41. Heinger U, Stehr K, Schmitt-Grobe S, Lorenz C, Rost R, et al. (1994) Clinical characteristics of illness caused by *Bordetella parapertussis* compared with illness caused by *Bordetella pertussis*. Pediatr. Infect. Dis. J. 13: 306–309. doi: 10.1097/00006454-199404000-00011
42. Preston A, Allen AG, Cadisch J, Thomas R, Stevens K, et al. (1999) Genetics basis for lipopolysaccharide O-antigen biosynthesis in *Bordetella*e. Infect. Immun. 67: 3763–3767. doi: 10.1097/00006454-199404000-00011
43. Harvill ET, Cotter PA, Miller JF (1999) Pregenomic comparative analysis between *Bordetella bronchiseptica* RB50 and *Bordetella pertussis* Tohama I in murine models of respiratory tract infection. Infect. Immun. 67: 6109–6118. doi: 10.1097/00006454-199404000-00011
44. Schneider CA, Rasband WS, Eliceiri KW (2012) NIH Image to ImageJ: 25 years of image analysis. Nature Methods 9: 671–675. doi: 10.1038/nmeth.2089
45. Buboltz AM, Nicholson TL, Weyrich LS, Harvill ET (2009) Role of the type III secretion system in a hypervirulent lineage of *Bordetella bronchiseptica*. Infect. Immun. 77(9): 3969–3977. doi: 10.1097/00006454-199404000-00011
46. Nicholson TL (2007) Construction and validation of a first-generation *Bordetella bronchiseptica* long-oligonucleotide microarray by transcriptional profiling the Bvg regulon. BMC Genomics 8: 220. doi: 10.1186/1471-2164-8-220
47. Nicholson TL, Buboltz AM, Harvill ET, Brockmeier SL (2009) Microarray and functional analysis of growth phase-dependent gene regulation in *Bordetella bronchiseptica*. Infect. Immun. 77(10): 4221–4231. doi: 10.1128/iai.00136-09
48. Tusher V G, Tibshirani R Chu G (2001) Significance analysis of microarrays applied to the ionizing radiation response. Proc. Natl. Acad. Sci. 98: 5116–5121. doi: 10.1128/iai.00136-09
49. Saeed AI, Sharov V, White J, Li, Liang W, et al. (2003) TM4: a free, open-source system for microarray data management and analysis. Biotechniques 34: 374–378. doi: 10.1128/iai.00136-09
50. Livak KJ, Schmittgen TD (2001) Analysis of relative gene expression data using real-time quantitative PCR and the 2-ΔΔCT method. Methods 25: 402–408. doi: 10.1006/meth.2001.1262
51. Cummings CA, Bootsma HJ, Relman DA, Miller JF (2006) Species- and Strain-Specific Control of a Complex, Flexible Regulon by *Bordetella* BvgAS. J. Bacteriol. 188: 1775–1785. doi: 10.1128/jb.188.5.1775-1785.2006
52. Yuk MH, Harvill ET, Cotter PA, Miller JF (2000) Modulation of host immune responses, induction of apoptosis and inhibition of NF-kappaB activation by the *Bordetella* type III secretion system. Mol. Microbiol. 35: 991–1004. doi: 10.1128/jb.188.5.1775-1785.2006
53. Cotter PA, Yuk MH, Mattoo S, Akerley BJ, Boschwitz J, et al. (1998) Filamentous hemagglutinin of *Bordetella bronchiseptica* is required for efficient establishment of tracheal colonization. Infect. Immun. 66(12): 5921–5929. doi: 10.1128/jb.188.5.1775-1785.2006

54. Mattoo S, Miller JF, Cotter PA (2000) Role of *Bordetella bronchiseptica* fimbriae in tracheal colonization and development of a humoral immune response. Infect. Immun. 68(4): 2024–2033. doi: 10.1128/jb.188.5.1775-1785.2006

There are several supplemental files that are not available in this version of the article. To view this additional information, please use the citation information cited on the first page of this chapter.

CHAPTER 9

CO_2 EFFLUX FROM CLEARED MANGROVE PEAT

CATHERINE E. LOVELOCK, ROGER W. RUESS, AND ILKA C. FELLER

9.1 INTRODUCTION

Mangroves are being cleared at a rapid rate, exceeding that of tropical forests [1], [2]. Clearing of above-ground biomass in mangrove forests results in changes in ecosystem processes [3] and losses of ecosystem services, including fisheries and storm protection [4], [5]. Additionally, clearing of forests reduces carbon sequestration and may lead to CO_2 emissions due to loss of aboveground carbon stocks and increased rates of soil decomposition [6]. In terrestrial ecosystems land-use change is one of the major sources of CO_2 emissions above the burning of fossil fuels [7]. In the tropics clearing of rainforests has led to high levels of CO_2 emissions [8] which have made these forests particularly valuable for conservation schemes developed to reduce emissions from deforestation and forest degradation, and to enhance carbon storage (REDD and REDD+) [9], [10], [11]. Similar schemes are proposed for carbon rich marine ecosystems, including mangroves, but there are many uncertainties around factors influencing carbon sequestration and carbon stocks in these coastal systems [12], [13].

There are few estimates of ecosystem carbon stocks in mangroves [14], [15]. The few that are available indicate a large proportion of carbon is in soils [14], [15], [16]. Carbon stocks in mangrove soils can be extremely high at some sites, as they contain accumulated peat (>20% carbon) derived mainly from roots as sea level has risen in the last interglacial period and anoxic conditions have slowed decomposition [17], [18]. The high levels of carbon in mangrove soils, the potential oxidation of peat deposits with land use change [6] indicate that once cleared mangrove forests on peat soils may become significant sources of CO_2.

In terrestrial tropical forest settings, clearing and draining of peat soils results in oxidation of carbon leading to peat collapse and the emission of CO_2 and other greenhouse gases [11]. Peat collapse and CO_2 emissions from cleared peat lands correlate with the level of the water table, increasing with the lowering of the water table and thus the exposure of peat to aerobic conditions [11], [19], [20]. Similarly, clearing of mangrove forests could result in significant CO_2 emissions due to oxidation of C in mangrove peat. In mangrove ecosystems that have been damaged by hurricanes peat collapse has been observed [21]. Additional oxidation may occur if peat is disturbed and contact with air is increased, as would be the case when shrimp ponds are constructed in peat soils and peat is pushed up on to banks or levees. The increase in CO_2 emissions with clearing of mangroves may be a major cost of disturbance of mangrove world-wide, and thus may contribute to the case for strengthening protection of these ecosystems through abating CO_2 emissions [12], [22].

In this study we measured CO_2 emissions from mangrove peat in Belize that had been cleared of vegetation over the last 2 decades in anticipation of tourism development. This mangrove peat at the site has a carbon (C) concentration of approximately 300 mg C g^{-1} [23]. We used a chronosequence of clearing to assess the potential change in CO_2 efflux over time since disturbance. Additionally we experimentally disturbed cleared peat to assess the potential enhancements in CO_2 efflux through increasing contact with air.

9.2 RESULTS

Over our chronosequence of sites representing time since clearing of mangroves, CO_2 efflux declined logarithmically with time, from 7.6 to 2.1 μmol m^{-2} s^{-1} over 20 years (Figure 1, F1, 30 = 40.50, P<0.0001). Soil temperature varied during the measurements, but there was no significant correlation between CO_2 efflux and soil temperature. At 4 years after clearing, CO_2 efflux had reached a relatively constant level of approximately $_2$ μmol m^{-2} s^{-1}. Extrapolation of CO_2 efflux rates to annual CO_2 loss indicates that CO_2 emissions from cleared peat would be ~10 600 tonnes km^{-2} $year^{-1}$ in the first year after clearing, falling to ~2900 tonnes CO_2 km^{-2} $year^{-1}$ (Table 1). Higher rates of CO_2 efflux were observed with acute disturbance of the peat, reaching a mean of 27 μmol m^{-2} s^{-1} when blocks of peat were cut from the soil (Figure 2, $F2,15 = 25.37$, $P<0.0001$). However this increase was transitory, as CO_2 efflux had returned to ambient levels within $_2$ days of disturbance.

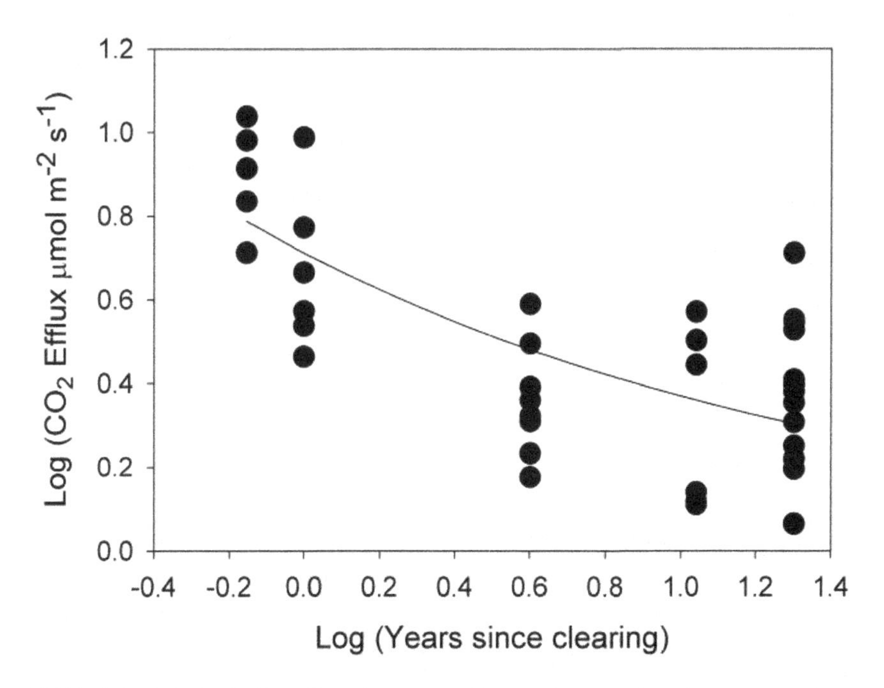

FIGURE 1: Variation in CO_2 efflux from peat soils over the time since the mangrove forest was cleared from Twin Cays Belize. The fitted line is of the form: Log CO_2 Efflux = a x exp (-b x time) where a = 0.712 and b = 0.656; $R^2 = 0.51$. The model is significant: $F1,30 = 40.4988$, $P<0.0001$.

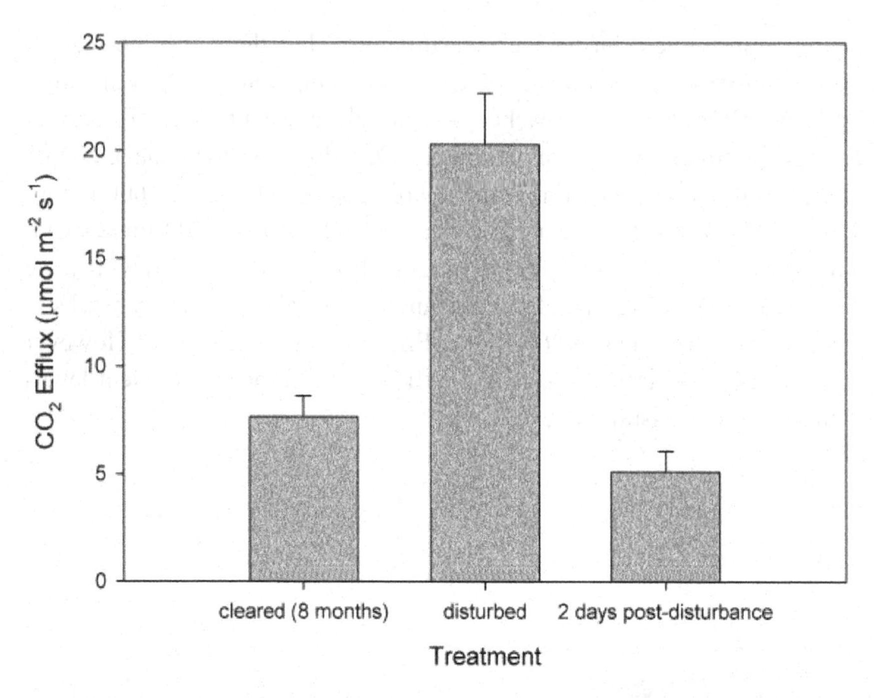

FIGURE 2. CO_2 efflux from peat soils that were cleared of forest (cleared 8 months) where peat was disturbed by cutting blocks from the soils (disturbed) and two days after the blocks of peat were cut (2 days post-disturbance). There was a significant effect of the disturbance treatment ($F2,15 = 25.37$, $P<0.0001$) but after two days there was no significant difference in soil CO_2 efflux between disturbed and undisturbed samples.

9.3 DISCUSSION

Based on short term measurements of CO_2 efflux from the soil surface of cleared mangrove forests, we found that CO_2 efflux is substantial, estimated to be approximately 2900 tonnes km^{-2} $year^{-1}$ (Table 1). This value is similar to CO_2 losses estimated for collapsing terrestrial peat soils in Indonesia [11], similar to that which can be estimated from peat collapse (losses in elevation) after hurricane damage in mangroves in Honduras [21], and greater than estimates of CO_2 emissions with permafrost thaw and decomposition of tundra peat [24]. In contrast, intact mangrove forests absorb approximately 5000 tonnes CO_2 km^{-2} $year^{-1}$ of which only ~20%

is respired as CO_2 [14], [15]. Carbon export from mangroves to adjacent systems (which could be up to 70% of total production) may potentially contribute to CO_2 emissions, but also support secondary production [14], [15]. Clearing mangroves from peat soils will clearly be unfavourable for regional and global carbon budgets [6] as well as reducing other ecosystem services offered by mangroves [5].

TABLE 1: Estimates of CO_2 efflux from modified mangrove and other habitats with peat soils.

Habitat	Modification	CO_2 effluc tonnes km^{-2} $year^{-1}$	Method	Reference
Mangrove, Belize	Cleared	2900	CO_2 efflux	This study
Mangrove, Honduras	Forest damaged by hurricane	1500	Inferred from peat collapse	Cahoon et al. 2003
Mangrove, Australia	Shrimp pond	1750 (220–5000)	CO_2 efflux	Burford and Longmore 2001
Rainforest, Indonesia	Drained for agriculture	3200	Infered from peat collapse and measured as CO_2 efflux	Couwenburg et al. 2010 and references therein
Tundra, Alaska	Thawed (vegetation intact)	150–430	Net CO_2 exchange	Schuur et al. 2009

While CO_2 efflux from intact forest soils is strongly associated with root respiration [25], CO_2 efflux from cleared and disturbed mangrove soils reflects microbial degradation of organic matter within soils [15]. The large, but transient increase in CO_2 efflux with disturbance of the peat (Figure 2) probably reflects oxidation of relative labile fractions (e.g. sugars and phenols) as they are exposed to enhanced oxygen concentrations [26]. However, this fraction is rapidly depleted before relatively slower decomposition of refractory pools (e.g. lignin) dominates CO_2 efflux. Short term high levels of CO_2 efflux from soil directly after clearing (Figure 1, 8 months) or from disturbing the peat are not included in our annual estimate of CO_2 emissions but may contribute a significant proportion to total emissions.

Once cleared, mangroves are often converted to shrimp ponds [1], [2]. Rates of CO_2 emissions from cleared mangroves are within the same range as those measured from shrimp ponds [27]. Thus, once established this alternative land-use, unlike conversion to agriculture [28] does not mitigate CO_2 emissions from clearing mangroves. Additionally, aquaculture and agriculture often increase nutrient availability of coastal waters [29]. Mangrove peat collapse has been observed to be enhanced by addition of nitrogen due to increases in decomposition and compaction [18]. Thus, increasing levels of nutrients in cleared mangrove areas may contribute to loss of habitat and possibly to increased CO_2 emissions associated with decomposition of peat.

Approximately half of Caribbean mangrove forests are anticipated to be growing on carbon rich peat soils [30]. The proportion of mangrove forests on peat soils is not known for the Indo-Pacific region and Africa, but could be substantial, particularly if mangrove peat is associated with upland peat forest soils which are common in the Indo-Pacific region [31]. The documentation of acid sulphate soils in shrimp farm developments from South East Asia and elsewhere [32] also indicates the presence of high concentrations of organic matter in many mangrove soils that have already been cleared for aquaculture. CO_2 emissions from cleared mangroves growing on mineral soils have not been assessed, but are needed in conjunction with improved soil mapping of soil carbon stocks within mangrove forest soils in order to estimate the global effects of clearing mangroves on CO_2 emissions.

Our annual estimate of CO_2 emissions of $_2$900 tonnes km^{-2} year^{-1} may be improved through measurement of CO_2 efflux over seasons which vary in tidal height, temperature and rainfall, however the timing of our measurements have probably lead to a underestimate of CO_2 emissions. Our measurements were made in the winter months in Belize when temperatures are relatively low and may limit bacterial activity. Although tidal variation is low in Belize (~0.5 m) in the winter months tides are higher than in summer [33] and thus we may have underestimated CO_2 flux compared to periods when tides are lower and peat maybe exposed to air at greater depth in the soil. Increases in sea level may also influence CO_2 emissions from cleared forest soils, changing oxidation status and potentially altering decomposition processes [26].

We conclude that the clearing of mangroves and the use of mangrove peat soils for alternative uses (e.g. cleared, shrimp ponds) results in increases in CO_2 emissions, in addition to resulting in losses in other ecosystem functions including fisheries and coastal protection. Incentive payments for maintaining intact forests, thus avoiding carbon emissions, as proposed by REDD and REDD+ [9], would be beneficial for conservation of mangroves in the tropics. There are significant gaps in our knowledge in: 1) the global extent of carbon currently stored in peat and mineral soils in mangrove forests, 2) the rate of CO_2 emissions from clearing mangroves growing on mineral soils, 3) the spatial and temporal variation in CO_2 emissions from cleared mangrove forests and alternative land-uses, and 4) the loss of carbon as dissolved organic and inorganic forms of carbon from intact and disturbed forest systems [12], [14], [15]. Filling these knowledge gaps will improve arguments for conservation of mangroves based on carbon stocks and sequestration.

9.4 MATERIALS AND METHODS

9.4.1 STUDY SITE

This study was conducted at Twin Cays, a peat-based, 92-ha archipelago of intertidal mangrove islands in a carbonate setting, just inside the crest of the Mesoamerican Barrier Reef System of central Belize, 12 km off shore (16°50′N, 88°06′W). These islands receive no terrigenous inputs of freshwater or sediments. Mangrove islands in this part of the reef, which originated approximately 8000 yr B.P. on a limestone base formed by a Pleistocene patch reef, have an underlying peat deposit ~7–10 m thick and have been mangrove communities throughout the Holocene [18]. Mangrove forests are dominated by Rhizophora mangle the roots of which the peat is derived [18]. Since 1980, this group of islands has been the primary study site for the Smithsonian Institution's National Museum of Natural History Field Station on nearby Carrie Bow Cay [34].

Illegal clearing of mangroves has occurred on Twin Cays over the last 20 years, primarily for housing and prospective tourism developments. Multiple clearing events allowed us to measure CO_2 efflux over soils that have been cleared of vegetation over 20 years. We measured CO_2 efflux from 3 5 ha patches that had been cleared for durations of 8 months, 12 months, 4 years, 11 years and 20 years. We measured efflux at 6–12 locations within each aged clearing.

In order to test whether disturbance of the peat increased soil CO_2 efflux we cut blocks of peat from the area that had been cleared 8 months previously. Six replicate blocks approximately 30×30×30 cm were cut with a shovel and placed on the soil surface. We measured CO_2 efflux from the soil, from the peat blocks directly after cutting them from the peat and then again after 2 days.

CO_2 efflux from soils was measured using a LiCor 6400 portable photosynthesis system configured with the LiCor Soil Respiration chamber (LiCor Corp, Lincoln, NE, USA). The chamber was set to penetrate 5 mm into the soil. Settings for measurement were determined at each site following the procedure described by the manufacturer. Soil temperature was measured at 2 cm depth simultaneously with CO_2 efflux. Soil temperatures varied from 28 to 34 C during the measurements. Measurements were made in February of 2004 and January of 2007.

9.4.2 DATA ANALYSIS

Differences in soil CO_2 flux over time among areas of differing time since clearing were assessed using linear models where time was considered a random, continuous variable in the model. Changes in CO_2 efflux with disturbance of peat was assessed using repeated measures ANOVA. Scaling instantaneous CO_2 efflux data was done by simply multiplying CO_2 efflux (μmol m^{-2} s^{-1}) by time to give tonnes CO_2 km^{-2} $year^{-1}$.

REFERENCES

1. Valiela I, Bowen JL, York JK (2001) Mangrove forests: one of the world's threatened major tropical environments. Bioscience 51: 807–815. doi: 10.1890/0012-9658(2002)083[1065:minacf]2.0.co;2

2. Alongi DM (2002) Present state and future of the world's mangrove forests. Environ Conserv 29: 331–349. doi: 10.3354/meps224187
3. Granek EF, Ruttenberg BI (2008) Changes in biotic and abiotic processes following mangrove clearing. Estuar Coastal Shelf Sci 80: 555–562. doi: 10.1007/978-1-4020-4271-3
4. Aburto-Oropeza O, Ezcurra E, Danemann G, Valdez V, Murray J, et al. (2008) Mangroves in the Gulf of California increase fishery yields. Proc Nat Acad Sci USA 105: 10456–10459. doi: 10.1073/pnas.0804601105
5. Barbier EB, Koch EW, Silliman B (2008) Coastal ecosystem-based management with nonlinear ecological functions and values. Science 319: 321–323. doi: 10.1126/science.1150349
6. Donato DC, Kauffman JB, Murdiyarso D, Kurnianto S, Stidham M, et al. (2011) Mangroves among the most carbon-rich tropical forests and key in land-use carbon emissions. Nature Geosci 4: 293–297. doi: 10.3354/meps224187
7. Houghton RA (1995) Land-use change and the carbon cycle. Glob Chan Biol 1: 275–287. doi: 10.1890/0012-9658(2002)083[1065:minacf]2.0.co;2
8. Kauffman JB, Hughes RF, Heider C (2009) Dynamics of carbon and nutrient pools associated with land conversion and abandonment in Neotropical landscapes. Ecol Appl 19: 1211–1222. doi: 10.1890/08-1696.1
9. Anglesen A (2009) Realising REDD+: National Strategy and Policy Options. Bogor, Indonesia: Center for International Forestry Research. 362 p.
10. Muradian R, Kumar P (2009) Payment for ecosystem services and valuation: challenges and research gaps. In: Kumar P, Muradian R, editors. Payment for Ecosystem Services. New Delhi: Oxford University Press. pp. 1–16.
11. Couwenberg J, Dommain R, Joosten H (2010) Greenhouse gas fluxes from tropical peatlands in south-east Asia. Glob Chan Biol 16: 1715–1732. doi: 10.1890/0012-9658(2002)083[1065:minacf]2.0.co;2
12. Alongi DM (2011) Carbon payments for mangrove conservation: ecosystem constraints and uncertainties in sequestration potential. Environ Sci Pol 14: 462–470.
13. McCleod E, Chmura GL, Bouillon S, Salm R, Bjork M, et al. (2011) A Blueprint for Blue Carbon: Towards an improved understanding of the role of vegetated coastal habitats in sequestering CO2. Front Ecol Environ. in press.
14. Alongi DM (2009) The Energetics of Mangrove Forests. Dordrecht, The Netherlands: Springer. 216 p.
15. Bouillon S, Borges AV, Castañeda-Moya E, Diele K, Dittmar T, et al. (2008) Mangrove production and carbon sinks: a revision of global budget estimates. Glob Biogeochem Cycl 22: doi:10.1029/2007GB003052.
16. Kauffman JB, Heider C, Cole TG, Dwire KA, Donato DC (2011) Ecosystem carbon stocks of Micronesian mangrove forests. Wetlands 31: 343–352. doi: 10.1016/j.chemgeo.2008.07.009
17. Chmura GL, Anisfeld SC, Cahoon DR, Lynch JC (2003) Global carbon sequestration in tidal, saline wetland soils. Glob Biogeochem Cycl 17: 1111–1120. doi: 10.1007/0-387-21657-x_11
18. McKee KL, Feller IC, Cahoon DR (2007) Caribbean mangroves adjust to rising sea level through biotic controls on change in soil elevation. Glob Ecol Biogeogr 16: 545–556. doi: 10.1111/j.1466-8238.2007.00317.x

19. DeLaune RD, Nyman JA, Patrick WH (1994) Peat collapse, ponding and wetland loss in a rapidly submerging coastal marsh. J Coast Res 10: 1021–1030. doi: 10.1007/978-1-4020-4271-3
20. Crow SE, Wieder K (2005) Sources of CO2 emission from a northern peatland: root respiration, exudation, and decomposition. Ecology 86: 1825–1834. doi: 10.1007/978-1-4020-4271-3
21. Cahoon DR, Hensel P, Rybczyk J, McKee KL, Proffitt CE, et al. (2003) Mass tree mortality leads to mangrove peat collapse at Bay Islands, Honduras after Hurricane Mitch. J Ecol 91: 1093–1105. doi: 10.1111/j.1466-8238.2007.00317.x
22. Emmett-Mattox S, Crooks S, Findsen J (2010) Wetland grasses and gases: Are tidal wetlands ready for the carbon markets? Nat Wetl Newslet 32: 6–10.
23. McKee KL, Feller IC, Popp M, Wanek W (2002) Mangrove isotopic fractionation (d15N and d13C) across a nitrogen versus phosphorus limitation gradient. Ecology 83: 1065–1075. doi: 10.1890/0012-9658(2002)083[1065:minacf]2.0.co;2
24. Schuur EAG, Vogel JG, Crummer KG, Lee H, Sickman JO, et al. (2009) The effect of permafrost thaw on old carbon release and net carbon exchange from tundra. Nature 459: 556–559. doi: 10.1038/nature08031
25. Lovelock CE (2008) Soil respiration in tropical and subtropical mangrove forests. Ecosystems 11: 342–354. doi: 10.1890/080090
26. Lallier-Vergès E, Marchand C, Disnar J-R, Lottier N (2008) Origin and diagenesis of lignin and carbohydrates in mangrove sediments of Guadeloupe (French West Indies): Evidence for a two-step evolution of organic deposits. Chem Geol 255: 388–398. doi: 10.1016/j.chemgeo.2008.07.009
27. Burford M, Longmore (2001) High ammonium production from sediments in hypereutrophic shrimp ponds. Mar Ecol Prog Ser 224: 187–195. doi: 10.3354/meps224187
28. Chimner RA, Ewel KC (2004) Differences in carbon fluxes between forested and cultivated micronesian tropical peatlands. Wetl Ecol Manag 12: 419–427. doi: 10.3354/meps224187
29. Burford M, Costanzo SD, Dennison WC, Jackson CJ, Jones AB, et al. (2003) A synthesis of dominant ecological processes in intensive shrimp ponds and adjacent coastal environments in NE Australia. Mar Poll Bull 46: 1456–1469. doi: 10.1007/0-387-21657-x_11
30. Ellison AM, Farnsworth EJ (1996) Anthropogenic disturbance of Caribbean mangrove ecosystems: past impacts, present trends, and future predictions. Biotropica 28: 549–565. doi: 10.1016/j.chemgeo.2008.07.009
31. Ewel KC (2010) Appreciating tropical coastal wetlands from a landscape perspective. Front Ecol Environ 8: 20–26. doi: 10.1890/080090
32. Boyd CE (1992) Shrimp pond bottom soil and sediment management. In: Wyban J, editor. Proceedings of the Special Session on Shrimp Farming. Baton Rouge, LA: World Aquaculture Society. pp. 166–181.
33. Lee RY, Porubsky WP, Feller IC, McKee KL, Joye SB (2008) Porewater biogeochemistry and soil metabolism in dwarf red mangrove habitats (Twin Cays, Belize). Biogeochemistry 87: 181–198. doi: 10.1890/080090
34. Rützler K, Feller IC (1996) Caribbean mangrove swamps. Sci Am 274: 94–99. doi: 10.1038/scientificamerican0696-94

CHAPTER 10

SOIL MICROBIAL RESPONSES TO ELEVATED CO_2 AND O_3 IN A NITROGEN-AGGRADING AGROECOSYSTEM

LEI CHENG, FITZGERALD L. BOOKER, KENT O. BURKEY, CONG TU, H. DAVID SHEW, THOMAS W. RUFTY, EDWIN L. FISCUS, JARED L. DEFOREST, AND SHUIJIN HU

10.1 INTRODUCTION

Soil microbes critically affect plant and ecosystem responses to climate change by modulating organic C decomposition and nutrient availability for plants. Experimental evidence accumulated over the last several decades has clearly shown that climate change factors such as CO_2 enrichment in the atmosphere can significantly alter plant growth [1], [2] and the availability of organic C, N and cation nutrients for microbes [3], [4], [5]. Ozone is a greenhouse gas with demonstrated inhibitory effects on plant growth and resource allocation belowground [6], [7]. Although less well-studied, O_3 is considered to have an impact on soil microbial processes [6]. Alterations in soil microbes can, in turn, profoundly influence soil C processes and the long-term potential of terrestrial ecosystems as a C sink to mitigate anthropogenic sources of atmospheric CO_2. However, predicting what these changes will be is hampered by our limited understanding

---

of the underlying mechanisms by which soil microbes respond to altered resource availability.

The current prevailing hypothesis, building on the assumption that soil microbes are generally C limited [8], predicts that elevated CO_2 increases soil microbial biomass and activities due to enhanced soil C availability [9], [10], [11], whereas O_3 reduces them due to lower C allocation belowground [6], [12], [13]. This broad hypothesis has been extensively tested over the past two decades for CO_2 but less so with O_3 [4], [10], [12], [13], [14], [15], [16]. Though C availability to microbes has been commonly reported to increase under elevated CO_2 [5], [17], [18] and to decrease under elevated O_3 [6], [13], [19], results of soil microbial responses to elevated CO_2 and O_3 have been inconsistent [6], [11], [20], [21]. In a meta-analysis study, de Graaff et al. (2006) found that elevated CO_2 increased microbial biomass C and microbial respiration by 7.7% and 17.1%, respectively, across 40 studies that mainly included herbaceous species. In the meantime, Hu et al. (2006) reviewed 135 studies examining elevated CO_2 effects on a suite of soil microbial parameters such as microbial biomass and respiration and found that microbial biomass C and microbial respiration increased under elevated CO_2 in 19 of 40 studies and 20 of 38 studies, respectively, but remained unchanged or even decreased in the remainder. Despite considerable efforts in the past two decades, there is a lack of conceptual understanding of why and how these inconsistencies in CO_2 and O_3 effects on microbes occur.

Soil microbial responses to elevated CO_2 can also be influenced by CO_2-induced alterations in soil moisture [9], [22], grazing activity of soil animals [23], and soil nutrient availability [4], [9]. Such mechanisms could operate either singly or in combination with changes in soil C availability. In particular, CO_2-induced alteration in the stoichiometery of available C and N has been proposed to be a primary control over microbial responses to elevated CO_2 [4], [9], [24]. Soil N availability may influence microbial responses to elevated CO_2 by affecting both physiological activities and the community structure composition of microbes [9], [25]. When soil N was limiting, competitive plant N uptake can significantly reduce soil N availability for microbes under elevated CO_2, limiting microbial decomposition over the short-term [4]. Conversely, high N availability in soil often increases microbial activities [26], [27] and favors bacteria over fungi

[28], [29]. Yet most studies that examined microbial responses to elevated CO_2 were conducted in N-limiting forest and grassland ecosystems [20], [21]. It remains unclear how soil microbes respond to elevated CO_2 and O_3 in N-rich or N-aggrading agroecosystems.

Many crop plants, particularly C_3 crops, are usually responsive to elevated CO_2 and O_3 [2], [30]. For instance, it has been estimated that elevated CO_2 alone increased the shoot biomass of soybean and wheat by 48% and 16%, respectively [2], [31], but elevated O_3 reduced them by 21% and 18% [32], [33]. Also, elevated CO_2 has been shown to ameliorate O_3 effects on plants by reducing O_3 uptake and increasing C assimilation rates [30], [32], [34]. However, whether the CO_2- and O_3-induced changes in plant biomass translate into alterations in soil C sequestration depends largely on the responses of soil microbial processes. Additionally, elevated CO_2 significantly increased symbiotic N_2 fixation in legumes such as soybean and peanut [35], [36], whereas elevated O_3 tended to reduce it [36]. It has been suggested that high N availability in agro- and grassland ecosystems can sustain plant responses to rising CO_2 over a long time frame and provide an opportunity for soil C sequestration in soil in a higher CO_2 world [2], [37], [38], [39]. Convincing evidence is still lacking, but soil microbial responses may be indicative for understanding the long-term soil C dynamics in high N or N-aggrading ecosystems [27], [40].

In a long-term study examining climate change effects on soil C dynamics in a wheat-soybean agroecosystem with no-till practice, we continually monitored a suite of soil microbial parameters in response to elevated CO_2 and O_3 for more than four years. Because soybean and its symbiotic N_2 fixation are sensitive to elevated CO_2 and O_3 [35], [36], we expected that elevated CO_2 would enhance both C and N inputs belowground through increasing residue returns, while elevated O_3 would offset this CO_2 effect. Also, we expected that the stoichiometry of available C and N for microbes might change over time as a portion of residue C was mineralized and released back to the atmosphere as CO_2 while a large proportion of residue N was retained in the system. Consequently, alterations in C and N availability for microbes induced by elevated CO_2 and O_3 may further affect microbial biomass and activities over time, and possibly induce a shift in the microbial community structure. Therefore, our specific objectives were to: 1) document the time-course of CO_2 and O_3 effects on microbial biomass, activities and community structure; and, $_2$) examine how changes in microbial parameters were related to residue inputs and soil C and N availability.

10.2 MATERIALS AND METHODS

10.2.1 SITE DESCRIPTION

We initiated a long-term field experiment in May 2005 to investigate the response of a wheat-soybean rotation agroecosystem to elevated atmospheric CO_2 and O_3 using open-top field chambers (OTC). The experimental site is located at the Lake Wheeler Experimental Station, 5 km south of North Carolina State University, Raleigh, North Carolina, USA (35°43′N, 78°40′W; elevation 120 m). Annual mean temperature is 15.2°C and annual mean precipitation is 1050 mm. The field had been left fallow for eight years prior to this study. Before CO_2 and O_3 treatments were initiated, the soil was repeatedly turned-over using a disc implement and rotovator. The soil is an Appling sandy loam (fine, kaolinitic, thermic Typic Kanhapludult), well drained with a pH of 5.5, and contained 9.0 g C and 0.86 g N kg^{-1} soil when the experiment started.

This experiment was a 2×2 factorial design with four treatments randomly assigned into four blocks. Four different trace-gas treatments were: (a) charcoal-filtered air and ambient CO_2 (CF); (b) charcoal-filtered air plus ambient CO_2 and 1.4 times ambient O_3 (+O_3); (c) charcoal-filtered air plus 180 μl l^{-1} CO_2 (+CO_2); and (d) charcoal-filtered air plus 180 μl l−1 CO_2 and 1.4 times ambient O_3 (+CO_2+O_3). The seasonal daily average concentrations of CO_2 and O_3 over the experimental duration are shown in Table 1. The purpose of filtration of ambient air with activated charcoal was to reduce the concentrations of ambient O_3 to levels considered non-phytotoxic to soybean and wheat plants. Ozone was deemed as a major air pollutant in this area, while other air pollutants such as NO_2 and SO_2 were below the phytotoxic levels at the experimental location [41].

TABLE 1: The seasonal daily average (12 h) CO_2 and O_3 concentrations at canopy height during the 4-year period.

Crop	CO_2 (μl l^{-1})		O_3 (nl l^{-1})	
	Ambient	Elevated	CF	Elevated
Soybean	376.0 ± 0.4	555.0 ± 0.7	19.9 ± 0.3	65.7 ± 0.4
Wheat	388.0 ± 0.4	547.0 ± 0.5	20.7 ± 0.2	49.8 ± 0.3

CF: charcoal filtered air. Values are mean ± s.e.m,

Soybean [cv. CL54 RR (Year 1), Asgrow 5605 RR (Years 2 and 3) and SS RT5160N RR (Year 4)] was planted each spring followed by soft red winter wheat (Coker 9486) in the fall using no-till practices. Plants were exposed to reciprocal combinations of CO_2 and O_3 within cylindrical OTCs (3.0 m diameter×2.4 m tall) from emergence to physiological maturity. Carbon dioxide was released from a 14-ton liquid-receiving tank 24 h daily and monitored at canopy height using an infrared CO_2 analyzer (model 6252, Li-Cor Inc. Lincoln, NE, USA). Ozone was generated by electrostatic discharge in dry O_2 (model GTC-1A, Ozonia North America, Elmwood Park, NJ, USA) and dispensed 12 h daily (08:00–20:00 hours Eastern Standard Time) in proportion to concentrations of ambient O_3. The O_3 concentration in the chambers was monitored at canopy height with a UV photometric O_3 analyzer (model 49, Thermo Environmental Instruments Co., Franklin, MA, USA). During wheat growing seasons, each plot initially received 48 g NH_4NO_3 (equivalent to 24 kg N ha^{-1}) in November each year, followed by an additional input of 19_2 g NH_4NO_3 (equivalent to 96 kg N ha^{-1}) in March. Plots were treated with lime, K and P in November during the experiment according to soil test recommendations. During soybean growing seasons, plants were irrigated with drip lines to prevent visible signs of water stress, but no additional N fertilizers were applied. Upon senescence of the plants, all aboveground plant biomass in each chamber was harvested. Soybean plants were divided into leaves, stems, husks and seeds, while wheat plants were separated as straw, chaff and seeds, then dried and quantified. Afterward, residues other than seeds were uniformly returned to their corresponding treatment plots and evenly distributed on the soil surface.

10.2.2 SOIL SAMPLING

The chamber plot was divided into two parts: the sampling area (an inner circular area with a diameter of 2.4 m) and the border area (for purpose of reducing chamber effects, 0.3 m in width). To avoid taking soil samples from the same location, the sampling area was divided into 448 small subplots (10×10 cm). Soil sampling locations were determined using a random number generator and each subplot was sampled only once. In June and November of each year (Year 1–Year 4), corresponding to harvest time for each crop, we used a 5-cm diameter soil corer to take three soil cores to 20 cm depth in the center of three pre-determined subplots from each chamber. Three additional soil cores were immediately taken from the border areas to fill holes in the sampling areas. Sample holes in border areas were refilled by soil cores taken just outside of each chamber. Soil cores were separated into 0–5 cm, 5–10 cm and 10–20 cm depth fractions. Core sections were then pooled by the depth fraction into three soil samples per chamber. Soil samples were also collected at the mid-growing season to check whether microbial parameters were significantly different from those obtained at the end of the growing season. Soil samples (0–5 and 5–10 cm) were collected in each April of the first two years and each August of the last two years, corresponding to the maximal physiological activity of wheat and soybean plants, respectively. All samples were sealed in plastic bags, stored in a cooler and transported to the laboratory.

Field moist soils were mixed thoroughly and sieved through a 4-mm mesh within 24 hours of the field sampling and all visible residues and plant roots were carefully removed. Subsamples (~20 g) were then taken immediately; frozen and stored at −20°C for the phospholipid fatty acid (PLFA) analysis and the rest of soils were stored at 4°C for other microbial and chemical analyses. A 10-g subsample was oven-dried at 105°C for 48 h and weighed for the determination of the water content. All the soil and microbial data were calculated on the dry weight basis of soils.

10.2.3 SAMPLE ANALYSES

10.2.3.1 C AND N CONTENTS IN PLANT RESIDUES AND SOILS.

Air-dried subsamples of aboveground plant components (stems, leaves, husks and seeds of soybean; straw, chaff and seeds of wheat) were ground in a Tecator Cyclotec mill fitted with a 1-mm screen (FOSS, Eden Prairie, MN, USA). Soil samples were ground into fine powder using an 8000-D Mixer Mill (SPEX CertiPrep Inc. Metuchen, NJ, USA). The C and N concentrations in various plant components and in soil were determined with a CHN elemental analyzer (Carla Erba and model 2400, Perkin Elmer Co., Norwalk, CT, USA). Aboveground residue C and N inputs to soil were calculated by adding up the C and N, respectively, in all aboveground plant components except for seeds.

10.2.3.2 SYMBIOTIC N_2 FIXATION IN SOYBEAN.

Soybean N_2 fixation was estimated using the conventional N accumulation method [42]. To estimate total N_2 fixation by soybean in each season, we first estimated total biomass N in wheat and soybean, respectively. Total aboveground plant N was calculated by directly adding up the N in all plant components. Root biomass was estimated by using the fixed root:shoot ratios of wheat (0.07) and soybean (0.22) according to the literature [2], [31], [43], [44], [45]. We also assumed that the C:N ratio of roots was the same as that of shoots [46]. Then, we used wheat in the following season as the nonfixing plant to estimate total N fixed by soybean in each season by subtracting total N in wheat plants from total N in soybean plants on a per chamber basis. Further, the CO_2 effect on N_2 fixation was estimated by subtracting total N in soybean in ambient CO_2 from elevated

CO_2. Although wheat has been often used as a non-fixing control plant [42], we realized that this method does not provide an exact estimate of N_2 fixation by soybean plants in the field. Using wheat plants as the non-fixing control in our system should provide a conservative underestimate of soybean N_2 fixation because: 1) N inputs to soil through soybean root exudates and fine root turnover were not considered; 2) inorganic N fertilizers (120 kg N ha^{-1}) were applied for wheat; and 3) wheat should have also obtained significantly higher amounts of N from the mineralization of soybean residues. In addition to estimating N_2 fixation, changes in total soil N over time were documented by comparing the soil N content (0–5 cm soil layer) at the end of the fourth year to the pretreatment soil N content (0–5 cm soil layer).

10.2.3.3 SOIL MICROBIAL BIOMASS C AND N.

Soil microbial biomass C (MBC) and biomass N (MBN) were determined using the fumigation-extraction method [47]. Twenty-g dry weight equivalent soil samples were fumigated with ethanol-free chloroform for 48 h and then extracted with 50 mL of 0.5 M K_2SO_4 by shaking for 30 min. Another 20-g sample of non-fumigated soil was also extracted with 50 mL of 0.5 M K_2SO_4. Soil extractable organic C in both fumigated and non-fumigated K_2SO_4 extracts was measured using a TOC analyzer (Shimadzu TOC-5050A, Shimadzu Co., Kyoto, Japan). Soluble inorganic N in the extracts of fumigated and non-fumigated soils was quantified on the Lachat flow injection analyzer (Lachat Instruments, Milwaukee, WI, USA) after digestion with alkaline persulfate [48]. The differences in extractable organic C and inorganic N between fumigated and non-fumigated soils were assumed to be from lysed soil microbes. The released C and N were used to calculate MBC and MBN using a conversion factor of 0.45 (k_{EC}) and 0.45 (k_{EN}), respectively [47], [49].

10.2.3.4 SOIL EXTRACTABLE C AND N

The concentration of organic C in non-fumigated soil extracts was used to represent soil extractable C. The extractable inorganic N referred to the sum of NH_4^+-N and NO_3^--N in non-fumigated soil extracts.

10.2.3.5 SOIL MICROBIAL RESPIRATION.

We determined soil heterotrophic respiration using an incubation-alkaline absorption method [50]. In brief, 20-g dry mass equivalent soil samples were adjusted to moisture levels of around 60% water holding capacity, placed in 1-L Mason jars, and then incubated at 25°C in the dark for 2 weeks. Respired CO_2 was trapped in 5 mL of 0.25 M NaOH contained in a beaker suspended in the jar. After the first week incubation, NaOH solutions were replaced with fresh solutions. The CO_2 captured in the NaOH solution was titrated with 0.125 M HCl to determine the amount of CO_2 evolved from the soil. Soil microbial respiration (SMR) rate was expressed as mg CO_2 kg^{-1} soil d^{-1} by averaging the data across two 1-wk incubations.

10.2.3.6 NET SOIL N MINERALIZATION.

Potential N mineralization was determined following a 4-wk incubation at 25°C in the dark. Soil NH_4^+ and NO_3^- in un-incubated and incubated subsamples (20-g each) were extracted with 50 mL of 0.5 M K_2SO_4 by shaking for 30 min. The concentrations of inorganic N were then measured on the Lachat flow injection analyzer. Net mineralized N (NMN) was determined by the difference in extractable total inorganic N (NH_4^+-N + NO_3^--N) between incubated and un-incubated soil samples.

10.2.3.7 PHOSPHOLIPID FATTY ACIDS.

PLFAs were extracted following a procedure described by Bossio et al. (1998). Briefly, 10 g of freeze-dried soils (0–5 cm soil layer) were extracted using a solution containing CH_3OH:CH_3Cl:PO_4^{3-} (vol/vol/vol 2:1:0.8). Solid phase extraction columns (Thermo Scientific, Vernon Hills, IL, USA) were used to separate phospholipids from neutral and glycollipids. The phospholipids were then subjected to an alkaline methanolysis to form fatty acid methyl esters (FAMEs). The resulting FAMEs were separated and measured using gas chromatography on a HP GC-FID

(HP6890 series, Agilent Technologies, Inc. Santa Clara, CA, USA); peaks were identified using the Sherlock Microbial Identification System (v. 6.1, MIDI, Inc., Newark, DE, USA). We chose the following fatty acids, i14:0, i15:0, a15:0, 15:0, i16:0, 16:1ω7c, i17:0, a17:0, 17:0cy, 17:0, 18:1ω7c, and 18:1ω5c, to represent the bacterial PLFAs [51], [5$_2$], [5$_3$], and the other three fatty acids (16:1ω5c, 18:$_2$ω6.9c and 18:1ω9c) as the fungal PLFAs [52], [53], [54]. We used the ratio of signature fungal and bacterial PLFAs as an indicator of soil microbial community structure [51], [53], [55]. The fatty acid profile of soil microbes was examined in soils collected in years 1, 3, 4, and 5.

10.2.4 STATISTICAL ANALYSIS

We examined results for the entire experimental duration from 2005–2009 (Year 1–Year 4), and used the linear mixed model [56] to test the main effects of CO_2, O_3 and the interaction of CO_2 and O_3, and whether these changed over time. We employed a set of covariance structures including compound symmetric model (CS), the first-order autoregressive model [AR (1)], and autoregressive with random effect to reduce autocorrelation. The P values for treatments and interaction terms were reported based on the covariance structure that minimized Akaike information criterion (AIC) and Bayesian information criterion (BIC) [56]. Data for soil and microbial parameters from mid-seasons (Appendix S1), plant and soil N contents, fungal and bacterial PLFAs and the fungi:bacteria ratios were subjected to the analysis of variance using the mixed model. To test for relationships between variables, we conducted a correlation analysis using all the data generated over the 4-year period. A Chi-square (χ^2) test was also conducted to examine whether the CO_2 effect on microbial biomass, respiration and the community structure were correlated with the CO_2 effect on N availability. We thus developed four contingency tables for SMR, MBC, MBN, and fungi:bacteria ratio, respectively. All statistical analyses were performed using the SAS 9.1 (SAS Institute, Inc., Cary, NC, USA). For all tests, $P \leq 0.05$ was considered to indicate a statistically significant difference.

TABLE 2: Effects of elevated CO_2 and O_3 on soybean N_2 fixation and total N in the surface soil.

	Year 1	Year 2	Year 3	Year 4
N inputs to soil derived from soybean N_2 fixation (g N m^{-2})				
Treatment				
CF	11.4 ± 0.8	2.9 ± 0.4	6.0 ± 0.5	7.5 ± 0.3
$+O_3$	10.6 ± 0.5	1.8 ± 0.4	5.0 ± 0.5	4.5 ± 0.5
$+CO_2$	15.1 ± 0.4	4.8 ± 0.2	8.3 ± 0.7	10.0 ± 0.5
$+CO_2xO_3$	14.7 ± 1.4	3.8 ± 0.2	7.8 ± 0.3	8.6 ± 0.7
Source				
O_3	NS	**	NS	**
CO_2	***	***	***	***
CO_2+O_3	NS	NS	NS	NS
Total soil N in the surface soil (0-5 cm) (g N m^{-2})				
Treatment				
CF	85. 7 ± 8.2	ND	ND	93.7 ± 11.7
$+O_3$	74.1 ± 16.0	ND	ND	90.0 ± 3.1
$+CO_2$	76.4 ± 10.7	ND	ND	95.6 ± 6.4
$+CO_2+O_3$	75.1 ± 4.7	ND	ND	97.5 ± 8.1

Values shown for N inputs to soil from N_2 fixation exclude seed.
*Value are mean ± s.e.m. ***(P<0.001) and **(P<0.0!) denote statistically significant main treatment effects, ANOVA mixed models. ND, not determined. NS, not significant. CF, charcoal-filtered ambient air. $+O_3$, elevated O_3 + CO_2. $+CO_2$, elevated CO_2. $+CO_2+O_3$, elevated CO_2+O_3. The main treatment effects of CO_2, O_3, and the CO_2xO_3 interaction on soil N were not statistically significant for any years.*

10.3 RESULTS

10.3.1 SOYBEAN N_2 FIXATION AND PLANT RESIDUE C AND N INPUTS TO SOIL

Elevated CO_2 significantly increased symbiotic N_2 fixation by soybean plants, while O_3 decreased it. Over the 4-year period, total N derived from symbiotic N_2 fixation in soybean was estimated at 92.5, 68.4, 119.1 and

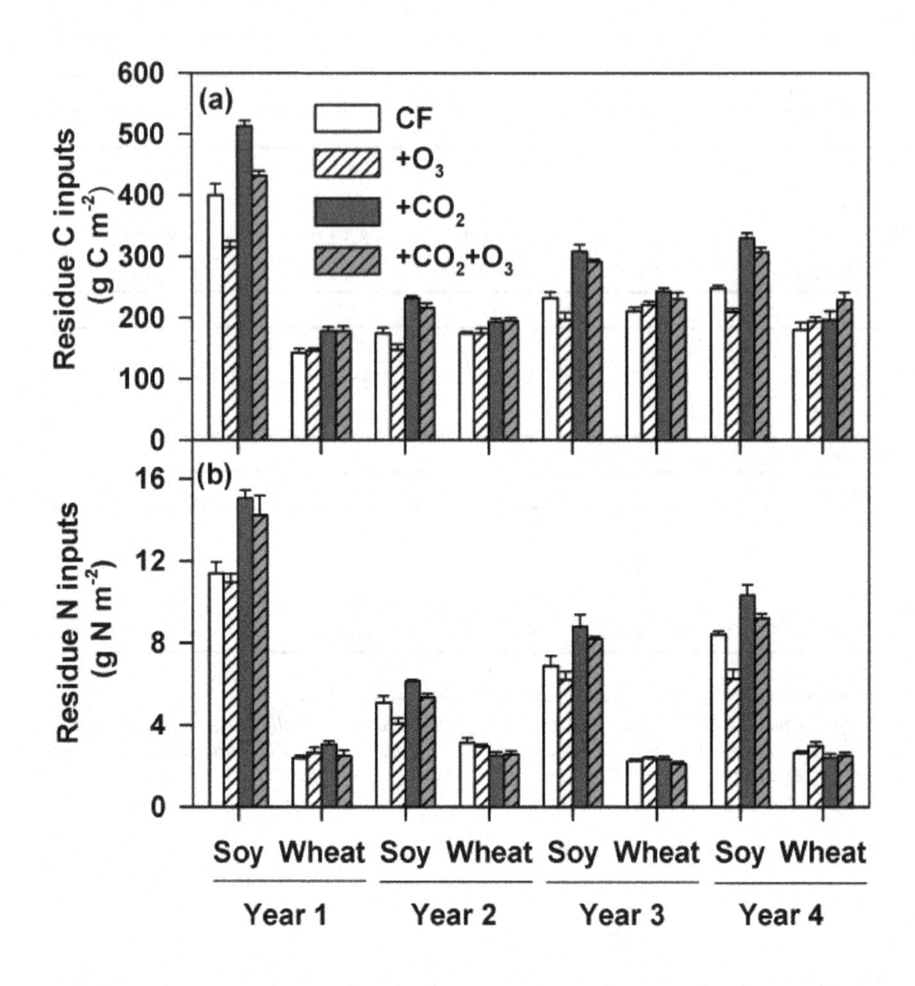

FIGURE 1: Aboveground residue C and N inputs under elevated CO_2 and O_3. Soy, soybean. CF, charcoal-filtered ambient air. $+O_3$, elevated O_3.$+CO_2$, elevated CO_2.$+CO_2+O_3$, elevated CO_2+O_3. Data represent means (n = 4) ± s.e.m. (a) Residue C inputs. Soybean residue C inputs: CO_2 effect, $P \leq 0.001$ for every year; O_3 effect, $P<0.05$ for every year; $CO_2 \times O_3$, $P>0.1$ for every year. Wheat residue C inputs: CO_2 effect, $P<0.05$ for every year; O_3 effect, $P>0.1$ for every year; $CO_2 \times O_3$, $P>0.1$ for every year, ANOVA mixed model. (b) Residue N inputs. Soybean residue N inputs: CO_2 effect, $P \leq 0.001$ for every year; O_3 effect, $P<0.01$ only for year $_2$ and 4; $CO_2 \times O_3$, $P>0.1$ for every year. Wheat residue N inputs: CO_2 effect, $P<0.05$ for year $_2$ (significantly decreased) but >0.1 for year 1, $_3$ and 4; O_3 effect, $P>0.1$ for every year; $CO_2 \times O_3$, $P>0.1$ for every year, ANOVA mixed model.

109.7 g N m^{-2} in the CF, $+O_3$, $+CO_2$ and $+CO_2+O_3$ treatments, respectively. Compared to the CF treatment, the $+CO_2$ treatment significantly increased the net N inputs to soil from symbiotic N_2 fixation (excluding seed harvests) on average by 43%, while the $+O_3$ treatment decreased it by 23% over the 4-year period (Table 2). However, there was no significant effect of $CO_2 \times O_3$ on the net N inputs derived from symbiotic N_2 fixation in any year (Table 2).

Elevated CO_2 significantly increased both soybean aboveground residue C and N inputs to the chambers in all years ($P \leq 0.001$ for each year; Fig. 1), leading to an average increase by 38% and 30%, respectively, over the experimental period. Elevated CO_2 also significantly increased wheat residue C inputs by 15%, but did not affect wheat residue N inputs (Fig. 1). Elevated O_3 had no significant effects on wheat residue C and N inputs, but reduced soybean C and N inputs by 12% (Fig. 1). No significant $CO_2 \times O_3$ interaction was observed on soybean and wheat residue C and N inputs ($P>0.1$). Additionally, the total amounts of C and N in soybean residues were significantly different ($P<0.01$) among four years, which primarily resulted from the differences in biomass production of three different cultivars as well as the variability among years.

10.3.2 SOIL ORGANIC C AND N, AND EXTRACTABLE C AND N

Over the experimental period, no significant CO_2 or O_3 effects on total soil organic C (data not shown) or total soil N (Table 2) were observed. However, there was a significant increase in total soil N over the experimental period in all treatments compared to the soil N before the treatments were applied in 2005. Although the magnitude of increase in soil N over time tended to be higher under elevated (26%) than ambient (17%) CO_2 plots, this effect was not statistically significant.

Neither CO_2 nor O_3 treatments had any significant effects on soil extractable C in the whole soil profile or on the interactions between time and gas treatments over the 4-year period (Appendix S2). In general, elevated CO_2 tended to increase concentrations of total extractable inorganic N (Appendix S3). Soil extractable N in elevated CO_2 plots increased by, on average, 17% ($P>0.1$), 18% ($P<0.01$) and 8% ($P>0.1$), respectively, in

0–5, 5–10, and 10–20 cm soil layers over the 4-year period. There were no significant effects of O_3 and the $CO_2 \times O_3$ interaction on soil extractable N (Appendix S2).

10.3.3 MICROBIAL BIOMASS C AND BIOMASS N

Elevated CO_2 significantly enhanced both MBC and MBN in the 0–5 cm soil layer, leading to an average increase of 8% ($P<0.05$; Fig. 2a) and 14% ($P<0.05$; Fig. 2d), respectively, over the 4-year period. However, these increases resulted primarily from CO_2-induced enhancement in the third and fourth years of the experiment (Fig. 2). Both MBC and MBN at elevated CO_2 remained unchanged in the top soil layer during year 1 and 2, but increased on average by 14% and 26%, respectively, within year $_3$ and 4 of the experiment. The CO_2 effects were also significant for MBN in the 10–20 cm soil layer ($P<0.05$), but not significant for MBC (Appendix S2). However, neither O_3 nor the $CO_2 \times O_3$ interaction had any significant impacts on MBC or MBN along the soil profile (Appendix S_2).

10.3.4 SOIL MICROBIAL RESPIRATION (SMR)

Over the 4-year period, atmospheric CO_2 enrichment increased SMR rates (Fig. 3). Compared to ambient CO_2, SMR under elevated CO_2 was 26% ($P<0.05$), 17% ($P>0.1$) and $_3$1% ($P<0.05$) higher in 0–5, 5–10 and 10–20 cm soil layers, respectively. Similar to microbial biomass, the observed increases in SMR were largely due to the CO_2 stimulation effects in the third and fourth years of the experiment (Fig. 3). In the 0–5 cm soil layer, for example, elevated CO_2 only increased SMR by 9% in the first two years, but by 43% over the subsequent two years. Neither the O_3 effect nor the $CO_2 \times O_3$ interaction resulted in significant effects on SMR in any soil layer (Appendix S_2).

Metabolic quotient of soil microbes (the respiration rate per unit of microbial biomass C) under elevated CO_2 increased on average by 16% (repeated measures mixed models; CO_2 effect: $P = 0.003$), 9% ($P = 0.2$), and 20% ($P = 0.02$) in 0–5, 5–10, and 10–20 cm soil layers, respectively,

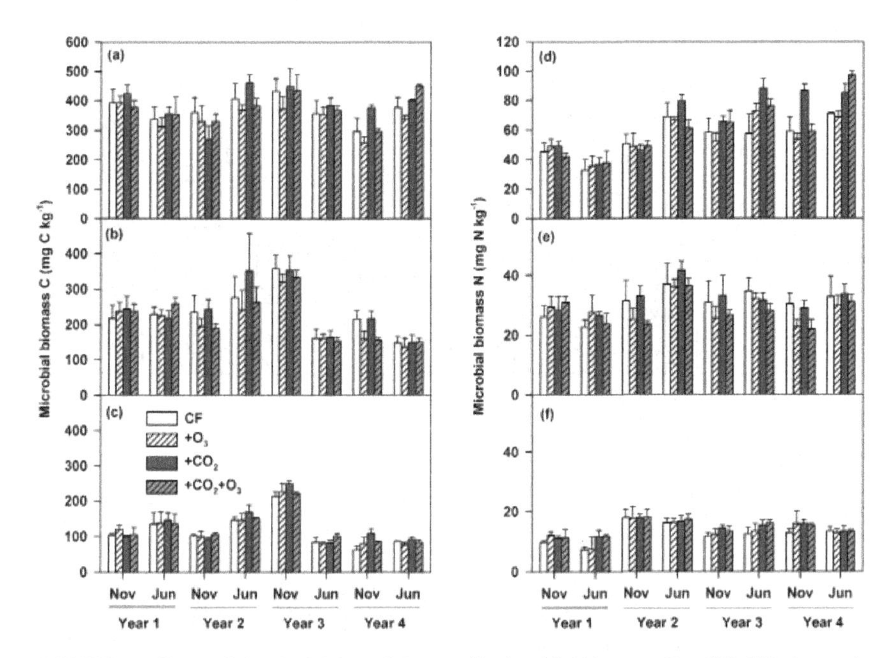

FIGURE 2: Effects of elevated CO_2 and O_3 on soil microbial biomass C and N. CF, charcoal-filtered ambient air. $+O_3$, elevated O_3.$+CO_2$, elevated CO_2.$+CO_2+O_3$, elevated CO_2+O_3. Microbial biomass C: (a) 0–5 cm soil layer (Repeated measures mixed model; CO_2 effect: P = 0.0,6; CO_2×Time: P>0.1), (b) 5–10 cm soil layer (Repeated measures mixed model; CO_2 effect: P>0.1; CO_2×Time: P>0.1) and (c) 10–20 cm soil layer (Repeated measures mixed model; CO_2 effect: P>0.1; CO_2×Time: P>0.1). Microbial biomass N: (d) 0–5 cm soil layer (Repeated measures mixed model; CO_2 effect: P = 0.0,5; CO_2×Time: P = 0.018), (e) 5–10 cm soil layer (Repeated measures mixed model; CO_2 effect: P>0.1; CO_2×Time: P>0.1) and (f) 10–,0 cm soil layer (Repeated measures mixed model; CO_2 effect: P = 0.040; CO_2×Time: P>0.1). The O_3 and CO_2×O_3 effects were not significant for all soil layers. Data represent means (n = 4) ± s.e.m.

over the 4-year period (Appendix S4). The CO_2 effect on the metabolic quotient also changed considerably over time. In the 0–5 cm soil layer, CO_2 enrichment slightly increased the metabolic quotient by 7% in the first two years, but significantly increased it by 25% within the following two years. Neither the O_3 effect nor the CO_2×O_3 interaction resulted in significant impacts on metabolic quotient of soil microbes in any soil layer (Appendix S2).

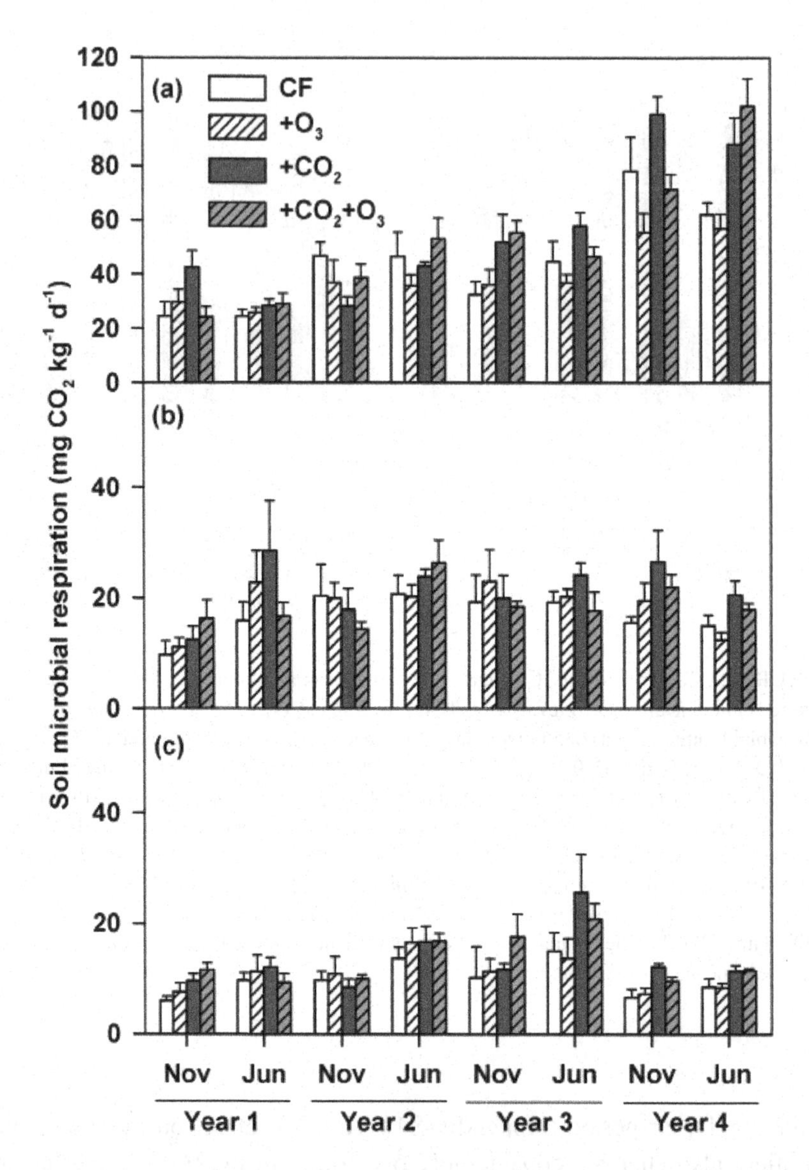

FIGURE 3: Effects of elevated CO_2 and O_3 on soil microbial respiration. CF, charcoal-filtered ambient air. $+O_3$, elevated O_3.$+CO_2$, elevated CO_2.$+CO_2+O_3$, elevated CO_2+O_3. (a) 0–5 cm soil layer (Repeated measures mixed model; CO_2 effect: $P = 0.01_2$; $CO_2 \times$Time: $P = 0.00_3$). (b) 5–10 cm soil layer (Repeated measures mixed model; CO_2 effect: $P>0.1$; $CO_2 \times$Time: $P>0.1$). (c) 10–20 cm soil layer (Repeated measures mixed model; CO_2 effect: $P = 0.044$; $CO_2 \times$Time: $P>0.1$). The O_3 and $CO_2 \times O_3$ effects were not significant for all soil layers. Data represent means (n = 4) ± s.e.m.

10.3.5 NET SOIL N MINERALIZATION

Similar to the effects on SMR and MBN, CO_2 enrichment significantly stimulated the rate of net soil N mineralization at both 0–5 ($P<0.01$; Fig. 4a) and 10–20 ($P<0.05$; Fig. 4c) cm soil layers. On average, net mineralizable N (NMN) in elevated CO_2 plots was 13%, 5%, and 26% higher than those in ambient CO_2 plots, respectively, in the 0–5, 5–10, and 10–$_2$0 cm soil layers. Again, these effects were mainly due to the CO_2-induced increases within the year 3 and 4 of the experiment (Fig. 4). Elevated CO_2 showed no impacts on net soil N mineralization in the first two years, but caused an average increase by 22%, 12% and 49%, respectively, in the 0–5, 5–10, and 10–20 cm soil layers during the third and fourth years. In contrast, neither the O_3 treatment effect nor the $CO_2 \times O_3$ interaction were statistically significant (Appendix S2).

10.3.6 STRATIFICATION OF SOIL MICROBIAL PARAMETERS UNDER ELEVATED CO_2

The time course of CO_2-effects on various parameters along the soil profile was significantly different. In the top 5-cm soil samples, MBC fluctuated over the whole period (Fig. 2a), but MBN, SMR and NMN started to increase by the third year (Fig. 2d, 3a and 4a). In the deeper soils (5–10 and 10–20 cm), MBC significantly decreased (Fig. 2b and 2c), MBN remained unchanged (Fig. 2e and 2f), but SMR and NMN increased (Fig. 3b, 3c, 4b and 4c) in years 3 and 4. Over the first two years of the experiment, all these parameters remained largely unaffected by CO_2 enrichment in the 5–10 and 10–20 cm soil layers (Figs. 2, 3, 4). By the third and fourth years, elevated CO_2 had no impacts on MBC in the deeper soil depths (Fig. 2b and 2c), but still increased SMR and NMN (Fig. 3b, 3c, 4b and 4c).

The microbial parameters from soil samples collected at the mid-seasons were similar with those at the harvest of the corresponding growing season and those results were shown in Appendix S1.

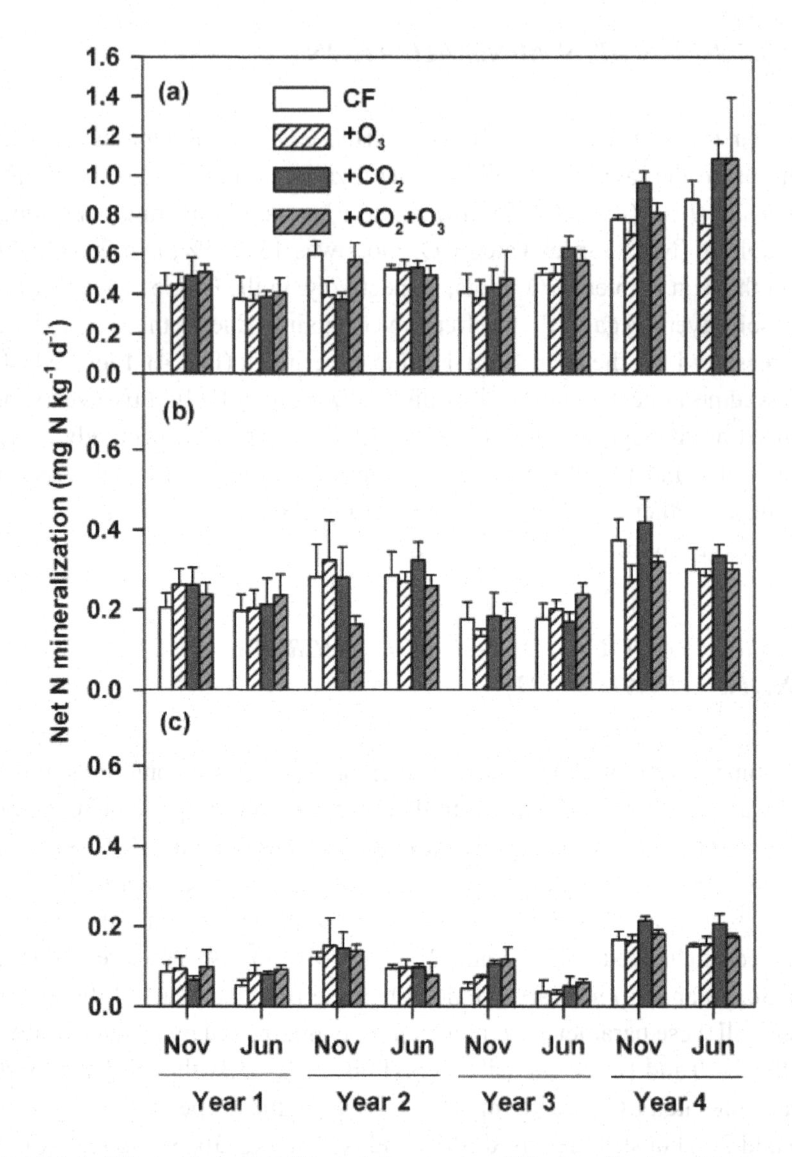

FIGURE 4: Effects of elevated CO_2 and O_3 on net soil N mineralization. CF, charcoal-filtered ambient air. $+O_3$, elevated O_3.$+CO_2$, elevated CO_2.$+CO_2+O_3$, elevated CO_2+O_3. (a) 0–5 cm soil layer (Repeated measures mixed model; CO_2 effect: P = 0.00$_2$; CO_2×Time: P = 0.011), (b) 5–10 cm soil layer (Repeated measures mixed model; CO_2 effect: P>0.1; CO_2×Time: P>0.1) and (c) 10–$_2$0 cm soil layer (Repeated measures mixed model; CO_2 effect: P = 0.019; CO_2×Time: P>0.1). The O_3 and CO_2×O_3 effects were not significant for all soil layers. Data represent means (n = 4) ± s.e.m.

10.3.7 PLFAS OF SOIL MICROBES AND THE MICROBIAL COMMUNITY STRUCTURE

Two trends in fungal and bacterial PLFAs emerged. First, the abundance of fungal and bacterial PLFAs and fungi:bacteria ratios remained largely unchanged in the first two years but decreased significantly in years 4 and 5 of the experiment (Fig. 5). On average, fungal and bacterial PLFAs decreased by 31% and 13%, respectively, from the years 1–3 to years 4–5. Second, elevated CO_2 significantly increased microbial PLFA biomass which was due only to increased fungal PLFA biomarkers starting from year 3 ($P<0.05$; Fig. 5a). As such, the fungi:bacteria ratio increased significantly due to elevated CO_2 ($P<0.05$; Fig. 5c). Neither the O_3 effect nor the $CO_2 \times O_3$ interaction had significant effects on fungal and bacterial PLFAs at any time points (Fig. 5a, 5b).

10.3.8 CORRELATION ANALYSIS

Correlation analysis, conducted among extractable C, extractable N, MBC, MBN, SMR and NMN, showed that all six parameters were significantly correlated with each other, though the coefficients varied considerably (Appendix S5). Net soil N mineralization can best explain the variation of MBC ($R^2 = 0.44$), MBN ($R^2 = 0.72$) and SMR ($R^2 = 0.72$). The χ^2 values for MBC vs N, MBN vs N, SMR vs N and the fungi:bacteria ratio vs N were 14.7 ($P<0.001$), 14.5 ($P<0.001$), 10.8 ($P<0.01$) and 4 ($P<0.05$) (Appendix S6), respectively, indicating that the CO_2 effects on microbial biomass, activity and the community structure were closely related to CO_2-induced alterations in N availability.

10.4 DISCUSSION

10.4.1 THE EFFECT OF ELEVATED CO_2 ON SOIL MICROBES

Results obtained in this study showed that elevated CO_2 influenced microbial processes over time likely through its impacts on C and N availability

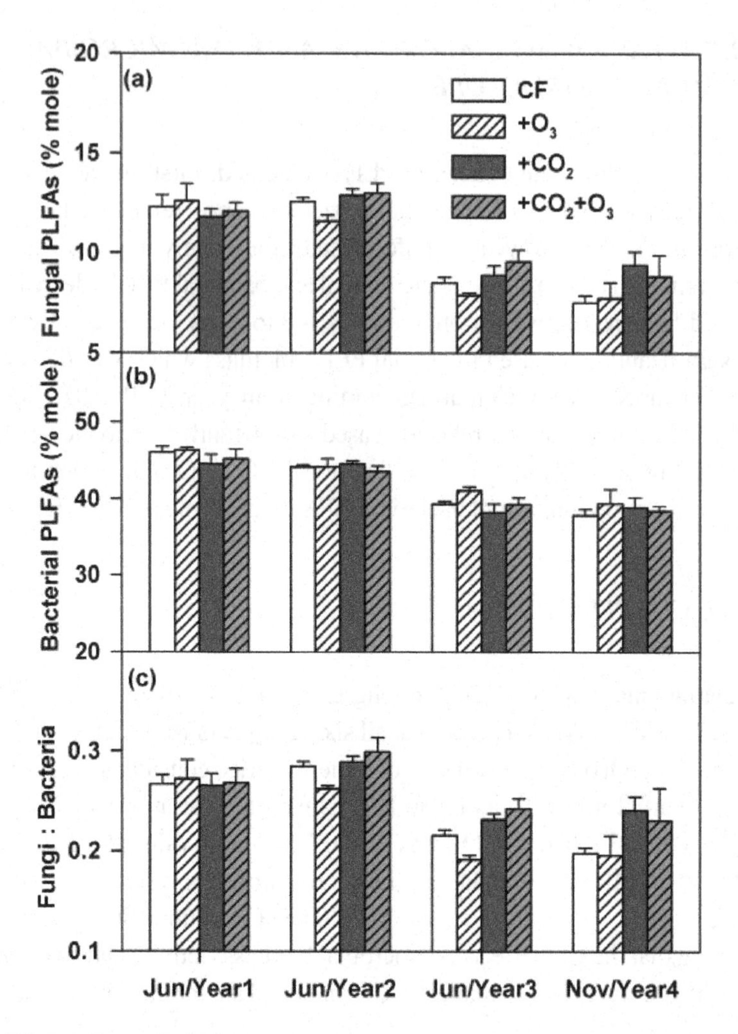

FIGURE 5: Effects of elevated CO_2 and O_3 on microbial community composition. CF, charcoal-filtered ambient air. $+O_3$, elevated O_3.$+CO_2$, elevated CO_2.$+CO_2+O_3$, elevated CO_2+O_3. (a) the abundance of fungal phospholipid fatty acids (PLFAs) [ANOVA mixed model; CO_2 effect: Jun/Year 1 ($P>0.1$), Jun/Year 2 ($P = 0.01$), Jun/Year 3 ($P = 0.0_29$), Nov/Year 4 ($P = 0.015$); O_3 effect: $P>0.1$ for all four time points; $CO_2 \times O_3$: $P>0.1$ for all four time points)], (b) the abundance of bacterial PLFAs (ANOVA mixed model; CO_2 effect: $P>0.1$ for all four time points, O_3 effect: $P>0.1$ for all four time points; $CO_2 \times O_3$: $P>0.1$ for all four time points) and (c) the ratio of fungal to bacterial PLFAs in top (0–5 cm) soils [ANOVA mixed model; CO_2 effect: Jun/Year 1 ($P>0.1$), Jun/Year 2 ($P = 0.006$), Jun/Year 3 ($P<0.001$), Nov/Year 4 ($P = 0.03$); O_3 effect: $P>0.1$ for all four time points; $CO_2 \times O_3$: Jun/Year 1 ($P>0.1$), Jun/Year $_2$ ($P = 0.019$), Jun/Year 3 ($P = 0.032$), Nov/Year 4 ($P>0.1$))]. Data represent means ($n = 4$) ± s.e.m.

(Figs. 2, 3, 4, 5, Appendix S5 and F). Microbial responses to elevated CO_2 have so far largely been considered in the context that soil microbes are C-limited [8], [11] while plant growth is N-limited [39], [57]. Many experiments have provided evidence that increased C availability induced by elevated CO_2 enhanced soil microbial biomass and/or activities [5], [10], [11], [20], [26], [58] and can alter the structure of soil microbial communities in favor of fungal growth [16], [25], [55], [59]. In the current study, elevated CO_2 significantly increased C and N availability for microbes by enhancing both aboveground (and presumably belowground) soybean and wheat residue C and N inputs in all four years (Fig. 1). However, microbial biomass, respiration and the community structure did not respond significantly until the third year (Fig. 2a, 2d, 3a, and 5c). Likely, it took a period of time for the increased residue inputs to accumulate to levels that affected soil microbial processes. It should be noted that the CO_2-stimulation in soybean residue inputs was lower in year 2 compared with in other years (Fig. 1), which may have contributed to the time-lag in elevated CO_2 effects on soil microbial processes observed in the present study.

Along with available C in soil, other factors such as soil moisture, soil food-web interactions and nutrient availability have also been suggested to affect microbial responses to elevated CO_2, either singly or in combination [9], [11]. The availability of soil N has so far received the most attention in studies of elevated CO_2 effects on soil microbes [9], [20], [21] because N is the most abundant nutrient element required for microbial growth [60]. The coincidence of higher microbial activities with increasing soil N availability and microbial biomass N in the CO_2 treatments during the third and fourth years of the experiment suggests a link between soil N availability and microbial responses to elevated CO_2 in this N-aggrading system. With the surface placement of residues in no-till systems, N existing in plant residues (mainly soybean) gradually moves into the soil profile, particularly the top soil layer, through leaching and decomposition processes. Higher N inputs from residues of plants grown under elevated CO_2 (Fig. 1b), which stemmed from both CO_2-stimulation of N_2 fixation (Table 2) [35], [36] and possibly plant N retention [4], [61], can in turn increase soil N availability for microbes. In a recent meta-analysis of 131 manipulation studies with tree species, Dieleman et al. (2010) found that the CO_2-enhancement of microbial activity and decomposition was positively correlated with increasing soil N availability. These results are similar to the CO_2-stimulation of microbial growth and activities along with increasing available soil N observed in our study. Also, it has been well documented that mineral N additions can stimulate decomposition of plant residues, particularly the non-lignin components [28], [62]; thus increased soil N availability can significantly facilitate decomposition of non-lignin components of crop residues. All exoenzymes responsible for disintegrating organic materials are N-rich proteins, and sufficient supplies of N for microbes may facilitate enzyme production [28], [63], though addition of inorganic N could also suppress lignin-degrading enzymes [64]. Other studies have also recently showed that N addition stimulated microbial respiration [26] and decomposition activities [25].

Changes in fungal and bacterial PLFAs, and their ratios observed in our study provide new insights into how alterations in the relative availability of C and N can modulate microbial activities and their responses to elevated CO_2. First, the coincidence between decreased fungi:bacteria ratios in years 4 and 5 in comparison with previous years and increased N availability and MBN over time is consistent with the general concept that high N availability favors bacteria over fungi [28], [29]. Evidently, high N inputs due to N fertilization of wheat and soybean N_2 fixation in our system gradually increased soil N availability and altered the soil microbial community composition over time. In a loblolly pine system, Feng et al. (2010) also observed that N fertilization reduced the fungi:bacteria ratio. Second, significantly higher fungi:bacteria ratios under elevated than ambient CO_2 indicate that CO_2-enhancement of C inputs may still play a major role in shaping the community structure in N-rich agroecosysystems, as shown in many forests and grasslands [4], [16], [25], [55], [59], [65]. What was surprising is that the significant increases in microbial respiration (Fig. 3a) concurred with decreased microbial PLFA biomass (Fig. 5a, 5b), leading to an increase in metabolic quotient over time as well as under elevated CO_2 (Appendix S4). Since both fungi and bacteria identified by PLFAs represent the most active part of soil microorganisms [52], [59], these results indicate that high N inputs may have stimulated microbial physiological activities and/or microbial biomass turnover. Taken together, our results suggest that CO_2-induced changes in soil N availability might be an important factor that concurrently mediated elevated CO_2 effects on soil microbes and microbial feedbacks in this N-aggrading agroecosystem.

The findings that the stimulation of soil microbes under elevated CO_2 over the course of the experiment may have significant implications for understanding residue turnover and soil C sequestration in agroecosystems under future climate change scenarios. In many natural and semi-natural ecosystems, the CO_2-induced stimulation of plant growth may not persist because of nutrient limitation [9], [39]. In agricultural ecosystems, however, N is typically not a limiting factor for plant growth due to the application of chemical N fertilizers and/or the incorporation of legume plants, and CO_2-stimulation of biomass production is expected to be sustained [2], [31], [66]. Therefore, it has been suggested that elevated CO_2 can increase long-term C storage in agroecosystems, particularly in combination with no-tillage management [38], [67], [68]. However, this assumption does not fully consider the C output from agroecosystems: unlike forest ecosystems where the standing biomass constitutes a major C pool, most agroecosystems must accumulate C in the soil for ecosystem C sequestration to occur. Consequently, the fate of returning residues will largely determine the potential of agroecosystem C sequestration. The close correlations between N availability and both microbial respiration and metabolic quotient under elevated CO_2 in our study (Fig. 3, Appendix S4 and Appendix S6) indicate that soil microbes became more active with CO_2 enrichment. Our results, along with other previous findings [27], suggest that high N availability may significantly increase soil organic C turnover in agroecosystems through stimulating residue decomposition under future CO_2 scenarios, highlighting the need to examine the interactive effect of soil N availability and atmospheric CO_2 on soil organic C dynamics.

It is also interesting to note that microbial parameters along the soil profile exhibited different patterns under elevated CO_2 (Figs. 2, 3, 4). No-till systems are characterized by vertical stratification of soil organic C and microbial biomass because of continuous residue surface placement [37], [69]. Rapid deceases in MBC and diminished CO_2 effects on MBC starting in the third year in deeper soil layers (Fig. 2b and 2c) seems to suggest that alteration in C availability caused by residue placement may dominate microbial responses. However, other parameters [MBN (Fig. 2f), SMR

(Fig. 3c) and NMN (Fig. 4c)] did not decrease correspondingly with MBC and continued to significantly respond to elevated CO_2 (Appendix S2), suggesting that other factors may significantly exert control. High correlations between MBN and SMR, and NMN in deeper soil depths (Appendix S5) suggest that N availability critically modulated microbial activities. In no-till systems, root-derived C is the primary source for deep soil C and CO_2-stimulation of both fine and deep roots has been proposed as a potential mechanism that facilitates C sequestration there [37]. However, higher metabolic quotient (Appendix S4), SMR (Fig. 3) and NMN (Fig. 4) under elevated CO_2 indicate that not only were microbes more active but also organic C turnover was more rapid in the deeper soil layers. Consequently, high root production under elevated CO_2 might stimulate C losses from deep soil layers by priming decomposition of indigenous organic matter [15], [58], [66], [70], [71]. Long-term experiments are critically needed to examine whether the stimulation of SMR and NMN in our study is transient or will be sustained over time.

10.4.2 EFFECTS OF ELEVATED O_3 AND $CO_2$4O_3 ON SOIL MICROBES

Elevated O_3 often leads to a substantial decline in the aboveground biomass of O_3-sensistive plants [7], [30], [72] and subsequent C allocation belowground [6], [19]. In the current study, the statistically significant decline in plant residue C primarily stemmed from O_3-reduction of soybean residue C (by 12% on average; $P<0.05$). The unresponsiveness of wheat to O_3 was likely due to the relatively low O_3 concentrations during the wheat growing season (Table 1), use of a relatively O_3-tolerant cultivar, and possibly other environmental conditions (for example, temperature and light levels). Ambient O_3 concentrations during winter wheat growing seasons are usually low due to the lower concentrations of precursors of O_3 formation and the lower temperatures during the winter and the early spring. The decrease in soybean residue N inputs under elevated O_3 (Fig. 1b) resulted from O_3-induced reduction in residue biomass and possibly symbiotic N_2 fixation in soybean plants (Table 2) [36]. However, no significant O_3 effects were detected on any soil microbial parameters in this study (Figs. 2, 3, 4, 5, Appendix S2). These results suggest that N inputs through both fertilization and N_2 fixation in our system might overtake O_3-induced reduction of residue N in affecting soil microbes. Alternatively, these results also suggest that the magnitude of reductions in both C and N under elevated O_3 were insufficient to substantially affect soil microbial activity in our experiment. In an OTC experiment under conventional tillage practice, Islam et al. (2000) also found that elevated O_3 had no significant impacts on soil microbial respiration.

Our results showed that elevated O_3 tended to reduce soybean residue C and N inputs under elevated CO_2 (Fig. 1). This indicated that added O_3 prevented a portion of the CO_2-induced stimulation in biomass production from occurring. Such a pattern, however, was not observed for microbial parameters over the course of the experiment (Figs. 2, 3, 4, 5). The lack of microbial responses to O_3 under elevated CO_2 suggests that, as noted above, the magnitude of the combination of elevated CO_2 and O_3 effect on residue C and N inputs was not enough to influence soil microbes in the current study. It is also possible that O_3 might not necessarily diminish the stimulation effect of elevated CO_2 on C allocation belowground through fine root biomass, root exudation and turnover during plant growth, as observed in the Rhinelander free-air CO_2 and O_3 enrichment study using tree species [73]. Regardless of the underlying causes, our results suggest that O_3 may have limited impact on soil microbial processes in agricultural systems under future CO_2 scenarios and that its effect will be dependent on the sensitivity of crop cultivars to O_3.

10.5 CONCLUSIONS

In summary, results obtained from this study showed that the responses of soil microbes and their community structure to elevated CO_2 significantly changed through time in the N-aggrading wheat-soybean rotation system, and that these may be largely related to CO_2-induced alterations in soil C and N availability. While soil microbial biomass, activities and the community structure composition were little affected by elevated CO_2 in the first two years, they significantly responded to CO_2 enrichment in the third and fourth years of the experiment as N availability increased.

However, O_3 effects on soil C and N availability were likely insufficient in magnitude to produce detectable changes in the soil microbial parameters measured. Together, these results highlight the urgent need for considering the interactive impact of C and N availability on microbial activities and decomposition when projecting soil C balance in N-rich systems under future CO_2 scenarios.

REFERENCES

1. Drake BG, GonzalezMeler MA, Long SP (1997) More efficient plants: A consequence of rising atmospheric CO2? Annual Review of Plant Physiology and Plant Molecular Biology 48: 609–639.
2. Kimball BA, Kobayashi K, Bindi M (2002) Responses of agricultural crops to free-air CO2 enrichment. In: Sparks DL, editor. Advances in Agronomy, Vol 77. San Diego: Academic Press Inc. pp. 293–368.
3. Cheng L, Zhu J, Chen G, Zheng X, Oh NH, et al. (2010) Atmospheric CO2 enrichment facilitates cation release from soil. Ecology Letters 13: 284–291. doi: 10.1111/j.1461-0248.2009.01421.x
4. Hu S, Chapin FS, Firestone MK, Field CB, Chiariello NR (2001) Nitrogen limitation of microbial decomposition in a grassland under elevated CO2. Nature 409: 188–191. doi: 10.1038/35051576
5. Hungate BA, Holland EA, Jackson RB, Chapin FS, Mooney HA, et al. (1997) The fate of carbon in grasslands under carbon dioxide enrichment. Nature 388: 576–579. doi: 10.1111/j.1469-8137.2004.01054.x
6. Andersen CP (2003) Source-sink balance and carbon allocation below ground in plants exposed to ozone. New Phytologist 157: 213–228. doi: 10.1046/j.1439-037x.1999.00325.x
7. USEPA (2006) Air Quality Criteria for Ozone and Related Photochemical Oxidants. Washinton, D.C.: U.S. Environmental Protection Agency. pp. 705–734.
8. Smith JL, Paul EA (1990) The significance of soil microbial biomass estimations. In: Bollag JM, Stotzky G, editors. Soil Biochemistry. New York: Marcel Dekker, Inc. pp. 357–396.
9. Hu S, Firestone MK, Chapin FS (1999) Soil microbial feedbacks to atmospheric CO2 enrichment. Trends in Ecology & Evolution 14: 433–437. doi: 10.1046/j.1439-037x.1999.00325.x
10. Zak DR, Pregitzer KS, Curtis PS, Teeri JA, Fogel R, et al. (1993) Elevated atmospheric CO2 and feedback between carbon and nitrogen cycles. Plant and Soil 151: 105–117. doi: 10.1007/s002489900087
11. Zak DR, Pregitzer KS, King JS, Holmes WE (2000) Elevated atmospheric CO2, fine roots and the response of soil microorganisms: a review and hypothesis. New Phytologist 147: 201–222.

12. Islam KR, Mulchi CL, Ali AA (2000) Interactions of tropospheric CO2 and O3 enrichments and moisture variations on microbial biomass and respiration in soil. Global Change Biology 6: 255–265.
13. Phillips RL, Zak DR, Holmes WE, White DC (2002) Microbial community composition and function beneath temperate trees exposed to elevated atmospheric carbon dioxide and ozone. Oecologia 131: 236–244. doi: 10.1111/j.1365-2486.2010.02207.x
14. Blagodatskaya E, Blagodatsky S, Dorodnikov M, Kuzyakov Y (2010) Elevated atmospheric CO2 increases microbial growth rates in soil: results of three CO2 enrichment experiments. Global Change Biology 16: 836–848. doi: 10.1007/s002489900087
15. Langley JA, McKinley DC, Wolf AA, Hungate BA, Drake BG, et al. (2009) Priming depletes soil carbon and releases nitrogen in a scrub-oak ecosystem exposed to elevated CO2. Soil Biology & Biochemistry 41: 54–60. doi: 10.1111/j.1365-3040.2005.01349.x
16. Rillig MC, Field CB, Allen MF (1999) Soil biota responses to long-term atmospheric CO2 enrichment in two California annual grasslands. Oecologia 119: 572–577. doi: 10.1111/j.1365-3040.2005.01349.x
17. Pendall E, Mosier AR, Morgan JA (2004) Rhizodeposition stimulated by elevated CO2 in a semiarid grassland. New Phytologist 162: 447–458. doi: 10.1111/j.1469-8137.2004.01054.x
18. Talhelm AF, Pregitzer KS, Zak DR (2009) Species-specific responses to atmospheric carbon dioxide and tropospheric ozone mediate changes in soil carbon. Ecology Letters 12: 1219–1228. doi: 10.1111/j.1461-0248.2009.01380.x
19. Grantz DA, Gunn S, Vu HB (2006) O3 impacts on plant development: a meta-analysis of root/shoot allocation and growth. Plant Cell and Environment 29: 1193–1209. doi: 10.1007/s11104-006-9093-4
20. Hu S, Tu C, Chen X, Gruver JB (2006) Progressive N limitation of plant response to elevated CO2: a microbiological perspective. Plant and Soil 289: 47–58. doi: 10.1007/s11104-006-9093-4
21. de Graaff MA, van Groenigen KJ, Six J, Hungate B, van Kessel C (2006) Interactions between plant growth and soil nutrient cycling under elevated CO2: a meta-analysis. Global Change Biology 12: 2077–2091.
22. Rice CW, Garcia FO, Hampton CO, Owensby CE (1994) Soil microbial responses in tall grass prairie to elevated CO2. Plant and Soil 165: 67–74. doi: 10.1016/j.soilbio.2003.10.002
23. Jones TH, Thompson LJ, Lawton JH, Bezemer TM, Bardgett RD, et al. (1998) Impacts of rising atmospheric carbon dioxide on model terrestrial ecosystems. Science 280: 441–443. doi: 10.1007/s002489900087
24. Barnard R, Le Roux X, Hungate BA, Cleland EE, Blankinship JC, et al. (2006) Several components of global change alter nitrifying and denitrifying activities in an annual grassland. Functional Ecology 20: 557–564.
25. Feng XJ, Simpson AJ, Schlesinger WH, Simpson MJ (2010) Altered microbial community structure and organic matter composition under elevated CO2 and N fertilization in the duke forest. Global Change Biology 16: 2104–2116. doi: 10.1111/j.1365-2486.2009.02080.x

26. West JB, Hobbie SE, Reich PB (2006) Effects of plant species diversity, atmospheric [CO2], and N addition on gross rates of inorganic N release from soil organic matter. Global Change Biology 12: 1400–1408. doi: 10.1111/j.1365-2486.2005.00939.x
27. Dieleman WIJ, Luyssaert S, Rey A, De Angelis P, Barton CVM, et al. (2010) Soil [N] modulates soil C cycling in CO2-fumigated tree stands: a meta-analysis. Plant, Cell & Environment 33: 2001–2011. doi: 10.1111/j.1365-3040.2010.02201.x
28. Fog K (1988) The effect of added nitrogen on the rate of decomposition of organic matter. Biological Reviews of the Cambridge Philosophical Society 63: 433–462. doi: 10.2307/2269568
29. Kaye JP, Hart SC (1997) Competition for nitrogen between plants and soil microorganisms. Trends in Ecology & Evolution 12: 139–143. doi: 10.1016/S0169-5347(97)01001-X
30. Fiscus EL, Booker FL, Burkey KO (2005) Crop responses to ozone: uptake, modes of action, carbon assimilation and partitioning. Plant Cell and Environment 28: 997–1011. doi: 10.1111/j.1365-3040.2005.01349.x
31. Ainsworth EA, Davey PA, Bernacchi CJ, Dermody OC, Heaton EA, et al. (2002) A meta-analysis of elevated [CO2] effects on soybean (Glycine max) physiology, growth and yield. Global Change Biology 8: 695–709. doi: 10.1111/j.1365-2486.2009.02080.x
32. Feng ZZ, Kobayashi K, Ainsworth EA (2008) Impact of elevated ozone concentration on growth, physiology, and yield of wheat (Triticum aestivum L.): a meta-analysis. Global Change Biology 14: 2696–2708. doi: 10.1007/s11104-006-9093-4
33. Morgan PB, Ainsworth EA, Long SP (2003) How does elevated ozone impact soybean? A meta-analysis of photosynthesis, growth and yield. Plant Cell and Environment 26: 1317–1328. doi: 10.1007/s11104-006-9093-4
34. Booker FL, Fiscus EL (2005) The role of ozone flux and antioxidants in the suppression of ozone injury by elevated CO2 in soybean. Journal of Experimental Botany 56: 2139–2151. doi: 10.1093/jxb/eri214
35. Rogers A, Ainsworth EA, Leakey ADB (2009) Will elevated carbon dioxide concentration amplify the benefits of nitrogen fixation in legumes? Plant Physiology 151: 1009–1016. doi: 10.1104/pp.109.144113
36. Tu C, Booker FL, Burkey KO, Hu S (2009) Elevated atmospheric carbon dioxide and O3 differentially alter nitrogen acquisition in peanut. Crop Science 49: 1827–1836. doi: 10.1111/j.1365-2486.2005.00939.x
37. Prior SA, Runion GB, Rogers HH, Torbert HA, Reeves DW (2005) Elevated atmospheric CO2 effects on biomass production and soil carbon in conventional and conservation cropping systems. Global Change Biology 11: 657–665. doi: 10.1111/j.1365-3040.2010.02201.x
38. Lal R (2004) Soil carbon sequestration impacts on global climate change and food security. Science 304: 1623–1627. doi: 10.1126/science.1097396
39. Reich PB, Hobbie SE, Lee T, Ellsworth DS, West JB, et al. (2006) Nitrogen limitation constrains sustainability of ecosystem response to CO2. Nature 440: 922–925. doi: 10.1038/nature04486
40. Liu LL, Greaver TL (2010) A global perspective on belowground carbon dynamics under nitrogen enrichment. Ecology Letters 13: 819–828. doi: 10.1111/j.1461-0248.2010.01482.x

41. Booker FL, Prior SA, Torbert HA, Fiscus EL, Pursley WA, et al. (2005) Decomposition of soybean grown under elevated concentrations of CO2 and O3. Global Change Biology 11: 685–698. doi: 10.1111/j.1365-2486.2005.00939.x
42. Warembourg FR (1993) Nitrogen fixation in soil and plant systems. In: Knowles R, Blackburn TH, editors. Nitrogen Isotope Techniques. San Diego: Academic Press. pp. 127–157.
43. Kimball BA, Pinter PJ, Garcia RL, LaMorte RL, Wall GW, et al. (1995) Productivity and water use of wheat under free-air CO2 enrichment. Global Change Biology 1: 429–442. doi: 10.1007/s11104-006-9093-4
44. McMaster GS, LeCain DR, Morgan JA, Aiguo L, Hendrix DL (1999) Elevated CO2 increases wheat CER, leaf and tiller development, and shoot and root growth. Journal of Agronomy and Crop Science 183: 119–128. doi: 10.1046/j.1439-037x.1999.00325.x
45. Nissen T, Rodriguez V, Wander M (2008) Sampling soybean roots: A comparison of excavation and coring methods. Communications in Soil Science and Plant Analysis 39: 1875–1883.
46. Hu S, Wu J, Burkey KO, Firestone MK (2005) Plant and microbial N acquisition under elevated atmospheric CO2 in two mesocosm experiments with annual grasses. Global Change Biology 11: 213–223. doi: 10.2307/2269568
47. Vance ED, Brookes PC, Jenkinson DS (1987) An extraction method for measuring soil microbial biomass-C. Soil Biology & Biochemistry 19: 703–707. doi: 10.1111/j.1469-8137.2004.01054.x
48. Cabrera ML, Beare MH (1993) Alkaline persulfate oxidation for determining total nitrogen in microbial biomass extracts. Soil Science Society of America Journal 57: 1007–1012. doi: 10.1021/cr00090a003
49. Jenkinson DS, Brookes PC, Powlson DS (2004) Measuring soil microbial biomass. Soil Biology & Biochemistry 36: 5–7. doi: 10.1016/j.soilbio.2003.10.002
50. Coleman DC, Anderson RV, Cole CV, Elliott ET, Woods L, et al. (1978) Trophic interactions in soils as they affect energy and nutrient dynamics .4. flows of metabolic and biomass carbon. Microbial Ecology 4: 373–380.
51. Bossio DA, Scow KM, Gunapala N, Graham KJ (1998) Determinants of soil microbial communities: Effects of agricultural management, season, and soil type on phospholipid fatty acid profiles. Microbial Ecology 36: 1–12. doi: 10.1007/s002489900087
52. Frostegard A, Baath E (1996) The use of phospholipid fatty acid analysis to estimate bacterial and fungal biomass in soil. Biology and Fertility of Soils 22: 59–65.
53. Zhang W, Parker KM, Luo Y, Wan S, Wallace LL, et al. (2005) Soil microbial responses to experimental warming and clipping in a tallgrass prairie. Global Change Biology 11: 266–277. doi: 10.1111/j.1365-2486.2010.02207.x
54. Balser TC, Treseder KK, Ekenler M (2005) Using lipid analysis and hyphal length to quantify AM and saprotrophic fungal abundance along a soil chronosequence. Soil Biology & Biochemistry 37: 601–604. doi: 10.1111/j.1365-3040.2005.01349.x
55. Carney KM, Hungate BA, Drake BG, Megonigal JP (2007) Altered soil microbial community at elevated CO2 leads to loss of soil carbon. Proceedings of the National Academy of Sciences of the United States of America 104: 4990–4995. doi: 10.1073/pnas.0610045104

56. Littell RC, Milliken GA, Strooup WW, Wolfinger RD, Schabenberger O (2006) SAS for Mixed Models. Cary, , NC, USA: SAS Institute Inc.
57. Oren R, Ellsworth DS, Johnsen KH, Phillips N, Ewers BE, et al. (2001) Soil fertility limits carbon sequestration by forest ecosystems in a CO2-enriched atmosphere. Nature 411: 469–472. doi: 10.1038/35078064
58 Phillips RP, Finzi AC, Bernhardt ES (2011) Enhanced root exudation induces microbial feedbacks to N cycling in a pine forest under long-term CO2 fumigation. Ecology Letters 14: 187–194. doi: 10.1111/j.1461-0248.2010.01570.x
59. Jin VL, Evans RD (2010) Microbial 13C utilization patterns via stable isotope probing of phospholipid biomarkers in Mojave Desert soils exposed to ambient and elevated atmospheric CO2. Global Change Biology 16: 2334–2344. doi: 10.1111/j.1365-2486.2010.02207.x
60. Duboc P, Schill N, Menoud L, Vangulik W, Vonstockar U (1995) Measurements of sulfur, phosphorus and other ions in microbial biomass: influence on correct determination of elemental composition and degree of reduction. Journal of Biotechnology 43: 145–158. doi: 10.1016/0168-1656(95)00135-0
61. Finzi AC, DeLucia EH, Hamilton JG, Richter DD, Schlesinger WH (2002) The nitrogen budget of a pine forest under free air CO2 enrichment. Oecologia 132: 567–578.
62. Knorr M, Frey SD, Curtis PS (2005) Nitrogen additions and litter decomposition: A meta-analysis. Ecology 86: 3252–3257. doi: 10.1007/s002489900087
63. Schimel JP, Weintraub MN (2003) The implications of exoenzyme activity on microbial carbon and nitrogen limitation in soil: a theoretical model. Soil Biology & Biochemistry 35: 549–563. doi: 10.1021/cr00090a003
64. DeForest JL, Zak DR, Pregitzer KS, Burton AJ (2004) Atmospheric nitrate deposition, microbial community composition, and enzyme activity in northern hardwood forests. Soil Science Society of America Journal 68: 132–138.
65. Zak DR, Ringelberg DB, Pregitzer KS, Randlett DL, White DC, et al. (1996) Soil microbial communities beneath Populus grandidentata grown under elevated atmospheric CO2. Ecological Applications 6: 257–262. doi: 10.2307/2269568
66. Peralta AL, Wander MM (2008) Soil organic matter dynamics under soybean exposed to elevated [CO2]. Plant and Soil 303: 69–81. doi: 10.1111/j.1365-3040.2010.02201.x
67. Paustian K, Andren O, Janzen HH, Lal R, Smith P, et al. (1997) Agricultural soils as a sink to mitigate CO2 emissions. Soil Use and Management 13: 230–244. doi: 10.1046/j.1439-037x.1999.00325.x
68. West TO, Post WM (2002) Soil organic carbon sequestration rates by tillage and crop rotation: A global data analysis. Soil Science Society of America Journal 66: 1930–1946. doi: 10.1021/cr00090a003
69. Beare MH, Parmelee RW, Hendrix PF, Cheng WX, Coleman DC, et al. (1992) Microbial and faunal interactions and effects on litter nitrogen and decomposition in agroecosystems. Ecological Monographs 62: 569–591.
70. Fontaine S, Barot S, Barre P, Bdioui N, Mary B, et al. (2007) Stability of organic carbon in deep soil layers controlled by fresh carbon supply. Nature 450: 277–280. doi: 10.1038/nature06275

71. de Graaff M-A, Classen AT, Castro HF, Schadt CW (2010) Labile soil carbon inputs mediate the soil microbial community composition and plant residue decomposition rates. New Phytologist 188: 1055–1064. doi: 10.1111/j.1469-8137.2010.03427.x
72. Booker F, Muntifering R, McGrath M, Burkey K, Decoteau D, et al. (2009) The Ozone Component of Global Change: Potential Effects on Agricultural and Horticultural Plant Yield, Product Quality and Interactions with Invasive Species. Journal of Integrative Plant Biology 51: 337–351. doi: 10.1111/j.1744-7909.2008.00805.x
73. Pregitzer KS, Burton AJ, King JS, Zak DR (2008) Soil respiration, root biomass, and root turnover following long-term exposure of northern forests to elevated atmospheric CO2 and tropospheric O3. New Phytologist 180: 153–161. doi: 10.1111/j.1469-8137.2008.02564.x

There are several supplemental files that are not available in this version of the article. To view this additional information, please use the citation information cited on the first page of this chapter.

PART IV

CURRENT RESEARCH TRENDS IN CO_2 CAPTURE USING IONIC LIQUIDS

CHAPTER 11

OVERVIEW OF IONIC LIQUIDS USED AS WORKING FLUIDS IN ABSORPTION CYCLES

MEHRDAD KHAMOOSHI, KIYAN PARHAM, AND UGUR ATIKOL

11.1 INTRODUCTION

With fast economic growth and constantly increasing energy consumption, the human kind is about to face a growing degradation of the environment if business continues as usual. For this reason, the utilization of low-grade energy has become one of the most attractive solutions to heating and cooling problems encountered in industrial and residential applications. It is possible to recover the low-grade heat wasted in many industries to use it in some processes in order to increase the energy efficiency.

Absorption cycles which include absorption heat pump (AHP), absorption chiller (AC), and absorption heat transformer (AHT) can use waste heat economically resulting in decreasing the consumption of primary energy and reducing the negative impact on the environment. The basic principle of an absorption cycle is shown in Figure 1. Since the invention of absorption cycles, the properties of working fluids have been a challenging issue, as performance of an absorption cycle critically depends on thermodynamic properties of working pairs composed of refrigerant and absorbent. Thus, searching for more beneficial working pairs with ex-

This chapter was originally published under the Creative Commons Attribution License. Khamooshi M, Parham K, and Atikol U. Overview of Ionic Liquids Used as Working Fluids in Absorption Cycles. Advances in Mechanical Engineering **2013** (2013). http://dx.doi.org/10.1155/2013/620592.

cellent thermal stability, no corrosion, and no crystallization has become the subject of research in recent years [1]. The basic components of the absorption cycles are the evaporator, the condenser, the generator, and the absorber, while the evaporator and generator are supplied with waste heat at the same temperature and the increased heat is delivered from absorber. The operating system of the basic absorption cycles is explained as follows.

Refrigerant vapor is produced at state 4 in the evaporator, by low- medium-grade heat source. The refrigerant vapor dissolves and reacts with the strong refrigerant-absorbent solution that enters the absorber from state 10, and weak solution returns back to generator at state 5. In the generator some refrigerant vapor is removed from the weak solution to be sent to the condenser and consequently the strong solution from the generator is returned to the absorber. After condensing the vaporized refrigerant in the condenser, it is pumped to a higher pressure level as it enters the evaporator. The waste heat delivered to the evaporator causes its vaporization. Again the absorber absorbs the refrigerant vapor at a higher temperature. Therefore, the absorption cycles have the capability of raising the temperature of the solution above the temperature of the waste heat [2].

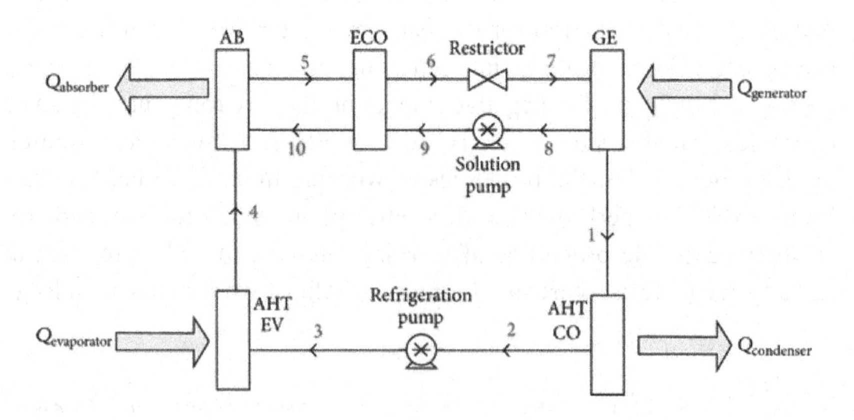

FIGURE 1: The basic absorption cycle.

The requirements of working fluids of absorption cycles areas are as follows [3].

1. The difference in boiling point between the pure refrigerant and the mixture at the same pressure should be as large as possible.
2. Refrigerant should have high heat of vaporization and high concentration within the absorbent in order to maintain low circulation rate between the generator and the absorber per unit of cooling capacity.
3. Transport properties that influence heat and mass transfer, for example, viscosity, thermal conductivity, and diffusion coefficient, should be favorable.
4. Both refrigerant and absorbent should be noncorrosive, environmentally friendly, and of low cost.

Ionic liquids (ILs) are a class of low-temperature molten salts, which are composed of an organic cation and an inorganic anion. During recent years, ILs have been used as organic green solvent in catalysis, separation process, electrochemistry, and many other industries for their unique physical and chemical properties, such as negligible vapor pressure [4], negligible flammability and thermal stability [5], low melting temperature and liquid state over a wide temperature range, and good solubility [6]. In particular, the low volatility of the ILs enables easy separation of the volatile working fluid from the ILs by thermal stratification with the minimum harmful impacts on environment [7]. ILs can be a substitute for some of the most used toxic working fluids (such as ammonia/water) in absorption cycles. Since many of the ionic liquids have melting points below the lowest solution temperature in the absorption system (~300) [8, 9], they also wipe out the crystallization and metal-compatibility problems of water/LiBr system.

Although a large number of ILs as the working fluids in absorption cycles are given in the literature, there is still a need for a complete review with comparison of all ILs for achieving higher performance of the absorption cycles. It seems to be necessary to have a review in this class of working pairs. So in the present work, a review of specific properties of ionic liquids as working fluids for absorption cycles has been carried out.

11.2 A REVIEW OF ILS AS WORKING FLUIDS

The thermodynamic performance of an absorption chiller operating with water + 1-ethyl-3-methylimidazolium dimethylphosphate [EMIM][DMP] and H_2O + LiBr mixtures has been compared under the same operating conditions by Zhang and Hu [10]. The coefficient of performance of the absorption chiller for working fluid H_2O + [EMIM][DMP] was lower than that of H_2O + LiBr by 7% but still higher than 0.7. Also for the same condensation and absorption temperatures, the generation temperature of refrigeration cycle for H_2O + [EMIM][DMP] is somewhat lower than that for H_2O + LiBr, allowing the operation of the absorption chiller at lower temperatures of heat supply. These features indicate that the binary ionic liquid solution, H_2O + [EMIM][DMP], has potential to be a new working pair for the absorption chiller driven by low-grade waste heat or hot water generated by common solar thermal collectors. The performance of absorption refrigeration cycle with methanol and 1,3-methylimidazolium dimethylphosphate (CH_3OH + [MMIM][DMP]) is investigated, and a COP of 0.82 was found with condensing, absorption, and evaporation temperatures of 40°C, 30°C, and 5°C, respectively. The requirements of operating pressure, condensing pressure, and vacuum for the system using CH_3OH + [MMIM][DMP] are lower, which makes the operation and the maintaining of the system more trouble free [11, 12]. Performance benchmarks and system-level simulations for refrigerant/[BMIM][PF_6] pairs are carried out by Kim et al. [13]. A nonrandom two-liquid model is built and used to predict the solubility of the mixture as well as the mixtures properties such as enthalpy and entropy. The evaporator and condenser saturation temperatures are 25°C and 50°C, respectively, with the operating temperature being set at 85°C. As R32, R134a, and R152a are compared with each other, and it is observed that R32 produces the highest performance with a maximum COP of 0.55, while R134a and R152a returned similar COPs of 0.4 at a generator outlet temperature of 80°C.

Kim et al. [14] in one of their other studies have obtained the thermo physical properties of H_2O + [EMIM][BF_4] with the saturation temperatures at the evaporator and condenser being 25°C and 50°C, respectively. A power dissipation of 100 W is estimated, while COP value of the system reaches 0.91. The suitable compatibility of water with [EMIM][BF4] and the superior properties of water as a heat transfer fluid, such as large latent heat of evaporation, followed by extremely small refrigerant (water) flow rate, resulted in its high performance. Higher viscosity ILs cause an increased pressure drop in the compression loop, which would result in larger pumping power or larger pipes and system volume. The viscosity increases with cation mass: EMIM < BMIM < HMIM. The viscosity is more dependent on the anion with the following order: $Tf_2N < BF_4 < PF_6$. The viscosity of [EMIM][Tf_2N] is only 31.3 MPa s at 294 K, which is 10 times smaller than that of [HMIM][PF_6] [14]. The COP of the [MMIM] DMP/methanol absorption refrigeration is lower than LiBr + H_2O in absorption refrigeration under the same temperature conditions, while higher than that of H_2O + NH_3 absorption refrigeration under most temperature conditions. [MMIM] DMP+methanol has excellent potential to be applied as the working pair of absorption refrigeration [15].

Ionic liquids with longer cation alkyl chain length cause a larger solubility but lower dependence of the solubility on temperature [16, 17]. Thus, Ionic liquids with shorter alkyl chains in cation are preferred ([EMIM] > [BMIM] > [HMIM]) due to more sensitive dependence of the solubility on temperature. R143a+[BMIM][PF_6] and R134a+[HMIM][PF_6] are remarkably less sensitive to temperature and show relatively low COPs [14]. The working pair H_2O + EMISE was considered as a potential working pair [18, 19]. The vapor pressure, heat capacity, and density of the H_2O + EMISE system were measured and correlated which verified the availability of being used in absorption cycles. Dong et al. [20] recommend H_2O + ([DMIM]DMP) as an alternative to H_2O + LiBr by comparing the coefficient of performance (COP). To predict the cycle performance, a single-effect absorption refrigeration cycle was simulated based on the models obtained from the studies of vapor pressure and heat capacity of the H_2O + ([DMIM]DMP) system. The simulation results show that the cycle performance of H_2O + ([DMIM]DMP) is close to that of conventional working pair . However, for the cycle using alternative working pair of the H_2O + ([DMIM]DMP) system, the operating temperature range has been extended and the disadvantages of crystallization and corrosion caused by H_2O + LiBr can be relieved. Ionic liquids have a large capacity of dissolving CO_2, so that the [BMIM][PF_6] can be used as a refrigerant in an absorption cycle. The COP of a system utilizing CO_2/bmimPF$_6$ is much lower than that of a traditional system using the $NH_3 + H_2O$ pair. However, it should be noted that serious questions remain regarding the accuracy of the ideal-gas heat capacity coefficients for the CO_2 bmimPF$_6$ mixture [21].

The permanent ion-dipole interaction between the ionic liquid and TFE evokes a considerable negative deviation from Raoult's law, which is normal in absorption pairs [22]. The Br anion shows a stronger interface with TFE than the BF_4 anion. This means that the required temperature for regeneration will be smaller and the vapor pressure lines will be steeper [23]. Novel working fluids for an absorption heat pump have been proposed by investigating ionic liquids and fluoroalcohol. In this study [BMIM][Br]+TFE and [BMIM][BF_4]+TFE have been considered as potential working pairs for the first time. The [BMIM][Br]+TFE system was found to be more favorable than the [BMIM][BF_4]+TFE from the results of vapor pressure. The excess volume, apparent molar volume, partial molar volume, and apparent molar expansibility of [MMIM] Cl aqueous solutions were investigated. Based on the results, it was found out that the water content has significant impact on the volumetric properties and the temperature dependence of the density is significantly less than the water content. It was proposed here that [MMIM] Cl has the potential to be used as a novel absorbent species of an absorption cycle working fluid [24].

High circulation ratio increases the energy requirements of heating and pumping processes. Circulation ratios in H_2O + LiBr and $NH_3 + H_2O$ systems usually are smaller, with typical values being around f = 10 [25, 26]. ILs circulation ratios compared to conventional absorption refrigerators are high. The lowest and therefore the best circulation ratios observed ranged from 20 to 25 [27]. The estimated energetic efficiency of the cycle with ionic liquids as absorbents with COP of 0.21 is lower than that of conventional pairs ($NH_3 + H_2O$ and LiBr + H_2O) in absorption refrigerators. However, CO_2 + [BMPYRR][Tf$_2$N] with COP of 0.55 is compared with $NH_3 + H_2O$ pairs operating in equivalent conditions. This is due to the necessity of operating the cycle with a relatively high circulation ratio (24 in the case of CO_2+ [BMPYRR/Tf$_2$N] pairs compared to 10 in conventional $NH_3 + H_2O$ systems), which increases the energy necessities of heating and pumping processes. The content of the ILs has a direct relation with the deviation from the Raoult's law meaning that these binary solutions have a negative deviation from Raoult's law. Solutions such as water, ethanol, and methanol containing [EMIM][DMP] have strong absorbing ability for coolant, which is a very important property of a working pair for absorption heat pump or absorption refrigeration [1].

The processes of [EMIM][DMP] mixed with water, alcohol, and methanol are exothermic under the temperature of the 298.15 K and pressure of 1 atm. They are ranked in the order of water, methanol, and alcohol according to the magnitude of mixing heat. Because of electron negativity of phosphorus contained in the [EMIM][DMP], hydrogen bonds are formed between [EMIM][DMP] and water, alcohol, and methanol, respectively. This property is especially important for the working pair of the absorption heat pump or refrigeration. They are ranked in the order of water, methanol, and alcohol according to the magnitude of mixing heat [1]. The three binary solutions, [MMIM][DMP] + water/ethanol/methanol systems, have exhibited some important characteristics which are needed for an absorption heat pump working pair. Their basic thermodynamic properties including vapor pressure, heat capacity, excess enthalpy, viscosity, and density were measured [18]. The viscosity of pure [MMIM][DMP] was very high, but it would decrease sharply when either [MMIM][DMP] was heated or the water/ethanol/methanol was added. As the absorption and generation temperatures of an absorption heat pump or absorption heat transformer are usually high, the viscosity of the [MMIM][DMP] + water/ethanol/methanol pairs is expected to be low. Consequently, the high viscosity of the ionic liquid would not limit its use as a heat pump absorbent. Zhao et al. [29] investigated the vapor pressure data for nine binary systems at varying temperature, and IL contents were measured using a quasi-static method. The effect of ILs on the vapor pressure lowering of solvent follows the order [MMIM][DMP] > [EMIM][DEP] > [BMIM][DBP] for water and [BMIM][DBP] > [EMIM][DEP] > [MMIM][DMP] for organic solvent methanol and ethanol. This suggests that the "ionic" is specifically dominant in water and "molecule" is prevailing in organic solvents.

Research on the solubility of ammonia in four ILs which contain [EMIM][Ac], [EMIM][SCN], [EMIM]-[$EtOSO_3$], and [DMEA][Ac] has been studied by Yokozeki and Shiflett [30]. The very high solubility behavior has been clearly demonstrated in terms of the thermodynamic excess functions based on the present EOS. Discussion shows that there is

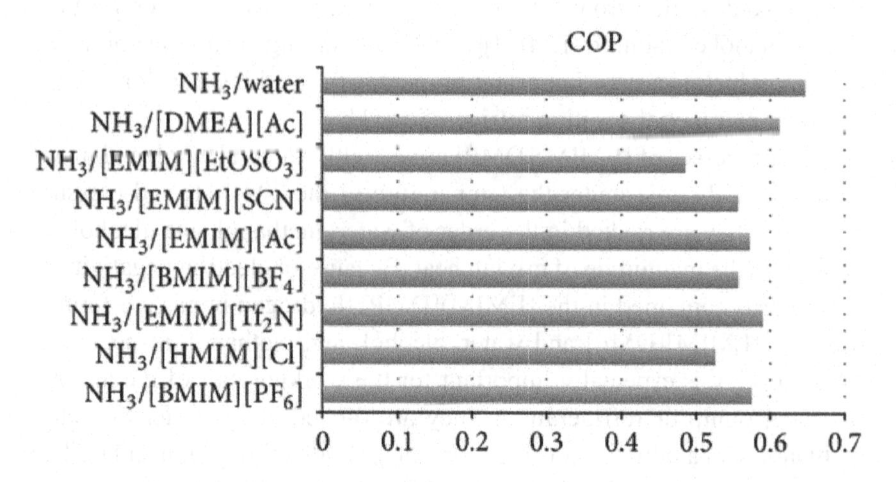

FIGURE 2: COP of series of and NH_3 + water [30–32] previously presented in [30].

opportunity for the absorption cycle application using ammonia-IL systems replacing the traditional ammonia–water system. Also, the COPs and flow ratios of series of ammonia-IL and NH_3 + water have been reported with the temperatures of 100, 40, 30, and 10°C corresponding to the generator, condenser, absorber, and the evaporator temperatures, respectively. These COPs and flow ratios obtained from studies [30–32] and compared in [30] are displayed in graphical form in Figures 2 and 3.

Among ammonia-ILs pairs studied, [DMEA][Ac] has shown the best result. The performance (COP) of the present ammonia-ILs systems is somewhat lower than that of the ammonia-water system. The functional capability of replacing LiBr with ILs as an absorbent for H_2O in absorption heat pumps and chillers has been demonstrated using theoretical absorption cycle analysis in some experimental works [10, 28, 33–35]. Yokozeki and Shiflett [28] used twelve ILs with H_2O in a simple cycle configuration analysis, using coefficient of performance and flow ratio as comparing parameters, which explains the efficiency and compactness of the system. In the analysis, [MMIM][$(CH_3)_2PO_4$] and [EMIM][$(CH_3)_2PO_4$] indicate the best results, with the highest COP and the lowest flow rate values. The COP result was about 85–88% of that of the H_2O + LiBr system. The

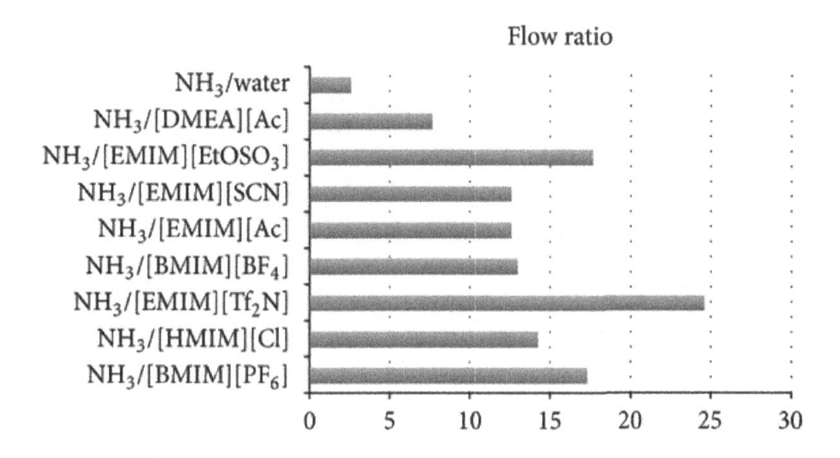

FIGURE 3: Flow ratio of series of and NH3 + water [30–32] previously presented in [30].

results of cycle simulations demonstrate that IL + H_2O systems could be competitive with H_2O + LiBr system particularly for optimized ILs.

The possible application of water and room-temperature ionic liquids (RTIL) mixtures for the absorption cooling cycle has been investigated using the present EOS and a simple absorption cycle model. It was found that an H_2O + RTIL system, when used with an optimized RTIL, could compete with an existing H_2O + LiBr system [28]. Also the COPs and flow ratios of series of H_2O + IL and NH_3 + water have been reported with temperatures of 100, 40, 30, and 10°C corresponding to the generator, condenser, absorber, and evaporator temperatures, respectively. The data used for the graphical representations in Figures 4 and 5 are taken from Yokozeki and Shiflett [28].

11.3 DISCUSSION

As illustrated in Table 1, a comparison was made among different ionic liquids in order to choose suitable working pairs. Among most of the ionic liquids, water was chosen as refrigerant due to its superior properties as

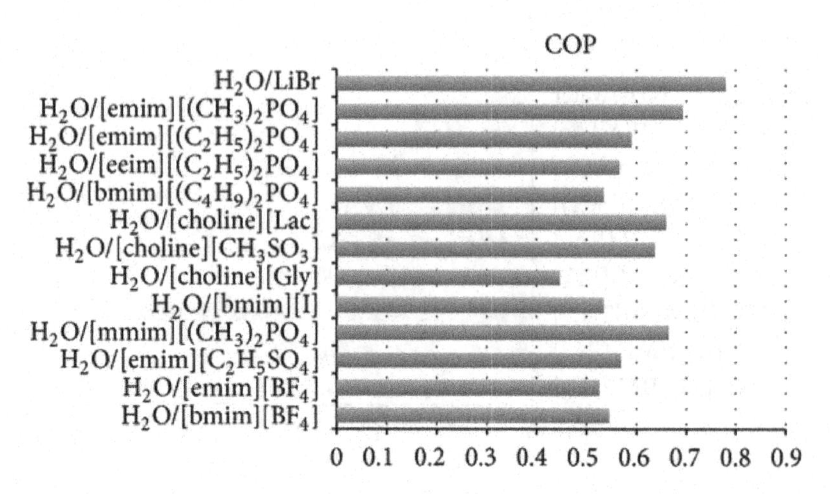

FIGURE 4: COP of series of H_2O + IL and H_2O + LiBr previously presented in [28].

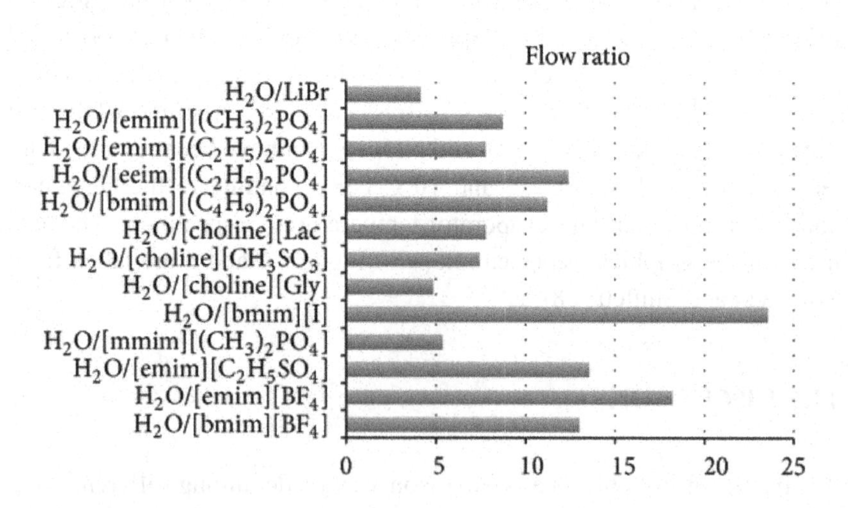

FIGURE 5: Flow ratio of series of H_2O + IL and H_2O + LiBr previously presented in [28].

a heat transfer fluid, such as large latent heat of evaporation followed by extremely small flow ratio that resulted in its high performance. Selecting working fluids for absorption cycles should be suitable for initial operating conditions and limitations, such as crystallization, corrosion problems, and environmental limits, and also material requirement should be considered.

TABLE 1: Characteristic comparison of different working fluids.

Working pair	References	Absorption cycle	Remarks
H_2O + [EMIM][DMP]	[10]	AC	COP lower than that of H_2O + LiBr by 7% but still higher than 0.7
H_2O + [EMIM][BF4]	[14]	AHP	High COP due to the following: (i) suitable compatibility of water with [EMIM][BF4] (ii) superior properties of water as a heat transfer fluid, such as large latent heat of evaporation (iii) extremely small refrigerant (water) flow rate
H_2O + [EMISE]	[18, 19]	AHP	Colorless ionic liquid; EMISE is easy to absorb the water vapor in the air.
H_2O + [DMIM]DMP	[20]	AC	Less crystallization and corrosion risk than those of H_2O + LiBr in air-cooled absorption chiller
CO_2 + [BMPYRR] [Tf2N]	[25–27]	AHP	Nontoxicity, nonflammability, and low cost, high circulation ratios in comparison wiht those of conventional absorption refrigerators, and increase in the energy necessities of heating and pumping processes
Methanol + [EMIM] [DMP] Ethanol + [EMIM] [DMP]	[1]	AHP	Strong absorbing ability for coolant
[MMIM][DMP] + water/ethanol/methanol	[18]	AHP	Suitable vapor pressure, heat capacity, excess enthalpy, and low viscosity
[MMIM][$(CH_3)_2PO_4$]	[28]	AC	High COP, low flow ratio
[EMIM][$(CH_3)_2PO_4$]	[28]	AC	High COP, low flow ratio
CH_3OH + [MMIM] [DMP]	[11, 12]	AHP	High COP, high refrigerant capacity, higher gas-emission scope and circulation ratio than those of NH_3 + H_2O, and fewer requirements of generating pressure and condensing pressure than those of conventional system.

11.4 CONCLUSIONS

In this paper a number of researches about working fluids of absorption cycles, which contain absorption heat pumps, absorption chillers, and absorption heat transformers including ionic liquids, were reviewed. Environmental issues and saving energy concerns have always been a major global problem. Absorption cycles can play an important role due to their capability of reducing CO_2 discharge and to reuse large amount of industrial waste heat. Since the performance of absorption cycles mainly depends on the working fluids, the research aspect of improving working fluids is quite essential and in progress. Using ionic liquids as the working fluids of absorption cycles can lead us to benefit from factors such as less crystallization, less corrosion, low toxicity, and nonflammability in comparison with conventional working fluids including (NH_3 + water and LiBr + water).

REFERENCES

1. J. Ren, Z. C. Zhao, and X. D. Zhang, "Vapor pressures, excess enthalpies, and specific heat capacities of the binary working pairs containing the ionic liquid 1-ethyl-3-methylimidazolium dimethylphosphate," Journal of Chemical Thermodynamics, vol. 43, no. 4, pp. 576–583, 2011.
2. I. Horuz and B. Kurt, "Absorption heat transformers and an industrial application," Renewable Energy, vol. 35, no. 10, pp. 2175–2181, 2010.
3. P. Holmberg and T. Berntsson, "Alternative, working fluids in heat transformers," Ashrae Transactions, vol. 96, part 1, pp. 1582–1589, 1990, Technical and Symposium Papers Presented at the 1990 Winter Meeting.
4. J. Sun, L. Fu, and S. G. Zhang, "A review of working fluids of absorption cycles," Renewable and Sustainable Energy Reviews, vol. 16, no. 4, pp. 1899–1906, 2012.
5. R. D. Rogers and K. R. Seddon, "Ionic liquids—solvents of the future?" Science, vol. 302, no. 5646, pp. 792–793, 2003.
6. K. R. Seddon, "Ionic liquids: a taste of the future," Nature Materials, vol. 2, no. 6, pp. 363–365, 2003.
7. S. Kim, Y. J. Kim, Y. K. Joshi, A. G. Fedorov, and P. A. Kohl, "Absorption heat pump/refrigeration system utilizing ionic liquid and hydrofluorocarbon refrigerants," Journal of Electronic Packaging, vol. 134, no. 3, Article ID 031009, 9 pages, 2012.

8. K. N. Marsh, J. A. Boxall, and R. Lichtenthaler, "Room temperature ionic liquids and their mixtures—a review," Fluid Phase Equilibria, vol. 219, no. 1, pp. 93–98, 2004.
9. M. E. van Valkenburg, R. L. Vaughn, M. Williams, and J. S. Wilkes, "Thermochemistry of ionic liquid heat-transfer fluids," Thermochimica Acta, vol. 425, no. 1-2, pp. 181–188, 2005.
10. X. D. Zhang and D. P. Hu, "Performance simulation of the absorption chiller using water and ionic liquid 1-ethyl-3-methylimidazolium dimethylphosphate as the working pair," Applied Thermal Engineering, vol. 31, no. 16, pp. 3316–3321, 2011.
11. S. Q. Liang, J. Zhao, L. Wang, and X. L. Huai, "Absorption refrigeration cycle utilizing a new working pair of ionic liquid type," Journal of Engineering Thermophysics, vol. 31, no. 10, pp. 1627–1630, 2010.
12. S. Q. Liang, W. Chen, K. Cheng, et al., "The latent application of ionic liquids in absorption refrigeration," in Applications of Ionic Liquids in Science and Technology, P. S. Handy, Ed., 2011.
13. Y. J. Kim, S. Kim, Y. K. Joshi, A. G. Fedorov, and P. A. Kohl, "Waste-heat driven miniature absorption refrigeration system using ionic-liquid as a working fluid," in Proceedings of the ASME 5th International Conference on Energy Sustainability, pp. 1299–1305, Washington, DC, USA, August 2011.
14. Y. J. Kim, S. Kim, Y. K. Joshi, A. G. Fedorovc, and P. A. Kohl, "Thermodynamic analysis of an absorption refrigeration system with ionic-liquid/refrigerant mixture as a working fluid," Energy, vol. 44, no. 1, pp. 1005–1016, 2012.
15. W. Chen, S. Q. Liang, Y. X. Guo, K. Cheng, X. Gui, and D. Tang, "Thermodynamic performances of mmim DMP/Methanol absorption refrigeration," Journal of Thermal Science, vol. 21, no. 6, pp. 557–563, 2012.
16. D. Kerlé, R. Ludwig, A. Geiger, and D. Paschek, "Temperature dependence of the solubility of carbon dioxide in imidazolium-based ionic liquids," Journal of Physical Chemistry B, vol. 113, no. 38, pp. 12727–12735, 2009.
17. W. Ren and A. M. Scurto, "Phase equilibria of imidazolium ionic liquids and the refrigerant gas, 1,1,1,2-tetrafluoroethane (R-134a)," Fluid Phase Equilibria, vol. 286, no. 1, pp. 1–7, 2009.
18. Z. B. He, Z. C. Zhao, X. D. Zhang, and H. Feng, "Thermodynamic properties of new heat pump working pairs: 1,3-dimethylimidazolium dimethylphosphate and water, ethanol and methanol," Fluid Phase Equilibria, vol. 298, no. 1, pp. 83–91, 2010.
19. G. Zuo, Z. Zhao, S. Yan, and X. Zhang, "Thermodynamic properties of a new working pair: 1-ethyl-3-methylimidazolium ethylsulfate and water," Chemical Engineering Journal, vol. 156, no. 3, pp. 613–617, 2010.
20. L. Dong, D. X. Zheng, N. Nie, and Y. Li, "Performance prediction of absorption refrigeration cycle based on the measurements of vapor pressure and heat capacity of H2O + DMIM DMP system," Applied Energy, vol. 98, pp. 326–332, 2012.
21. W. H. Cai, M. Sen, and S. Paolucci, "Dynamic modeling of an absorption refrigeration system using ionic liquids," in Proceedings of the ASME International Mechanical Engineering Congress and Exposition (IMECE '07), pp. 227–236, November 2007.

22. M. Ishikawa, H. Kayanuma, and N. Isshiki, "Absorption heat pump using new organic working fluids," in Proceedings of the International Sorption Heat Pump Conference, 1999.
23. K. S. Kim, B. K. Shin, H. Lee, and F. Ziegler, "Refractive index and heat capacity of 1-butyl-3-methylimidazolium bromide and 1-butyl-3-methylimidazolium tetrafluoroborate, and vapor pressure of binary systems for 1-butyl-3-methylimidazolium bromide + trifluoroethanol and 1-butyl-3-methylimidazolium tetrafluoroborate + trifluoroethanol," Fluid Phase Equilibria, vol. 218, no. 2, pp. 215–220, 2004.
24. L. Dong, D. X. Zheng, Z. Wei, and X. H. Wu, "Synthesis of 1,3-dimethylimidazolium chloride and volumetric property investigations of its aqueous solution," International Journal of Thermophysics, vol. 30, no. 5, pp. 1480–1490, 2009.
25. M. I. Karamangil, S. Coskun, O. Kaynakli, and N. Yamankaradeniz, "A simulation study of performance evaluation of single-stage absorption refrigeration system using conventional working fluids and alternatives," Renewable and Sustainable Energy Reviews, vol. 14, no. 7, pp. 1969–1978, 2010.
26. O. Kaynakli and R. Yamankaradeniz, "Thermodynamic analysis of absorption refrigeration system based on entropy generation," Current Science, vol. 92, no. 4, pp. 472–479, 2007.
27. A. Martín and M. D. Bermejo, "Thermodynamic analysis of absorption refrigeration cycles using ionic liquid + supercritical CO2 pairs," Journal of Supercritical Fluids, vol. 55, no. 2, pp. 852–859, 2010.
28. A. Yokozeki and M. B. Shiflett, "Water solubility in ionic liquids and application to absorption cycles," Industrial and Engineering Chemistry Research, vol. 49, no. 19, pp. 9496–9503, 2010.
29. J. Zhao, X. C. Jiang, C. X. Li, and Z. H. Wang, "Vapor pressure measurement for binary and ternary systems containing a phosphoric ionic liquid," Fluid Phase Equilibria, vol. 247, no. 1-2, pp. 190–198, 2006.
30. A. Yokozeki and M. B. Shiflett, "Vapor-liquid equilibria of ammonia + ionic liquid mixtures," Applied Energy, vol. 84, no. 12, pp. 1258–1273, 2007.
31. A. Yokozeki, "Theoretical performances of various refrigerant-absorbent pairs in a vapor-absorption refrigeration cycle by the use of equations of state," Applied Energy, vol. 80, no. 4, pp. 383–399, 2005.
32. A. Yokozeki and M. B. Shiflett, "Ammonia solubilities in room-temperature ionic liquids," Industrial and Engineering Chemistry Research, vol. 46, no. 5, pp. 1605–1610, 2007.
33. C. H. Romich, N. Merkel, K. Schaber, et al., "A comparison between lithiumbromide—water and ionic liquid—water as working solution for absorption refrigeration cycles," in Proceedings of the 23rd IIR International Congress of Refrigeration, Congres International du Froid-International Congress of Refrigeration, pp. 941–948, 2011.
34. A. Kuhn, O. Buchin, M. Seiler, P. Schwab, and F. Ziegler, "Ionic liquids—a promising solution for solar absorption chillers," in Proceedings of the 3rd Solar Air-Conditioning Conference, Sicily, Italy, 2009.
35. M. Radpieler and C. Scweigler, "Experimental investigation of ionic liquid EMIM+EtSO4 as solvent in a single effect cycle with adiabatic absorption and desorption," in Proceedings of the ISHPC, Padova, Italy, 2011.

CHAPTER 12

CO_2 CAPTURE IN IONIC LIQUIDS: A REVIEW OF SOLUBILITIES AND EXPERIMENTAL METHODS

ELENA TORRALBA-CALLEJA, JAMES SKINNER, AND DAVID GUTIÉRREZ-TAUSTE

12.1 INTRODUCTION

In recent years, increasing attention has been paid towards the worldwide climate change. Moreover, the exponential increase of carbon dioxide emissions into the atmosphere from the combustion of fossil fuels, making up the 86% of greenhouse gases [1], does not reflect a sustainable energy model. Entry into the Kyoto protocol has brought about the need to reduce anthropogenic emissions of CO_2. Thus carbon capture and storage (CCS) proves to be one of the most important initiatives to mitigate this global warming effect.

CCS is a concept based on the reduction of CO_2 emissions into the atmosphere from industrial processes, such as ammonia production, natural gas processing, or cement manufacture, to name a few. This review however will focus on CO_2 emissions from fossil fuel power plants, which is seen to be the main contributor to this effect [2]. It has been approximated that, if CCS is fully implemented, its potential by 2050 could be the total

This chapter was originally published under the Creative Commons Attribution License. Torralba-Calleja E, Skinner J, and Gutiérrez-Tauste D. CO_2 Capture in Ionic Liquids: A Review of Solubilities and Experimental Methods. Journal of Chemistry ***2013*** *(2013). http://dx.doi.org/10.1155/2013/473584*

capture and storage of 236 billion tons of CO_2 [3]. An approach to CCS that holds the greatest promise is the sequestration of captured carbon dioxide, in suitable deep sedimentary formations, for example, oil and depleted gas reservoirs, coal beds, and saline deposits [4–7]. The challenge is to develop a technology which will allow us to accomplish this task in an environmental, economic, and efficient way in the next years [8–10]. However the need to assess the environmental impact is great. The potential risks of geological storage to humans and ecosystems are abundant and need to be carefully monitored. Leakage of sequestered CO_2 would be the main concern. This could happen along fault lines, ineffective confining layers, abandoned wells, and so forth. The pollution of groundwater and mineral deposits is also a problem and could have lethal effects on plant life and animals. A recent review by Manchao et al. [11] offers a detailed risk assessment of the CO_2 injection process and storage in geological formations, with a main focus on abandoned coal mines and coal seams.

An alternative to geological storage of CO_2 would be the direct conversion of CO_2 into a high-valued product after the initial capture; this is sometimes referred to as carbon capture and usage (CCU). CO_2 is used in many industries such as the food industry (carbonation of beverages), electronics industry (surface cleaning and semiconductor manufacture),

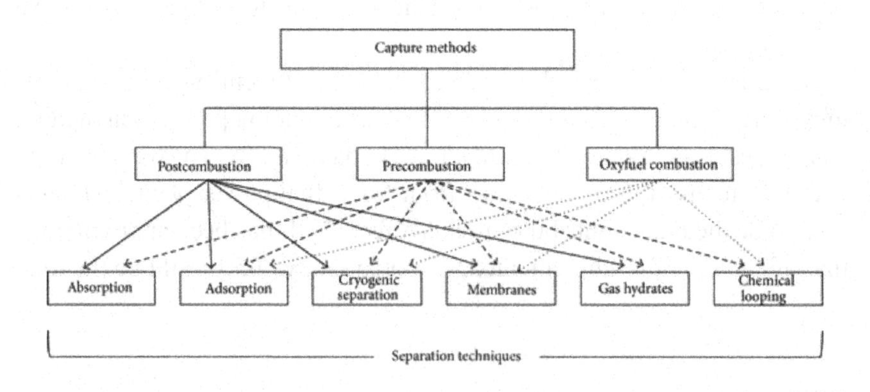

FIGURE 1: Possible techniques that can be used in conjunction with the processes of postcombustion, precombustion, and oxyfuel combustion.

and the chemical industry (polymers, plastics, and fertilizers). CCU is yet to be a mainstream technology so that many process aspects and methods are being published and reviewed [12, 13].

12.2 STATE-OF-THE-ART CO_2 CAPTURE TECHNOLOGIES

The capture of CO_2 is achieved through the use of specific materials that interact with the gas in one form or another. The materials that are used depend on the processes in which the flue gas is conditioned (Figure 1) [14].

There are three processes, each of which conditions the CO_2 for capture in different ways.

- Postcombustion. The separation of CO_2 from the flue gas after the combustion of fuel. Air is typically used as the oxidant in this process; therefore the flue gas becomes largely diluted with nitrogen.
- Precombustion. The hydrocarbon fuel (in this case gasified coal) is converted into carbon monoxide (CO) and hydrogen (H_2). This forms a synthesis gas. By using water shift conversion, CO is converted into CO_2. Finally the CO_2 is then separated from the H_2.
- Oxyfuel CO_2 Combustion. It uses pure oxygen as the oxidant instead of air, creating a flue gas mainly consisting of high-concentrated CO_2 and steam.

Although CO_2 capture and separation is a well-known technology, this technology is just applied in a small scale, so that right now it is not commercially available for being used in large power stations. The most challenging obstacle to overcome in CCS and CCU is finding an effective technique that satisfies environmental and economic factors. Some of the currently studied techniques for capturing CO_2 from the three conditioning processes are as follows (Figure 1).

Absorption occurs within the bulk of the material via a chemical or physical interaction. Chemical absorbents react with the CO_2, forming covalent bonds between the molecules. The solvent can be habitually regenerated through heating and captured CO_2 is released. This mechanism can also be made highly selective by the introduction of specific chemical complexes. Typical compounds used in this process are amines, or ammonia-based solutions. Physical absorbents obey Henry's law, where gas solubility is directly proportional to the partial pressure of the said gas in equilibrium, at a constant temperature. Typically this is at high CO_2 partial pressures and low temperatures. The interaction between CO_2 and the solvent is by nonchemical surface forces, that is, Van der Waals interaction. Regeneration of the solvent is achieved by increasing the temperature and lowering the pressure of the system [15]. Selexol and Rectisol are examples of physical absorbents that have been used in natural gas sweetening and synthesis gas treatment.

Adsorption, as opposed to absorption, takes place at the surface of the material. This interaction can also occur chemically (covalent bonding) or physically (Van der Waals). Typical adsorbers are solid materials with large surface areas, such as zeolites, activated carbons, metal oxides, silica gel, and ion-exchange resins. These can be used to capture CO_2 by separation, so that flue gas is put in contact with a bed of these adsorbers, allowing the CO_2 capture from the other gases which pass through. When the bed is fully saturated with CO_2, the flue gas is directed to a clean bed and the saturated bed is regenerated [16]. Three techniques can be employed to the adsorption mechanism: pressure swing adsorption (PSA) introduces the flue gas at high pressure until the concentration of CO_2 reaches equilibrium, then the pressure is lowered to regenerate the adsorbent, temperature swing adsorption (TSA) increases the temperature to regenerate the adsorbent, and electric swing adsorption (ESA) is where a low-voltage electric current is passed through the sorbent to regenerate. Adsorption is not yet considered practical for large-scale applications as the CO_2 selectivity in current sorbents is low. However, recently new sorbents are being investigated such as metal-organic frameworks and functionalised fibrous matrices that show some promise for the future of this particular technique.

Membrane separation technology is based on the interaction of specific gases with the membrane material by a physical or chemical interaction. Through modifying the material, the rate at which the gases pass through can be controlled. There are wide varieties of membranes available for gas separation, including polymeric membranes, zeolites, and porous inorganic membranes, some of which are used in an industrial scale and have the possibility of being implemented into the process of CO_2 capture. However achieving high degrees of CO_2 separation in one single stage has so far proved to be difficult; therefore, having to rely on multiple stages has led to increasing energy consumption and cost. An alternative approach is to use porous membranes as platforms for absorption and stripping. Here a liquid (typically aqueous amine solutions) provides the selectivity towards the gases. As the flue gas moves through the membrane, the liquid selects and captures the CO_2 [17].

Cryogenic Separation is a technique based on cooling and condensation. This has the advantage of enabling the direct production of liquid CO_2, benefiting transportation options. Although a major disadvantage of cryogenic technology in this respect is the high amounts of energy required to provide cooling for the process, this is especially prominent in low-concentration gas streams [18]. This technique is more suited to high-concentration and high-pressure gases, such as in oxyfuel combustion and precombustion.

Within these techniques lie the materials with which research pathways aim to develop more effective CO_2 capture mechanisms. Currently the postcombustion process is the most widely researched area for reducing CO_2 emissions from power stations. This is mainly because it can be retrofitted to existing combustion systems without a great deal of modification, unlike the other two processes. The flue gas emitted, from the postcombustion of fossil fuels in power stations, has a total pressure of 1-2 bars with a CO_2 concentration of approximately 15%. As this process creates low CO_2 concentration and partial pressures, strong solvents have to be used to capture the CO_2, resulting in a large energy input to regenerate the solvent for further use. This creates the technical challenge of finding an efficient, costeffective, and low-energy-demanding capture mechanism using novel materials.

12.2.1 AQUEOUS AMINES USED IN POSTCOMBUSTION

The conventional technologies used in this postcombustion process are solvent-based chemical absorbers. The common chemical solvents used for separation are aqueous amines, which are ammonia derivatives, where one or more of the hydrogen atoms have been replaced by alkyl groups. Some common amines used in this process are (Table 1) monoethanolamine (MEA) [19], methyldiethanolamine (MDEA) [20], and diethanolamine (DEA) [21]. Aqueous amines are stated as "conventional absorbers"

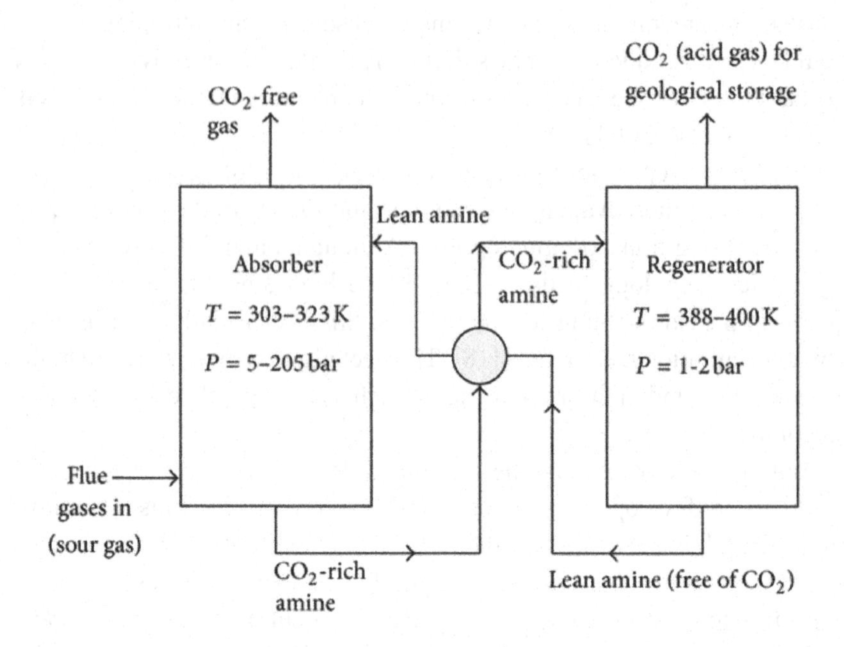

FIGURE 2: Conceptual scheme for CO_2 absorption using amine-based chemical absorption.

because they are well-known solvents used in the oil and gas industries, dating back to the 1930s; for example, Gregory and Scharmann investigated the implementation of amine CO_2 scrubbers in a hydrogenation plant of the Standard Oil Company of Louisiana in 1937. Today the aqueous amine absorption technology is still used in natural gas sweetening (removal of acidic gases, for example, hydrogen sulphide and carbon dioxide) and has also been applied to some small-scale fossil fuel power plants [22, 23], for example, Fundación Ciudad de la Energía (CIUDEN), Alstom power plant, and so forth.

Briefly, post-combustion capture with amines, seen in Figure 2, involves the CO_2 being removed by circulating a flue gas stream into a chamber containing an aqueous amine solution. In the case of primary amines like MEA, the CO_2 is captured by a chemical absorption process in which the CO_2 reacts with the amine in the form of a carbamate [24]. With secondary and tertiary amines, which do not possess a hydrogen atom attached to a nitrogen atom, they react with CO_2 in the form of bicarbonate through hydrolysis. This is a reversible reaction, and at high temperatures the captured CO_2 is released and the amine solution recycled. Piperazine (PIPA) is commonly used to improve reaction kinetics of secondary and tertiary amines in the form of an additive; this is because the heat of reaction to form a bicarbonate is low, causing more heat being needed for regeneration and thus higher costs [25].

TABLE 1: Chemical structures of commonly used amines.

Amine	Acronym	Structure
Monoethanolamine	MEA	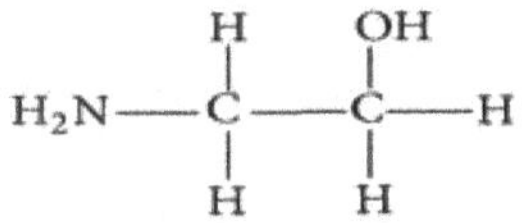 $H_2N-CH_2-CH(OH)-H$
Methyldiethanolamine	MDEA	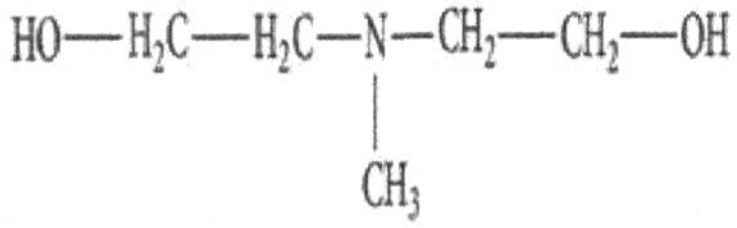 $HO-H_2C-H_2C-N(CH_3)-CH_2-CH_2-OH$
Diethanolamine	DEA	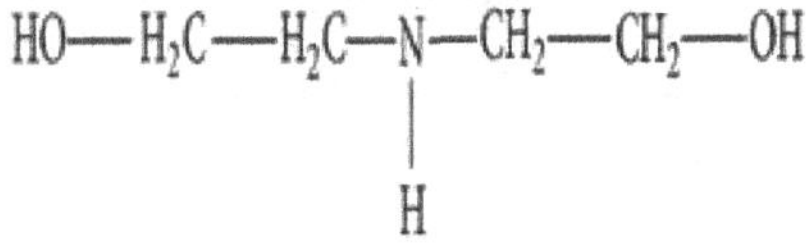 $HO-H_2C-H_2C-N(H)-CH_2-CH_2-OH$
Piperazine	PIPA	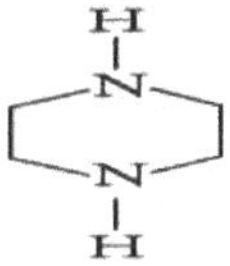 H–N(CH₂CH₂)₂N–H (six-membered ring)

Amines are so effective for CO_2 capture thanks to some of their properties such as high reactivity with CO_2, high absorbing capacity (in terms of mass of CO_2), relatively high thermal stability, and CO_2 selectivity [26]. However there are inherent disadvantages linked with amines, which need to be addressed in order to make a valid and efficient process for CO_2 capture. These disadvantages come in the form of high vapour pressure, corrosive nature, and high-energy input for regeneration. The high vapour pressure allows emission of amine gases into the air upon heating. These gases are unstable in nature thus giving them the possibility of producing dangerous toxins such as nitrosamines, nitramines, and amides. Nitrosamines are of the most concern as they are carcinogenic and toxic to humans even at low levels [27]. Amines are also corrosive, especially MEA. They take part in reactions in which waste forms and can eventually corrode the equipment, Kittel et al. [28] investigated the effects of MEA operating pilot plants and found that areas made of carbon steel had corrosion rates of 1 mm $year^{-1}$; so besides environmental impacts, expense on a large industrial scale is another issue. The recycling/regeneration process leads to high-energy consumption in order to break the chemical bonds formed between the CO_2 and amine [29]. This process also causes degradation of the amine which limits its CO_2 capture rate, causing them to be replaced frequently.

Much research has gone into developing new solvents with the foresight of being superior to amines. The factors that would allow new solvents to perform better than amines are lower cost, lower volatility, better thermal stability, less degradation, low corrosive nature, and low energy needed for regeneration and adaptability to an existing system. Although amines have high CO_2 solubility and selectivity, environmental and economic effects are taken into consideration when selecting the criteria for the most suited CO_2 capture mechanism. While continued research into improving the performance of these mature technologies is expected, research into novel materials and technologies could produce the significant breakthroughs required to minimise the environmental and energy penalties of capture.

12.2.2 IONIC LIQUID MEDIA FOR CO_2 CAPTURE

One of these advanced R&D pathways currently conveying great potential in the field of alternative technologies is ionic liquids (ILs). ILs are commonly defined as materials that are comprised of large organic cations and organic/inorganic anions, which demonstrate melting points below 100°C [30]. To date a wide range of ILs has been synthesised through different combinations of anions and cations. It has been stated that the theoretical number of potential ILs is to the order of 1018 [31]. An example of some of the common cations and anions used in IL synthesis can be seen in Table 2.

TABLE 2: Structures of common IL cations and anions.

Cation	Structure	Anion	Structure
Imidazolium		Tetrafluoroborate	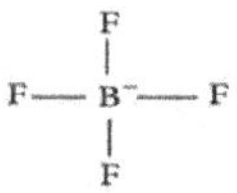
Pyrrolidinium	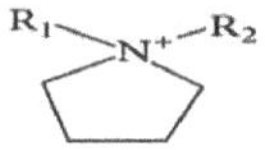	Hexafluoroborate	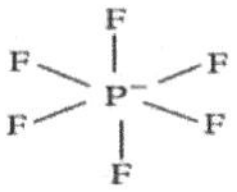
Pyridinium	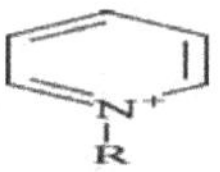	Bis(trifluorophoshate) imide	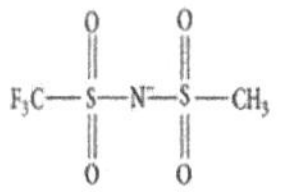
Quaternary Ammonium	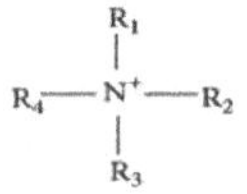	Nitrate	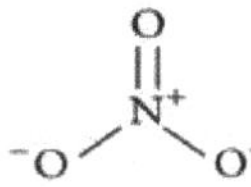
Tetra Alkyl Phosphonium	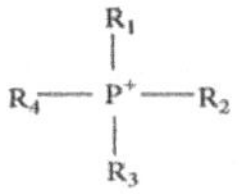	Acetate	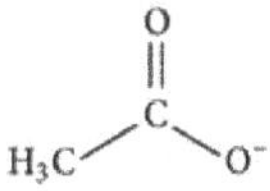

ILs possess several unique and diverse characteristics such as high thermal and chemical stability, low vapour pressure, large electrochemical window, tuneable/designer nature, and excellent solvent properties for a range of polar and nonpolar compounds. It is due to these characteristics that research into developing and implementing ILs over the past decade has spanned into many sectors of industry [32], for applications such as electrolytes [33], solar cells [34], lubricants [35], electropolishing and electroplating [36], and biomass processing [37], to name a few. This has become possible due to the large number of ILs that can be synthesised in the lab [38–40] and purchased commercially.

Companies like BASF, Merck, Sigma-Aldrich, Solvionic, Sachem, and IoLitec provide basic ILs and can also aid in the design and development of ILs for specific tasks. Therefore these compounds have created exciting new media for emerging technological applications already commercially available.

12.2.3 IONIC LIQUIDS IN THE SCOPE OF CO_2 CAPTURE

The aforementioned characteristics are particularly advantageous when applying ILs as solvents for CO_2 capture in comparison to current aqueous amine technology:

1. less energy is required when regenerating ILs to remove the captured CO_2 [41] due to their physical absorption mechanism,
2. further efficiency is attained by their low vapour pressure, which allows them to be regenerated and reused with no appreciable losses into the gas stream [42, 43],
3. ILs have a high thermal and chemical stability; typically they degrade at temperatures >300°C [44] avoiding their reaction with impurities and causing corrosion to the equipment,
4. the tuneable and designer nature of ILs offers many options concerning the physicochemical properties (viscosities and densities [45–47], heat capacities [48], thermal decomposition temperatures [49], surface tension [50], toxicity and health issues [51, 52], and corrosion [53, 54]) in the sense that the anions and cations can be manipulated to create an IL for a specific task.

This designer aspect can also be applied to the anion or cation in the sense that various chemical functionalities and structures can be attached, allowing properties such as absorption and viscosity to be controlled. These are commonly referred to as task-specific ionic liquids (TSILs). Generally ILs fulfil many of the major requirements stated in the green-chemistry principles stated by Anastas and Warner [55], in that they offer a new approach to industrial/chemical processes whereby steps are taken to eliminate hazardous waste in a system before a by-product is formed, thus neglecting the use of volatile organic solvent.

Most development concerning ILs for CO_2 capture is at present conducted at laboratory scale, while other technological applications are already in use as it was mentioned before. Conversely, their industrial application and implementation is being constantly investigated in areas of post-combustion [56]. For industrial-scale integration, it is necessary to achieve extensive knowledge of their physical and chemical properties. Therefore the need for experimental techniques and data is critical in enabling the ionic liquid to be the green, viable, and economic carbon capture technique of the future.

The solubility of CO_2 in ILs compared to other gases such as methane and nitrogen enables ILs to separate CO_2 from the source, be it a power plants' flue gas or natural gas. Even when there are low concentrations of CO_2 in a mixed gas, the IL can be designed to incorporate a functional group, such as an amine, thus rendering it task specific. The capacity for CO_2 solubility in ILs originates from the asymmetrical combination of the anion and cation, which results from short-range repulsive forces between their ionic shells. Therefore the more incompatible the ionic constituents are the greater the solubility is.

12.2.4 CONVENTIONAL IONIC LIQUIDS

Over the past decade, and at present, research has been built upon measuring the effects of variables such as pressure, temperature, and anion/cation choice. Results have shown high carbon dioxide solubility in what have become known as conventional ionic liquids. They are defined as ILs that do not possess an attached functional group and have been reported by many as portraying the typical behaviour of physical solvents [57–59]. This is evident when low-pressure CO_2 (1-2 bars) is put in contact with the IL, resulting in low CO_2 concentrations in the liquid phase. As the increment of pressure increases, typically to up to 100 bar, the concentration of absorbed CO_2 increases. Thus displaying the general characteristics of a physical absorber. As a rule, the solubility of CO_2 in ILs increases with increasing pressure and decreases with increasing temperature. The physical absorption mechanism is a result of the interaction between the CO_2 molecules and the IL, in which the CO_2 occupies the "free space" within the ILs structure through a large quadrupole moment and Van der Waals forces.

12.2.4.1 ANION AND CATION EFFECTS

In order to create an optimal process for capturing CO_2 in ILs, assessment of the essential building blocks, that is, cation/anion combinations, needs to be investigated. Synthesising ILs that encompass CO_2-philic groups on the anion such as carbonyls or fluorines has proven to increase CO_2 capture [60]. In the past decade studies have shown that the origin of high solubility is strongly dependent on the choice of anion [61]. Aki et al. [62] investigated the influence of the anion with seven ILs. They all contained the 1-butyl-3-methylimidazolium [Bmim] cation. The results are shown in Table 3.

TABLE 3: Influence of anions in different ionic liquids.

Anion	Nomenclature	Classification	Solubility of CO_2 in IL
Dicyanamide	$[DCA]^-$	Nonfluorinated anions	Low
Nitrate	$[NO_3]^-$		
Tetraflouroborate	$[BF_4]^-$		
Hexafluorophosphate	$[PF_6]^-$		
Trifuoromethanesulfonate	$[TfO]^-$	Fluorinated anions	Relatively high
Bis(trifluoromethylsulfonyl)imide	$[Tf_2N]^-$		
Tris(trifluoromethylsulfonyl)methide	$[methide]^-$		

Aki and coworkers also systematically investigated the effects of the cation on CO_2 solubility; they found that, in general, the increase of the alkyl chain on the cation resulted in a slight increase in solubility, which became more apparent at higher pressures. The effect of increasing the alkyl chain results in the increased volume available for CO_2 interaction. Muldoon et al. [60] concluded that adding partially fluorinated alkyl chains on the imidazolium cation does increase CO_2 solubility. They compared [hmim][Tf_2N] directly to [$C_6H_4F_9$mim][Tf_2N] and found that this increased solubility was due to fluorinating the last four carbons of the alkyl chain. Research on IL CO_2 solubility, in general, has focused intensively on imidazolium-based structures. However some groups have focused on using different cations. Recently Carvalho et al. [63] reported CO_2 solubilities in two phosphonium-based ILs, [THTDP][Tf_2N] and [THTDP][Cl]. They found exceptionally high solubility measurements exceeding those of current imidazolium-based ILs; they go on to conclude that their study shows the highest recorded solubility observed without chemical interactions in the absorption process. Although imidazolium is the most stable and commercially available cation of choice, it is evident that there are further enhancements and possibilities that can be developed from other bases.

To provide further insight into the interactions between CO_2 and the constituent anions and cations of RTILs, researches using spectroscopic approaches and molecular simulations have been made. Of which has broadened our understanding of absorption mechanisms and structure-property relationships, Kazarian et al. [64] used ATR-FTIR spectroscopy to analyse the specific interactions of CO_2 and ILs [Bmim][BF4] and [Bmim][PF6]. They saw evidence of chemical interactions between the anion $[PF6]^-$ and CO_2. They concluded that they observed weak Lewis acid-base interactions, where the anion acts as a Lewis base. ILs by their nature have intrinsic acid-base properties. These properties can be enhanced with the addition of acidic functions like carbonic or halide acids; likewise, basic functions like amino and fluorine groups can be added. This has shown to create specific Lewis acid-base chemical interactions between CO_2 and the IL.

As it can be seen in Tables 4 and 5, fluorination of the anion and in some cases the cation can improve CO_2 solubility in RTILs. However the associated disadvantages are cost increase, poor degradability, and a negative environmental impact [65]. Therefore paths to develop ILs with enhanced CO_2 solubility without fluorination are also being investigated.

TABLE 4: CO_2 solubility data for imidazolium-based ionic liquids.

Ionic liquid	Acronym	(K)	(bar)	χCO_2	References
1-N-Octyl-3-methylimidazolium hexafluorophosphate	C8mim[PF_6]	313	92.67	0.7550	Blanchard et al. 2001 [86]
1-N-Butyl-3-nethylimidazolium nitrate	Bmim[NO_3]	323	92.62	0.5300	Blanchard et al. 2001 [86]
1-N-Octyl-3-methylimidazolium tetrafluoroborate	C8mim[BF_4]	313	92.90	0.7080	Blanchard et al. 2001 [86]
1-Ethyl-3-methylimidazolium ethyl sulfate	EmimEt[SO_4]	333	94.61	0.4570	Blanchard et al. 2001 [86]
1-Butyl,3-methyl-imidazolium hexafluorophosphate	Bmim[PF_6]	313	96.67	0.7290	Blanchard et al. 2001 [86]
1-Butyl-3-methylimidazolium acetate	C_4mim[Ac]	333.3	12.75	0.2510	Carvalho et al. 2009 [87]
		323.09	755.26	0.5990	Carvalho et al. 2009 [87]
1-Butyl-3-methylimidazolium trifluoroacetate	C_4mim[TFA]	293.43	9.79	0.2250	Carvalho et al. 2009 [87]
		293.59	436.25	0.6790	Carvalho et al. 2009 [87]
1-Butyl,3-methyl-imidazolium tetrafluoroborate	Bmim[BF_4]	303	10	0.1461	Galan-Sanchez 2008 [88]
		333	10	0.0895	Galan-Sanchez 2008 [88]
1-Octyl,3-methyl-imidazolium tetrafluoroborate	Omim[BF_4]	303	10	0.1873	Galan-Sanchez 2008 [88]
		333	10	0.1213	Galan-Sanchez 2008 [88]
1-Butyl,3-methyl-imidazolium dicyanamide	Bmim[DCA]	303	10	0.1434	Galan-Sanchez 2008 [88]
		333	10	0.0997	Galan-Sanchez 2008 [88]
1-Butyl-3-methylimidazolium thiocyanate	Bmim[SCN]	303	10	0.0978	Galan-Sanchez 2008 [88]
		333	10	0.0664	Galan-Sanchez 2008 [88]
1-Butyl,3-methyl-imidazolium hexafluorophosphate	Bmim[PF_6]	303	10	0.1662	Galan-Sanchez 2008 [88]
1-Butyl,3-methyl-imidazolium hexafluorophosphate	Bmim[PF_6]	333	10	0.1012	Galan-Sanchez 2008 [88]

TABLE 4: *Cont.*

Ionic liquid	Acronym	(K)	(bar)	χCO_2	References
1-Butyl-3-methylimidazolium methylsulfate	Bmim[MeSO$_{4]}$	303	10	0.1190	Galan-Sanchez 2008 [88]
		333	10	0.0733	Galan-Sanchez 2008 [88]
1-N-Ethyl-3-mehylimidazolium bis(trifluoromethylsulfonyl) Imide	Emim[NTf2]	303	10	0.2257	Galan-Sanchez 2008 [88]
		333	10	0.1446	Galan-Sanchez 2008 [88]
1-Butyl,3-methyl-imidazolium hexafluorophosphate	Bmim[PF_6]	298.15	6.66	0.122	Kim et al. 2005 [89]
1-Hexyl-3-methylimidazolium hexafluorophosphate	C_6mim[PF_6]	298.15	9.27	0.167	Kim et al. 2005 [89]
1-Ethyl-3-methylimidazolium tetrafluoroborate	Emim[BF_4]	298.15	8.75	0.106	Kim et al. 2005 [89]
1-Hexyl-3-methylimidazolium tetrafluoroborate	C6mim[BF_4]	298.15	8.99	0.163	Kim et al. 2005 [89]
1-Ethyl-3-methylimidazolium bis(trifluoromethylsulfonyl) imide	Emim[Tf_2N]	298.15	9.03	0.209	Kim et al. 2005 [89]
1-Hexyl-3-methylimidazolium bis(trifluoromethylsulfonyl) imide	C_6mim[Tf_2N]	298.15	8.59	0.236	Kim et al. 2005 [89]
1-Ethyl-3-methylimidazolium trifluoromethanesulfonate	C_2mim[TfO]	303.85	149	0.6260	Shin and Lee 2008 [90]
		303.85	15	0.2610	Shin and Lee 2008 [90]
1-Butyl-3-methylimidazolium trifluoromethanesulfonate	C_4mim[TfO]	303.85	160	0.6720	Shin and Lee 2008 [90]
		303.85	11.5	0.2730	Shin and Lee 2008 [90]
1-Hexyl-3-methylimidazolium trifluoromethanesulfonate	C_6mim[TfO]	303.85	180	0.7170	Shin and Lee 2008 [90]
		303.85	12.5	0.2880	Shin and Lee 2008 [90]

TABLE 4: *Cont.*

Ionic liquid	Acronym	(K)	(bar)	χCO_2	References
1-Octyl-3-methylimidazolium trifluoromethanesulfonate	C_8mim[TfO]	303.85	180	0.7410	Shin and Lee 2008 [90]
		303.85	15.8	0.3440	Shin and Lee 2008 [90]
1,3-Dimethylimidazolium methylphosphonate	Dmim[MP]	313.35	95	0.4750	Revelli et al. 2010 [91]
		313.45	34	0.1620	Revelli et al. 2010 [91]
1-Butyl,3-methyl-imidazolium tetrafluroborate	Bmim[BF_4]	293.65	73	0.6100	Revelli et al. 2010 [91]
		293.25	10.5	0.1410	Revelli et al. 2010 [91]
1-Butyl-3-mcthylimidazolium thiocyanate	Bmim[SCN]	313.65	99	0.4300	Revelli et al. 2010 [91]
		292.35	10.5	0.1260	Revelli et al. 2010 [91]
1-Ethyl-3-methylimidazolium trifluoroacetate	Emim[TFA]	298.1	19.99	0.2820	Yokozeki et al. 2008 [67]
1-Ethyl-3-methylimidazolium acetate	Emim[Ac]	298.1	19.99	0.4280	Yokozeki et al. 2008 [67]
1-Butyl-3-methylimidazolium trifluoroacetate	Bmim[TFA]	298.1	19.99	0.3010	Yokozeki et al. 2008 [67]
1-Butyl-3-methylimidazolium acetate	Bmim[Ac]	298.1	19.99	0.4550	Yokozeki et al. 2008 [67]
1-Ethyl-3-methyl-imidazolium bis(trifluoromethylsulfonyl) imide	Emim[Tf_2N]	298.1	19.99	0.3900	Yokozeki et al. 2008 [67]
1-Hexyl-3-methylimidazolium tris(pentafluoroethyl) trifluoro-phosphate	Hmim[FAP]	298.1	19.99	0.4930	Yokozeki et al. 2008 [67]
1-Hexyl-3-methylimidazolium bis(trifluoromethylsulfonyl) imide	Hmim[Tf_2N]	298.1	19.74	0.4330	Yokozeki et al. 2008 [67]
1-Butyl-3-methylimidazolium 1,1,2,2-tetrafluoroethanesulfonate	Bmim[TFES]	298	19.9	0.2850	Yokozeki et al. 2008 [67]
1-Butyl-3-methylimidazolium propionate	Bmim[PRO]	298.2	19.9	0.3900	Yokozeki et al. 2008 [67]

TABLE 4: *Cont.*

Ionic liquid	Acronym	(K)	(bar)	χCO_2	References
1-Butyl-3-methylimidazolium isobutyrate	Bmim[ISB]	298.2	20	0.4030	Yokozeki et al. 2008 [67]
1-Butyl-3-methylimidazolium trimethylacetate	Bmim[TMA]	298.1	19.9	0.4310	Yokozeki et al. 2008 [67]
1-Butyl-3-methylimidazolium levulinate	Bmim[LEV]	298.1	19.9	0.4600	Yokozeki et al. 2008 [67]
1-Butyl-3-methylimidazolium succinamate	Bmim[SUC]	298.1	19.9	0.2320	Yokozeki et al. 2008 [67]
Bis(1-butyl-3-methylimidazolium) iminodiacetate	Bmim[2IDA]	298.1	19.9	0.3950	Yokozeki et al. 2008 [67]
1-Butyl-3-methylimidazolium iminoacetic acid acetate	Bmim[IAAc]	298.1	19.9	0.1910	Yokozeki et al. 2008 [67]
1-Hexyl-3-methylimidazolium tris(pentafluoroethyl) trifluorophosphate	Hmim[FEP]	283.5	17.99	0.5170	Zhang et al. 2008 [92]

TABLE 5: CO_2 solubility for ammonium ionic liquids.

Ionic liquid	Acronym	(K)	(bar)	χCO_2	References
Bis(2-hydroxyethyl)-ammonium acetate	(BHEAA)	298.15	15.15	0.1076	Kurnia et al. 2009 [78]
		298.15	5.48	0.0391	Kurnia et al. 2009 [78]
2-Hydroxy-N-(2-hydroxyethyl)-N-methylethanaminium acetate	(HHEMEA)	298.15	15.42	0.0761	Kurnia et al. 2009 [78]
		298.15	6.15	0.0300	Kurnia et al. 2009 [78]
Bis(2-hydroxyethyl)-ammonium lactate	(BHEAL)	298.15	15.12	0.0835	Kurnia et al. 2009 [78]
		298.15	3.46	0.0192	Kurnia et al. 2009 [78]
2-Hydroxy-N-(2-hydroxyethyl)-N-methylethanaminium lactate	(HHEMEL)	298.15	15.23	0.0776	Kurnia et al. 2009 [78]
		298.15	3.48	0.0179	Kurnia et al. 2009 [78]
2-Hydroxy ethyl ammonium formate	(HEF)	303	78.9	0.3083	Yuan et al. 2007 [93]
		303	4.4	0.0340	Yuan et al. 2007 [93]

TABLE 5: *Cont.*

Ionic liquid	Acronym	(K)	(bar)	χCO_2	References
2-Hydroxy ethyl ammonium acetate	(HEA)	303	90.1	0.4009	Yuan et al. 2007 [93]
		303	8.9	0.0687	Yuan et al. 2007 [93]
2-Hydroxy ethyl ammonium lactate	(HEL)	303	82	0.2422	Yuan et al. 2007 [93]
		303	7.8	0.0410	Yuan et al. 2007 [93]
Tri-(2-hydroxyethyl)-ammonium acetate	(THEAA)	303	82.5	0.2561	Yuan et al. 2007 [93]
		303	10.3	0.0534	Yuan et al. 2007 [93]
Tri-(2-hydroxyethyl)-ammonium lactate	(THEAL)	303	70.9	0.4617	Yuan et al. 2007 [93]
		303	9.6	0.1006	Yuan et al. 2007 [93]
2-(2-Hydroxyethoxy)-ammonium formate	(HEAF)	303	72.8	0.1907	Yuan et al. 2007 [93]
		303	6.6	0.0300	Yuan et al. 2007 [93]
2-(2-Hydroxyethoxy)-ammonium acetate	(HEAA)	303	65.7	0.4860	Yuan et al. 2007 [93]
		303	7.6	0.0889	Yuan et al. 2007 [93]
2-(2-Hydroxyethoxy)-ammonium lactate	(HEAL)	303	73.2	0.2640	Yuan et al. 2007 [93]
2-(2-Hydroxyethoxy)-ammonium lactate	(HEAL)	303	12.4	0.0704	Yuan et al. 2007 [93]

Due to certain limitations of conventional ionic liquid systems, where physical absorption takes place and high solubility is only seen at high pressures, numerous research groups have been developing the ILs designer character, by covalently tethering a functional group to either or both anion or cation. This resulting functionalized IL is capable of chemically binding to CO_2, adding chemical absorption to the capture mechanism.

[Bmim][Ac] has been found to be one of these RTILs in which a chemical complexion with CO_2 occurs [66]. In 2008, Yokozeki et al. [67] completed

CO_2 solubility tests for 18 RTILs, eight of which showed chemical absorption mechanisms. They found that RTILs that show strong chemical absorption with CO_2 all contain the anion $[X\text{-}COO]^-$, that is, [Bmim][Ac], [Emim][Ac], [Bmim][PRO], [Bmim][IBS], [Bmim][TMA], and [Bmim][LEU]. Their results can be seen in Table 4. In general it is assumed that conventional RTILs with acidic or basic functionalities strongly influence the absorption of CO_2.

As discussed, RTILs sufficiently absorb CO_2 especially those containing CO_2-philic groups like fluorine. These are known as TSILs (task-specific ionic liquids). Widely researched TSILs are those with appended amine group, examples of which can be seen in Table 5. Bates and coworkers [68] synthesized the amine functionalized IL [pNH_2Bim][Pf_6] and found it to chemically react with the CO_2. The CO_2 reacts with the amine on the IL, this then reacts with another amine and forms an ammonium carbamate double salt. This form of capture results in one CO_2 captured for every two ILs. This 1 : 2 capture mechanism is also observed on the molecular level with traditional aqueous amines. It is theoretically suggested that, when amines are tethered to the anion only, a 1 : 1 ratio can be met allowing a more efficient process.

Evidence has shown that TSILs have the ability to absorb CO_2 both chemically and physically. At low pressures (typically below 2 bars) chemical absorption takes place, in the same way as aqueous amines. After the majority of the chemical bonding have taken place, physical absorption dominates the capture mechanism; this is especially relevant at high pressures, whereas aqueous amines reach their absorption limits at low pressure. This shows how the absorption performances of TSILs with amine functionalities merge the characteristics of physical solvents with the attractive features of chemical solvents. In spite of TSILs showing greater CO_2 solubility than conventional RTILs, they tend to exhibit high viscosity in comparison to other commercially available absorbents. This poses a large problem for their implementation into large-scale platforms, as the heat required for absorption and regeneration would be a lot larger and energy intensive. In order to reduce the viscosity, some groups have combined mixtures of TSIL and RTIL. Bara et al. [69] dissolved their TSIL in a common RTIL, [C_6mim][Tf_2N]. Although the solution was stable and capable of absorbing in a 1 : 2 molar ratio, the viscosity was still high. As a whole TSILs and TSILs + RTILs are robust and have a high absorption capacity; however, they are limited by the intensive synthesis that is required, high viscosity, and the fact that the TSIL serves as both the capture material and the dispersant.

Instead of the direct incorporation of amino-functionalized anions and cations, some recent groups have reported using imidazolium-based RTILs with amines added in solution to act as the capture reagent. Camper and coworkers [70] first investigated this concept. They synthesized an [Rmim][Tf_2N] RTIL solution containing 16% v/v of MEA and found that this is capable of rapid and reversible capture of one mole of CO_2 per two moles of MEA at low CO_2 partial pressures. An MEA-carbamate was found to precipitate from the RTIL solution; this helps to drive the capture reaction. They have currently seen that this MEA-carbamate seems to be a consequence of the [Tf_2N] anion and does not occur in other [mim][X] RTILs.

TABLE 6: CO_2 solubility for phosphonium, pyridinium and pyrrolidinium ionic liquids.

Ionic liquid	Acronym	(K)	(bar)	χCO_2	References
N-Butylpyridinium tetrafluoroborate	N-BuPy[BF_4]	323	92.35	0.5810	Blanchard et al. 2001 [86]
Trihexyltetradecylphosphonium chloride	THTDP[Cl]	302.55	149.95	0.8000	Carvalho et al. 2010 [94]
		313.27	5.17	0.2000	Carvalho et al. 2010 [94]
Trihexyltetradecylphosphonium bis(trifluoromethylsulfonyl) imide	THTDP[NTf_2]	296.58	721.85	0.8790	Carvalho et al. 2010 [94]
		293.2	6.12	0.3080	Carvalho et al. 2010 [94]
N-Butyl-4-methylpyridinium tetrafluoroborate	MeBuPy[BF_4]	303	10	0.1443	Galan-Sanchez 2008 [88]
		333	10	0.0961	Galan-Sanchez 2008 [88]
N-Butyl-3-Methylpyridinium dicyanamide	MeBuPy[DCA]	303	10	0.1436	Galan-Sanchez 2008 [88]
		333	10	0.0683	Galan-Sanchez 2008 [88]

TABLE 6: *Cont.*

Ionic liquid	Acronym	(K)	(bar)	χCO_2	References
N-Butyl-4-Methylpyridinium thiocyanate	MeBuPy[SCN]	303	10	0.0962	Galan-Sanchez 2008 [88]
		333	10	0.0632	Galan-Sanchez 2008 [88]
1-Butyl-1-methylpyrrolidinium dicyanamide	MeBuPyrr[DCA]	303	10	0.1204	Galan-Sanchez 2008 [88]
		333	10	0.0613	Galan-Sanchez 2008 [88]
1-Butyl-1-Methylpyrrolidinium thiocyanate	MeBuPyrr[SCN]	303	10	0.0971	Galan-Sanchez 2008 [88]
		333	10	0.0608	Galan-Sanchez 2008 [88]
1-Butyl-1-methylpyrrolidinium trifluoroacetate	MeBuPyrr[TFA]	303	10	0.1674	Galan-Sanchez 2008 [88]
		333	10	0.1030	Galan-Sanchez 2008 [88]
Tetrabutylphosphonium formate	TBP[FOR]	298.1	19.9	0.3480	Yokozeki et al. 2008 [67]
1-Butyl-1-methylpyrrolidinium tri(pentafluoroethyl)trifluoro-phosphate	BmPyrr[FEP]	283.5	18.00	0.4980	Zhang et al. 2008 [92]

12.2.5 CO_2 SOLUBILITY RESULTS REPORTED BY VARIOUS EXPERIMENTAL GROUPS

Tables 4, 5, 6, and 7 aim to provide a range of experimental data cited by various experimental groups, for peak CO_2 absorption values for different cation-based ILs. This can then be used to characterise an experimental system to ensure correct implementation and method. (K) represents the system's temperature when measurements were recorded. (bar) is the corresponding pressure of CO_2. XCO_2 is the solubility of CO_2 expressed as a mole fraction, that is, moles of CO_2 to moles of IL. The tables also attempt to show the effects of temperature and pressure on ILs as well as different cation and anion combinations.

TABLE 7: CO_2 Solubility data for functionalized ionic liquids (TSILs).

Acronym	Functionalization	Anion	Cation	(K)	(bar)	χCO_2	References
APMim[NTf_2]	NH_2-cation	NTf2	Im	303	10.00	0.27	Galan-Sanchez 2008 [88]
				343	10.00	0.18	
APMim[DCA]	NH_2-cation	DCA	Im	303	10.00	0.29	Galan-Sanchez 2008 [88]
APMim[BF_4]	NH_2-cation	BF4	Im	303	10.00	0.32	Galan-Sanchez 2008 [88]
				343	10.00	0.36	
AEMPyrr[BF_4]	NH_2-cation	BF4	Pyrr	303	10.00	0.28	Galan-Sanchez 2008 [88]
				333	10.00	0.24	
MeImNet2[BF_4]	NR_3-cation	BF4	Im	303	4.00	0.09	Galan-Sanchez 2008 [88]
Bmim[Tau]	NH_2-Anion	Taureate	Im	333	10.00	0.43	Galan-Sanchez 2008 [88]
Bmim[Gly]	NH_2-Anion	Glycinate	Im	333	10.00	0.39	Galan-Sanchez 2008 [88]

12.3 EXPERIMENTAL AND MEASUREMENT TECHNIQUES

In order to integrate CO_2 separation techniques into large industrial systems, one needs to experimentally determine the ILs gas solubility in order to characterise the carrying capacity and selectiveness. These measurements can be accumulated via a number of experimental techniques, in which factors such as pressure and temperature can be controlled. The variety of techniques used for measuring solubility for high- and low-pressure phase equilibrium is vast and the naming of these techniques tends to vary from author to another. However all the techniques fall into two categories both of which are dependent on the equilibrium phases and mixture composition. If these two factors are unknown, measurements can be carried out analytically (analytical method); if the mixture is prepared with a precisely known composition, the synthetic method can be used. The experimental and measurement techniques reviewed here are gravimetric analysis, pressure drop method, and view-cell method and gas chromatography. All of which are being specifically applied to pure CO_2 solubility in ionic liquids. It is important to remember that impurities can occur in the gas and liquid, affecting the accuracy and precision of the results. Therefore degassing the liquid fully before analysis allows an accurate determination of the true solubility of the gas. This also relies on allowing true equilibrium conditions to be met between the gas and the liquid.

12.3.1 GRAVIMETRIC ANALYSIS

Gravimetric analysis is an analytical method which describes the quantitative determination of, in this case, gas solubility by measuring the overall

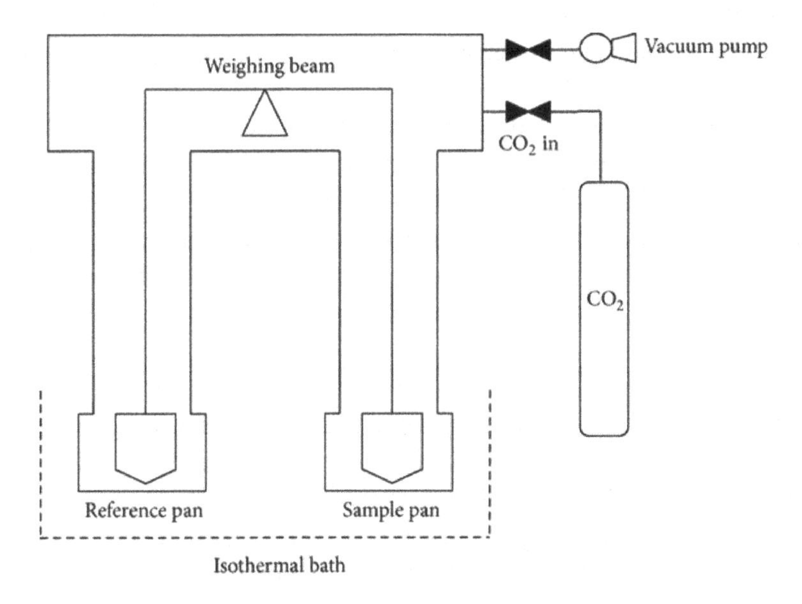

FIGURE 3: Basic components of a gravimetric system.

weight change of a sample during absorption. The gravimetric method is most commonly applied when the analyte is converted into a solid; however, as ILs are nonvolatile in nature and exhibit properties such as low vapour pressure, they can be used to a great effect with this method of analysis. Gravimetric gas analysers are used in laboratories conducting both fundamental studies into the physical properties of ILs and applications where the ability to measure gas solubility is of interest. The basic components of a typical gravimetric instrument can be seen in Figure 3.

High-precision gravimetric instruments are commercially available, in the form of thermogravimetric microbalances. These allow in situ measurements of gas absorption that record the mass gain of a sample with a high-precision electrobalance, which is capable of taking readings at high temperature and pressure. Also available are analysers which use magnetic suspension balances rather than an electrobalance [71]. The main difference between these two weighing systems is that, in magnetic suspension, the sample is weighed from the outside. Therefore the balance is not in physical contact with the high temperature and pressures subjected to the sample. This particular system is helpful when working with samples under extreme conditions. Petermann et al. [72] show the use and experimental setup of a magnetic suspension balance in conjunction with a volumetric determination method. The advantages of using magnetic suspension balances are also discussed by Dreisbach and Lösch [73].

Gravimetric analysis systems often measure gas solubility by recording isotherms, isobars, and kinetic sorption data, which can be output through a computer from which the system can be controlled. Hence when a sample is loaded, the operation of the instrument can be fully automated and programmed to carry out isothermal absorption and desorption measurements.

Due to gravimetric balances undergoing constant changes in temperature and pressure during measurements and the high sensitivity in which they operate, readings must be corrected for the changes in buoyant forces on the sample. In some apparatus, a counterweight side, which is symmetrical to the sample side, is used to minimise these effects. However they still need to be considered. Liu et al. [74] show a concise approach to calculate this.

A detailed experimental procedure using the gravimetric balance can be seen in [67]. Also measurements of CO_2 solubility for two imidazolium-based ILs using a thermogravimetric microbalance can be found in [75].

12.3.2 THE PRESSURE DROP METHOD

The pressure drop method is a synthetic technique that is widely used in this scientific community and is also known as the isochoric method. In this instance the volume of the system is held constant, as well as the temperature, and the pressure difference is recorded during gas absorption into the sample. This method for working out gas solubility is practically suited for ILs as they have negligible vapour pressure, therefore ensuring that the gas phase remains pure, and therefore the assumption can be made that changes in pressure are due to gas sorption. From an initial measurement of pressure, temperature, and volume, and a final measurement of these variables at equilibrium, the amount of gas absorbed by the IL can

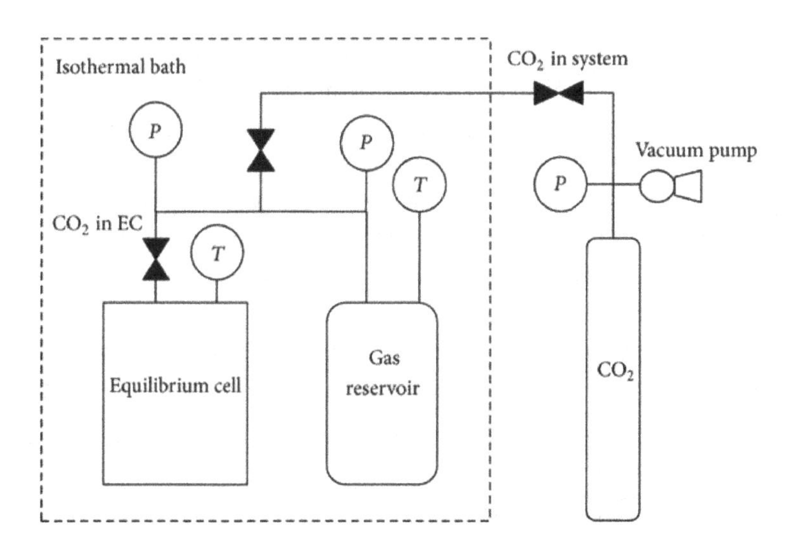

FIGURE 4: The pressure drop apparatus, where P and T correspond to pressure and temperature sensors.

be calculated. This calculation can be performed using an equation of state to convert all three variables into moles of gas.

The basic principles of this method are as follows: CO_2 gas is transferred into a reservoir of known volume and brought to a constant system temperature. An initial reading of pressure is measured. By using a PVT relation, the moles of CO_2 in the reservoir are calculated. The IL is loaded into an equilibrium cell/stainless steel reactor and equalized to system temperature. The CO_2 is then introduced to the ionic liquid and the pressure drop is recorded when the cell's pressure remains stable; this is the equilibrium point. From the pressure drop measured, the number of moles of CO_2 left in the gas phase can be calculated. The difference between CO_2 mole values corresponds to the amount of gas absorbed in the IL. A typical setup for the pressure drop method can be seen in Figure 4.

The moles of dissolved CO_2 in the ionic liquid can be calculated by (1).

Number of CO_2 moles dissolved in the ionic liquid

$$n_{CO_2} = \frac{P_{initial} V_{GR}}{Z_{CO_2}(P_{initial}, T_{initial}) R T_{initial}} - \frac{P_{eq}(V_{tot} - V_{IL})}{Z_{CO_2}(P_{eq}, T_{eq}) R T_{eq}} \quad (1)$$

$P_{initial}$ and $T_{initial}$ are the initial pressure and temperature in the gas reservoir. P_{eq} and T_{eq} are the pressure and temperature at equilibrium in the equilibrium cell. V_{tot} is the total volume of the entire apparatus. V_{IL} is the volume of the ionic liquid, assumed to be constant. R is the ideal gas constant. Z_{CO2} is the compressibility factor for CO_2; this modifies the ideal gas to account for real gas behaviour. A detailed experimental procedure and full calculations for CO_2 solubility measurements using the pressure drop method can be seen in [76, 77].

Further investigations that utilize this pressure drop method to derive gas solubility can be found where alternative experimental setups are shown [78–80].

12.3.3 VIEW-CELL METHODS

These involve the preparation of a mixture with a precisely known composition and then the observation of phase behavior inside an equilibrium cell, where measurements are recorded in the equilibrium state, that is, temperature and pressure. Synthetic methods consist of two main techniques, one being with a phase transition, and the other without. In synthetic methods with a phase transition a known amount of gas and IL is loaded into the equilibrium cell. The pressure is then varied at a constant temperature (or vice versa) until a second phase is formed, where the gas dissolves in the ionic liquid causing the vapor phase to diminish, whereby using different gas pressures, solubility can be worked out at various pressure, and temperatures.

In synthetic methods without a phase transition, equilibrium properties like temperature, pressure, density, cell volume, and gas/liquid phase volumes are measured, and the composition of the phase mixtures can be calculated in terms of moles or by a mass balance equation.

As can be seen in Figure 5, a pump releases CO_2 at a constant selected pressure and monitors the volume of CO_2 flowing into the system. The CO_2 is also heated to a constant temperature. By monitoring the volume, a known amount of CO_2 is then introduced to the high-pressure view cell, which contains a known amount of IL. In the case of non-phase transition, the amount of CO_2 absorbed is calculated by the difference in the amount of gas delivered to the cell and the amount of gas in the vapor phase. The amount of gas in the vapor phase can be calculated using a mass balance, shown in (2), coupled with an equation of state.

Equation to calculate amount of gas in the vapor phase:

$$m_g = m_{pump} - m_{lines} - m_{headspace} + m_{lines}^0 + m_{headspace}^0 \tag{2}$$

where m_g is the mass of CO_2 in the liquid phase, m_{pump} is the mass of CO_2 injected into the system, m_{lines} is the mass of CO_2 in the gas lines, connecting the pump to the equilibrium cell, $m_{headspace}$ is the mass of the gas in the headspace of the cell, $m_{lines}{}^0$ is the mass of the gas in the lines after venting the system, and $m_{headspace}{}^0$ is the mass of gas in the headspace, initially in the system after venting.

A full experimental procedure using a synthetic method without phase transition and demonstrating the use of mass balancing to determine gas solubility is explained in the literature [30, 81, 82].

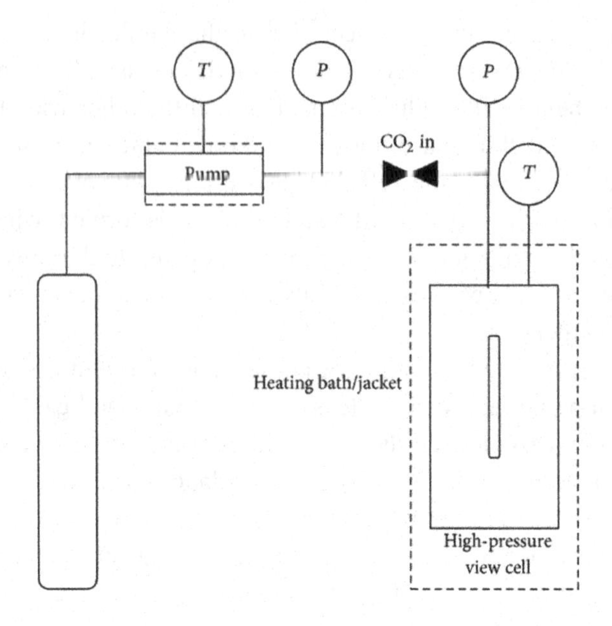

FIGURE 5: Scheme of a synthetic method setup

12.3.4 FURTHER TECHNIQUES

Gas chromatography is an analytical method that boosts high precision and accuracy. When applied to measuring gas solubility in absorption media, the gas chromatograph is usually coupled with a high-pressure reactor cell in which a synthetic or pressure drop method is applied, and at equilibrium, a sample is taken and analyzed [83]. Solubility data from gas chromatography can be achieved by using an extractive technique; here the solvent (IL) is saturated with the solute (CO_2) and then coated on a column. Nitrogen, or any other nonabsorbing carrier gas, is directed on to the column in order to extract the CO_2 from the IL. The nitrogen is then analyzed in the gas chromatograph. This determines the amount of CO_2 removed (per amount of coating). A detailed method for applying gas chromatography can be found in the thesis by Wilbanks [84].

Other analytical techniques can be used in some cases to determine the solubility of specific gases; this may be in the form of a titration; this was demonstrated by Shen and Li [85] with aqueous amine solutions. However this has so far not been applied to ILs. Inline gas sensors also have the potential to be used. A possible scenario could involve linking an electrochemical sensor to measure the difference in CO_2 concentration of the vapor phase before and after equilibrium conditions.

Many advantages come from using gravimetric microbalances for solubility measurements. The ability to measure mass change to a high precision is helpful for a variety of reasons. When initially degassing the ionic liquid sample, being able to measure mass decrease allows the experimentalist to see when a constant mass value has been reached, thus allowing the assumption that full degassing has occurred. Also, the mass reading is very important to ensure equilibrium conditions once the CO_2 has been introduced. Equilibrium is reached when the mass change is zero. In situ gravimetric balances, that is, when the balance is enclosed in the measuring gas, are limited to lower pressures and temperatures. Disadvantages of gravimetric systems are mainly due to their high retail price, making them impractical for small projects where the funding is restricted.

In comparison to the gravimetric analysis, the pressure drop and synthetic methods are much simpler in design. As samples of any size can be investigated with these methods, a high sensitivity can generally be achieved, however, not a high accuracy. The most significant errors in the pressure drop and synthetic methods are the error calculations of dead space; for gravimetric methods, it is the determination of buoyancy forces. In pressure drop and synthetic methods, the two variables, pressure and gas absorbed, are determined by the pressure sensors and calibrated volumes; this can result in measuring error which is added on each step of the absorption isotherm. With the gravimetric method, all of the variables, temperature, gas pressure, and absorbed gas, are measured independently and the absorption pressure is monitored at each step of the isotherm.

12.4 CONCLUSION

The versatility and inherent advantages of ionic liquids in the process of CO_2 capture are giving rise to a promising and expansive field. Their potential as physical absorbents is highly attractive, although at present their capture rate is not to the same scale as current aqueous amine technologies;

the fact that amines for CO_2 capture have been developed through many years and that ILs are a new research field leaves room for further research and improvement.

Solubility data of CO_2 in different imidazolium-based ionic liquids are the most often found in the literature. This is especially the case for bmim[BF_4] and bmim[PF_6], because these ionic liquids were among the first ones commercially available. Therefore an abundant amount of previous data is available and allows the validation of subsequent experimental procedures. Although commercially available, the price of these ionic liquids remains high. Quaternary ammonium and tetra alkylphosphonium bases provide a cheaper alternative. In comparison the synthesis process of these ionic liquids is simpler and the raw materials are accessible. However the lack of experimental data with these solvents means that they are constantly overshadowed.

Although experimental data on CO_2 solubility in ionic liquids is available in the literature, more is needed for process design. Here several different methods have been presented in order to obtain this data. These include gravimetric analysis, pressure drop, and synthetic methods, all of which are particularly well suited for the measurement of gases in non-volatile liquids. In terms of solubility data measurements, gravimetric balances offer the simplest and most precise route; however, their general high prices make them impractical for small research groups conducting initial experiments with ionic liquids. Pressure drop and synthetic methods provide a cheaper alternative and do not need sampling. However these methods depend on the models used to calculate the thermodynamic properties and phase equilibrium. It is important to observe that for some thermodynamic properties, such as excess molar enthalpy, research groups use a test system to check their equipment and methods accuracy. In the case of gas-liquid solubility, however, there is no test system, especially at elevated temperatures and pressures.

The main challenges affecting ionic liquids as green solvent for CO_2 capture are availability, cost, purity, and compatibility. These challenges are faced at present on a laboratory scale and must have solutions before expanding to industry. At present the advantages and disadvantages of ionic liquids and amines seem to be equally balanced. The main criteria for ideal CO_2 capture mechanisms are high CO_2 solubility, low energy input for regeneration, low cost, long-term reusability, and being environmentally benign. At the moment amines have the advantage of having high CO_2 solubility and being of low cost. However due to the vast number of ionic liquids that can be developed and different ways in which they can be synthesized, the potential is there. Moreover through increasing research and commercialization of ionic liquids in other areas of industry, the cost is set to decrease.

REFERENCES

1. B. Metz, O. Davidson, H. deConinck, M. Loos, and L. Meyer, IPCC special report on carbon dioxide capture and storage, prepared by working group III of the intergovernmental panel on climate change, Cambridge University Press, New York, NY, USA, 2005.
2. Carbon Capture and Storage in Industrial Applications: Technology Synthesis Report Working Paper—November 2010, United Nations Industrial Development Organization.
3. A. Stangeland, "A model for the CO2 capture potential," International Journal of Greenhouse Gas Control, vol. 1, no. 4, pp. 418–429, 2007.
4. S. T. Brennan, R. C. Burruss, M. D. Merrill, P. A. Freeman, and L. F. Ruppert, "A probabilistic assessment methodology for the evaluation of geologic carbon dioxide storage," U.S. Geological Survey Open-File Report 2010-1127, 2010.
5. R. P. Hepple and S. M. Benson, "Geologic storage of carbon dioxide as a climate change mitigation strategy: performance requirements and the implications of surface seepage," Environmental Geology, vol. 47, no. 4, pp. 576–585, 2005.
6. S. Holloway, "An overview of the Joule II project: the underground disposal of carbon dioxide," Energy Conversion and Management, vol. 37, no. 6–8, pp. 1149–1154, 1996.
7. E. T. Sundquist, R. C. Burruss, S. P. Faulkner et al., "Carbon sequestration to mitigate climate change," U.S. Geological Survey, Fact Sheet 2008-3097, 2008.
8. M. Finkenrath, Cost and Performance of Carbon Dioxide Capture from Power Generation, International Energy Agency, 2011.
9. R. S. Haszeldine, "Carbon capture and storage: how green can black be?" Science, vol. 325, no. 5948, pp. 1647–1652, 2009.
10. T. Kuramochi, A. Faaij, A. Ramírez, and W. Turkenburg, "Prospects for cost-effective post-combustion CO2 capture from industrial CHPs," International Journal of Greenhouse Gas Control, vol. 4, no. 3, pp. 511–524, 2010.
11. H. Manchao, L. Sousa, R. Sousa, A. Gomes, E. Vargas Jr., and Z. Na, "Risk assessment of CO2 injection processes and storage in carboniferous formations: a review," Journal of Rock Mechanics and Geotechnical Engineering, vol. 3, no. 1, pp. 39–56, 2011.

12. S.-E. Park, J.-S. Chang, and K.-W. Lee, Eds., Carbon Dioxide Utilization For Global Sustainability, Proceedings of 7th the International Conference on Carbon Dioxide Utilization, vol. 153, 2004.
13. I. Omae, “Aspects of carbon dioxide utilization,” Catalysis Today, vol. 115, no. 1–4, pp. 33–52, 2006.
14. D. M. D'Alessandro, B. Smit, and J. R. Long, “Carbon dioxide capture: prospects for new materials,” Angewandte Chemie, vol. 49, no. 35, pp. 6058–6082, 2010.
15. X. Gui, Z. Tang, and W. Fei, “CO2 capture with physical solvent dimethyl carbonate at high pressures,” Journal of Chemical and Engineering Data, vol. 55, no. 9, pp. 3736–3741, 2010.
16. A. L. Chaffee, G. P. Knowles, Z. Liang, J. Zhang, P. Xiao, and P. A. Webley, “CO2 capture by adsorption: materials and process development,” International Journal of Greenhouse Gas Control, vol. 1, no. 1, pp. 11–18, 2007.
17. C. A. Scholes, S. E. Kentish, and G. W. Stevens, “Carbon dioxide separation through polymeric membrane systems for flue gas applications,” Recent Patents on Chemical Engineering, vol. 1, no. 1, pp. 52–66, 2008.
18. S. Burt, A. Baxte, and L. Baxter, Cryogenic CO2 Capture to Control Climate Change Emissions, Brigham Young University, Provo, Utah, USA, 84602 Sustainable Energy Solutions Orem, 84058, 2010.
19. L. Faramarzi, G. M. Kontogeorgis, M. L. Michelsen, K. Thomsen, and E. H. Stenby, “Absorber model for CO2 capture by monoethanolamine,” Industrial and Engineering Chemistry Research, vol. 49, no. 8, pp. 3751–3759, 2010.
20. P. J. G. Huttenhuis, N. J. Agrawal, E. Solbraa, and G. F. Versteeg, “The solubility of carbon dioxide in aqueous N-methyldiethanolamine solutions,” Fluid Phase Equilibria, vol. 264, no. 1-2, pp. 99–112, 2008.
21. M. L. Kennard and A. Melsen, “Solubility of carbon dioxide in aqueous diethanolamine solutions at elevated temperatures and pressures,” Journal of Chemical and Engineering Data, vol. 29, no. 3, pp. 309–312, 1984.
22. H. Lepaumier, S. Martin, D. Picq, B. Delfort, and P. Carrette, “New amines for CO2 capture. III. Effect of alkyl chain length between amine functions on polyamines degradation,” Industrial and Engineering Chemistry Research, vol. 49, no. 10, pp. 4553–4560, 2010.
23. G. T. Rochelle, “Amine scrubbing for CO2 capture,” Science, vol. 325, no. 5948, pp. 1652–1654, 2009.
24. G. Puxty, R. Rowland, A. Allport et al., “Carbon dioxide postcombustion capture: a novel screening study of the carbon dioxide absorption performance of 76 amines,” Environmental Science and Technology, vol. 43, no. 16, pp. 6427–6433, 2009.
25. F. Closmann, T. Ngugen, and G. T. Rochelle, “MDEA/Piperazine as a solvent for CO2 capture,” Energy Procedia, vol. 1, no. 1, pp. 1351–1357, 2009.
26. P. W. F. Riemer and W. G. Ormerod, “International perspectives and the results of carbon dioxide capture disposal and utilisation studies,” Energy Conversion and Management, vol. 36, no. 6–9, pp. 813–818, 1995.
27. M. Karl, R. F. Wright, T. F. Berglen, and B. Denby, “Worst case scenario study to assess the environmental impact of amine emissions from a CO2 capture plant,” International Journal of Greenhouse Gas Control, vol. 5, no. 3, pp. 439–447, 2011.

28. J. Kittel, R. Idem, D. Gelowitz, P. Tontiwachwuthikul, G. Parrain, and A. Bonneau, "Corrosion in MEA units for CO2 capture: pilot plant studies," Energy Procedia, vol. 1, no. 1, pp. 791–797, 2009.
29. A. B. Rao and E. S. Rubin, "A technical, economic, and environmental assessment of amine-based CO2 capture technology for power plant greenhouse gas control," Environmental Science and Technology, vol. 36, no. 20, pp. 4467–4475, 2002.
30. G. Mahrer and D. Tuma, "Gas solubility (and related high-pressure phenomena) in systems with ionic liquids," in Ionic Liquids: From Knowledge to Application, N. V. Plechkova, R. D. Rogers, and K. R. Seddon, Eds., chapter 1, pp. 1–20.
31. J. D. Holbrey and K. R. Seddon, "Ionic liquids," Clean Products and Processes, vol. 1, no. 4, pp. 223–236, 1999.
32. K. R. Seddon, "Ionic liquids: a taste of the future," Nature Materials, vol. 2, no. 6, pp. 363–365, 2003.
33. M. Galiński, A. Lewandowski, and I. Stepniak, "Ionic liquids as electrolytes," Electrochimica Acta, vol. 51, no. 26, pp. 5567–5580, 2006.
34. P. Wang, S. M. Zakeeruddin, J.-E. Moser, and M. Grätzel, "A new ionic liquid electrolyte enhances the conversion efficiency of dye-sensitized solar cells," The Journal of Physical Chemistry B, vol. 107, no. 48, pp. 13280–13285, 2003.
35. A. E. Jimenez, M. D. Bermudez, F. J. Carrion, and G. Martınez-Nicolas, "Room temperature ionic liquids as lubricant additives in steel-aluminium contacts: influence of sliding velocity, normal load and temperature," Wear, vol. 261, no. 3-4, pp. 347–359, 2006.
36. A. P. Abbott, K. J. McKenzie, and K. S. Ryder, "Electropolishing and electroplating of metals using ionic liquids based on choline chloride," in Ionic Liquids IV, vol. 975 of ACS Symposium Series, chapter 13, pp. 186–197, 2007.
37. S. S. Y. Tan and D. R. Macfarlane, "Ionic liquids in biomass processing," Topics in Current Chemistry, vol. 290, pp. 311–339, 2009.
38. S. V. Dzyuba, K. D. Kollar, and S. S. Sabnis, "Synthesis of imidazolium room-temperature ionic liquids exploring green chemistry and click chemistry paradigms in undergraduate organic chemistry laboratory," Journal of Chemical Education, vol. 86, no. 7, pp. 856–858, 2009.
39. A. Stark, D. Ott, D. Kralisch, G. Kreisel, and B. Ondruschka, "Ionic liquids and green chemistry: a lab experiment," Journal of Chemical Education, vol. 87, no. 2, pp. 196–201, 2010.
40. T. Welton, "Room-temperature ionic liquids. Solvents for synthesis and catalysis," Chemical Reviews, vol. 99, no. 8, pp. 2071–2083, 1999.
41. M. J. Earle and K. R. Seddon, "Ionic liquids. Green solvents for the future," Pure and Applied Chemistry, vol. 72, no. 7, pp. 1391–1398, 2000.
42. C. M. Gordon, "New developments in catalysis using ionic liquids," Applied Catalysis A, vol. 222, no. 1-2, pp. 101–117, 2001.
43. J. D. Holbrey, "Industrial applications of ionic liquids," Chimica Oggi, vol. 22, no. 6, pp. 35–37, 2004.
44. Q. Gan, D. Rooney, and Y. Zou, "Supported ionic liquid membranes in nanopore structure for gas separation and transport studies," Desalination, vol. 199, no. 1–3, pp. 535–537, 2006.

45. C. P. Fredlake, J. M. Crosthwaite, D. G. Hert, S. N. V. K. Aki, and J. F. Brennecke, "Thermophysical properties of imidazolium-based ionic liquids," Journal of Chemical and Engineering Data, vol. 49, no. 4, pp. 954–964, 2004.
46. K. R. Seddon, A. Stark, and M. J. Torres, "Viscosity and density of 1-alkyl-3-methylimidazolium ionic liquids," in Clean Solvents, M. Abraham and L. Moens, Eds., pp. 34–49, American Chemical Society, Washington, DC, USA, 2002.
47. R. A. Mantz and P. C. Trulove, "Viscosity and density of ionic liquids. Physicochemical properties," in Ionic Liquids in Synthesis, P. Wasserscheid and T. Welton, Eds., pp. 56–68, Wiley-VCH Verlag GmbH & Co. KGaA, Weinheim, Germany, 2nd edition, 2008.
48. J. W. Magee, "Heat capacity and enthalpy of fusion for 1-butyl-3-methyl-imidazolium hexafluorophosphate," in Proceedings of the 17th IUPAC Conference on Chemical Thermodynamics (ICCT '02), Rostock, Germany, 2002.
49. M. Kosmulski, J. Gustafsson, and J. B. Rosenholm, "Thermal stability of low temperature ionic liquids revisited," Thermochimica Acta, vol. 412, no. 1-2, pp. 47–53, 2004.
50. G. Law and P. R. Watson, "Surface tension measurements of N-alkylimidazolium ionic liquids," Langmuir, vol. 17, no. 20, pp. 6138–6141, 2001.
51. A. S. Wells and V. T. Coombe, "On the freshwater ecotoxicity and biodegradation properties of some common ionic liquids," Organic Process Research and Development, vol. 10, no. 4, pp. 794–798, 2006.
52. T. D. Landry, K. Brooks, D. Poche, and M. Woolhiser, "Acute toxicity profile of 1-butyl-3-methylimidazolium chloride," Bulletin of Environmental Contamination and Toxicology, vol. 74, no. 3, pp. 559–565, 2005.
53. I. Perissi, U. Bardi, S. Caporali, and A. Lavacchi, "High temperature corrosion properties of ionic liquids," Corrosion Science, vol. 48, no. 9, pp. 2349–2362, 2006.
54. M. Uerdingen, C. Treber, M. Balser, G. Schmitt, and C. Werner, "Corrosion behaviour of ionic liquids," Green Chemistry, vol. 7, no. 5, pp. 321–325, 2005.
55. P. T. Anastas and J. C. Warner, Green Chemistry: Theory and Practice, Oxford University Press, New York, NY, USA, 1998.
56. P. Scovazzo, J. Kieft, D. A. Finan, C. Koval, D. DuBois, and R. Noble, "Gas separations using non-hexafluorophosphate [PF6]-anion supported ionic liquid membranes," Journal of Membrane Science, vol. 238, no. 1-2, pp. 57–63, 2004.
57. R. E. Baltus, B. H. Culbertson, S. Dai, H. Luo, and D. W. DePaoli, "Low-pressure solubility of carbon dioxide in room-temperature ionic liquids measured with a quartz crystal microbalance," The Journal of Physical Chemistry B, vol. 108, no. 2, pp. 721–727, 2004.
58. Á. P.-S. Kamps, D. Tuma, J. Xia, and G. Maurer, "Solubility of CO2 in the ionic liquid [bmim][PF6]," Journal of Chemical and Engineering Data, vol. 48, no. 3, pp. 746–749, 2003.
59. B.-C. Lee and S. L. Outcalt, "Solubilities of gases in the ionic liquid 1-n-butyl-3-methylimidazolium bis(trifluoromethylsulfonyl)imide," Journal of Chemical and Engineering Data, vol. 51, no. 3, pp. 892–897, 2006.
60. M. J. Muldoon, S. N. V. K. Aki, J. L. Anderson, J. K. Dixon, and J. F. Brennecke, "Improving carbon dioxide solubility in ionic liquids," The Journal of Physical Chemistry B, vol. 111, no. 30, pp. 9001–9009, 2007.

61. C. Cadena, J. L. Anthony, J. K. Shah, T. I. Morrow, J. F. Brennecke, and E. J. Maginn, "Why is CO2 so soluble in imidazolium-based ionic liquids?" Journal of the American Chemical Society, vol. 126, no. 16, pp. 5300–5308, 2004.
62. S. N. V. K. Aki, B. R. Mellein, E. M. Saurer, and J. F. Brennecke, "High-pressure phase behavior of carbon dioxide with imidazolium-based ionic liquids," The Journal of Physical Chemistry B, vol. 108, no. 52, pp. 20355–20365, 2004.
63. P. J. Carvalho, V. H. Álvarez, I. M. Marrucho, M. Aznar, and J. A. P. Coutinho, "High carbon dioxide solubilities in trihexyltetradecylphosphonium-based ionic liquids," Journal of Supercritical Fluids, vol. 52, no. 3, pp. 258–265, 2010.
64. S. G. Kazarian, B. J. Briscoe, and T. Welton, "Combining ionic liquids and supercritical fluids: in situ ATR-IR study of CO2 dissolved in two ionic liquids at high pressures," Chemical Communications, no. 20, pp. 2047–2048, 2000.
65. E. J. Beckman, "A challenge for green chemistry: designing molecules that readily dissolve in carbon dioxide," Chemical Communications, vol. 10, no. 17, pp. 1885–1888, $_2$004.
66. M. B. Shiflett and A. Yokozeki, "Solubilities and diffusivities of carbon dioxide in ionic liquids: [bmim][PF6] and [bmim][BF4]," Industrial and Engineering Chemistry Research, vol. 44, no. 12, pp. 4453–4464, 2005.
67. A. Yokozeki, M. B. Shiflett, C. P. Junk, L. M. Grieco, and T. Foo, "Physical and chemical absorptions of carbon dioxide in room-temperature ionic liquids," The Journal of Physical Chemistry B, vol. 112, no. 51, pp. 16654–16663, 2008.
68. E. D. Bates, R. D. Mayton, I. H. Ntai, et al., "CO2 2 capture by a task-specific ionic liquid," Journal of the American Chemical Society, vol. 124, no. 6, pp. 926–927, 2002.
69. J. E. Bara, D. E. Camper, D. L. Gin, and R. D. Noble, "Room-temperature ionic liquids and composite materials: platform technologies for CO2 capture," Accounts of Chemical Research, vol. 43, no. 1, pp. 152–159, 2010.
70. D. Camper, J. E. Bara, D. L. Gin, and R. Noble, "Room-temperature ionic liquid-amine solutions: tunable solvents for efficient and reversible capture of CO2," Industrial & Engineering Chemistry Research, vol. 47, no. 21, pp. 8496–8498, 2008.
71. R. Kleinrahm and W. Wagner, "Measurement and correlation of the equilibrium liquid and vapour densities and the vapour pressure along the coexistence curve of methane," The Journal of Chemical Thermodynamics, vol. 18, no. 8, pp. 739–760, 1986.
72. M. Petermann, T. Weissert, S. Kareth, H. W. Lösch, and F. Dreisbach, "New instrument to measure the selective sorption of gas mixtures under high pressures," Journal of Supercritical Fluids, vol. 45, no. 2, pp. 156–160, 2008.
73. F. Dreisbach and H. W. Lösch, "Magnetic suspension balance for simultaneous measurement of a sample and the density of the measuring fluid," Journal of Thermal Analysis and Calorimetry, vol. 62, no. 2, pp. 515–521, 2000.
74. H. Liu, J. Huang, and P. Pendleton, "Experimental and modelling study of CO2 absorption in ionic liquids containing Zn (II) ions," Energy Procedia, vol. 4, pp. 59–66, 2011.
75. A. N. Soriano, B. T. Doma Jr., and M.-H. Li, "Solubility of carbon dioxide in 1-ethyl-3-methylimidazolium tetrafluoroborate," Journal of Chemical and Engineering Data, vol. 53, no. 11, pp. 2550–2555, 2008.

76. J. Palgunadi, J. E. Kang, D. Q. Nguyen et al., "Solubility of CO2 in dialkylimidazolium dialkylphosphate ionic liquids," Thermochimica Acta, vol. 494, no. 1-2, pp. 94–98, 2009.
77. J. Palgunadi, J. E. Kang, M. Cheong, H. Kim, H. Lee, and H. S. Kim, "Fluorine-free imidazolium-based ionic liquids with a phosphorous-containing anion as potential CO2 absorbents," Bulletin of the Korean Chemical Society, vol. 30, no. 8, pp. 1749–1754, 2009.
78. K. A. Kurnia, F. Harris, C. D. Wilfred, M. I. Abdul Mutalib, and T. Murugesan, "Thermodynamic properties of CO2 absorption in hydroxyl ammonium ionic liquids at pressures of (100–1600) kPa," The Journal of Chemical Thermodynamics, vol. 41, no. 10, pp. 1069–1073, 2009.
79. D. Camper, P. Scovazzo, C. Koval, and R. Noble, "Gas solubilities in room-temperature ionic liquids," Industrial and Engineering Chemistry Research, vol. 43, no. 12, pp. 3049–3054, 2004.
80. F. C. Gomes, "Solubility of carbon dioxide, ethane, methane, oxygen, nitrogen, hydrogen, argon, and carbon monoxide in 1-butyl-3-methylimidazolium tetrafluoroborate between temperatures 283 K and 343 K and at pressures close to atmospheric," The Journal of Chemical Thermodynamics, vol. 38, no. 4, pp. 490–502, 2006.
81. W. Ren and A. M. Scurto, "High-pressure phase equilibria with compressed gases," Review of Scientific Instruments, vol. 78, no. 12, Article ID 125104, 7 pages, 2007.
82. A.-L. Revelli, F. Mutelet, and J.-N. Jaubert, "High carbon dioxide solubilities in imidazolium-based ionic liquids and in poly(ethylene glycol) dimethyl ether," The Journal of Physical Chemistry B, vol. 114, no. 40, pp. $1_2$908–$1_2$913, $_2$010.
83. Z. Lei, J. Yuan, and J. Zhu, "Solubility of CO_2 in propanone, 1-ethyl-3-methylimidazolium tetrafluoroborate, and their mixtures," Journal of Chemical and Engineering Data, vol. 55, no. 10, pp. 4190–4194, $_2$010.
84. K. E. Wilbanks, Phase behavior of carbon dioxide and oxygen in the ionic liquid 1-hexyl-3-methylimidazolium bis(trifluoromethylsulfonyl) imide [M.S. thesis], $_2$007.
85. K. P. Shen and M. H. Li, "Solubility of carbon dioxide in aqueous mixtures of monoethanolamine with methyldiethanolamine," Journal of Chemical and Engineering Data, vol. 37, no. 1, pp. 96–100, 199_2.
86. L. A. Blanchard, Z. Gu, and J. F. Brennecke, "High-pressure phase behavior of ionic liquid/CO_2 systems," The Journal of Physical Chemistry B, vol. 105, no. 1_2, pp. $_2$437–$_2$444, $_2$001.
87. P. J. Carvalho, V. H. Álvarez, B. Schröder et al., "Specific solvation interactions of CO_2 on acetate and trifluoroacetate imidazolium based ionic liquids at high pressures," The Journal of Physical Chemistry B, vol. 113, no. 19, pp. 6803–681_2, $_2$009.
88. L. M. Galan-Sanchez, Functionalised ionic liquids, absorption solvents for CO_2 and olefin separation [Ph.D. thesis], $_2$008.
89. Y. S. Kim, W. Y. Choi, J. H. Jang, K.-P. Yoo, and C. S. Lee, "Solubility measurement and prediction of carbon dioxide in ionic liquids," Fluid Phase Equilibria, vol. $_{22}$8-$_{22}$9, pp. 439–445, $_2$005.
90. E.-K. Shin and B.-C. Lee, "High-pressure phase behavior of carbon dioxide with ionic liquids: 1-alkyl-3-methylimidazolium trifluoromethanesulfonate," Journal of Chemical and Engineering Data, vol. 53, no. 1_2, pp. $_2$7$_2$8–$_2$734, $_2$008.

91. A.-L. Revelli, F. Mutelet, and J.-N. Jaubert, "High carbon dioxide solubilities in imidazolium-based ionic liquids and in poly(ethylene glycol) dimethyl ether," The Journal of Physical Chemistry B, vol. 114, no. 40, pp. 12908–12913, 2010.
92. X. Zhang, Z. Liu, and W. Wang, "Screening of ionic liquids to capture CO2 by COS-MO-RS and experiments," AIChE Journal, vol. 54, no. 10, pp. 2717–2728, 2008.
93. X. Yuan, S. Zhang, J. Liu, and X. Lu, "Solubilities of CO2 in hydroxyl ammonium ionic liquids at elevated pressures," Fluid Phase Equilibria, vol. 257, no. 2, pp. 195–200, 2007.
94. P. J. Carvalho, V. H. Alvarez, I. M. Marrucho, M. Aznar, and J. A. P. Coutinho, "High carbon dioxide solubilities in trihexyltetradecylphosphonium-based ionic liquids," The Journal of Supercritical Fluids, vol. 252, no. 3, pp. 258–265, 2010.

CHAPTER 13

CAPTURING CARBON DIOXIDE FROM AIR

KLAUS S. LACKNER, PATRICK GRIMES, AND HANS-J. ZIOCK

13.1 INTRODUCTION

The economic stakes in dealing with climate change are big and costs could escalate dramatically, if the transition to a zero emission economy would have to happen fast. Abandoning existing infrastructure is prohibitively expensive and as long as new technology is not yet ready to be phased in, improvements and additions to the existing infrastructure will tend to perpetuate the problem. For this reason alone it is important to consider the possibility of capturing carbon dioxide directly from the air [1-4]. If capture from air would prove feasible, one would not have to wait for the phasing out of existing infrastructure before addressing the greenhouse gas problem. Technology for extracting CO_2 from the air could be deployed as soon as it is developed; it could deal with all sources of CO_2, and it even could be scaled up to reduce present levels of atmospheric CO_2. Deployment of air extraction technology need not interfere with other approaches to the problem. Avoidance of emissions, either through capture at a plant or switching to non-carbon based energy sources would still make sense, but one would not have to abandon existing infrastructure or construct a complex CO_2 pipelining system in order to get started. For

---

This chapter was originally published by the U.S. Department of Energy. U.S. Department of Energy. Capturing Carbon Dioxide from Air. *By Lackner KS, Grimes P, and Ziock H-J. Available at: http://www.netl.doe.gov/publications/proceedings/01/carbon_seq/7b1.pdf.*

the portion of the CO_2 that is emitted from small and distributed sources, capture of CO_2 from the air may always the best solution.

In this paper we argue that capture of CO_2 from natural airflow is technically feasible at a rate far above the rate at which trees capture CO_2. The photosynthesis by plants seems to be more limited by sunlight than capture of CO_2. We will provide a rough estimate of the expected cost and the scale of operation required to deal with the world's CO_2 emissions. Finally we will discuss the benefits of the approach and how this approach would fit into a no-regret strategy.

Until recently, the world has been concerned exclusively with the first half of the fossil fuel carbon cycle, i.e. with bringing the energy resource to the energy user. The waste CO_2 was simply abandoned to the atmosphere. With the growing understanding that the atmosphere is not an infinite sink comes the realization that carbon has not only to be moved from the well to the wheel but on from there to an appropriate sink, i.e. from well to a disposal site. Utilizing the air as a temporary buffer makes this process easier and avoids the need for developing specific capture processes for each and every emitter.

13.2 OBJECTIVE

If fossil fuels are to play a significant role through the 21st century, thc accumulation of carbon dioxide in the air must be prevented. Current rates of fossil fuel consumption introduce an amount of carbon into the surface pool that over 100 years would match the size of the entire biomass. Unless painful actions are taken to reduce consumption, it is likely that world carbon consumption will grow rather than shrink. Natural processes are unlikely to absorb all this carbon, and CO_2 levels in the air will keep rising, unless CO_2 emissions are virtually stopped. To stabilize CO_2 levels, it is necessary to not only deal with CO_2 emissions from power plants, but from all sources in an industrial economy. While it is generally agreed that the reductions demanded by the Kyoto Treaty would be far less than what would ultimately be required to stabilize CO_2 levels in the atmosphere [5], it is also clear that even this goal would be too ambitious to be achieved by exclusively eliminating emissions from power plants. Since the economy and with it energy consumption have grown substantially since 1990, the reduction required in the United States is far more than the nominal seven percent reduction which is measured relative to 1990 emissions. The economy of 2010 would most likely have to reduce carbon emissions by more than 30% relative to business-as-usual. This is equivalent to eliminating all emissions from power plants. However, in the long-term carbon reductions will have to go far below 1990 emission levels and thus it is necessary to address all carbon dioxide emissions including those from small and mobile sources.

A portion of the desired reductions will be achieved by improved energy efficiency and energy savings, and another part might be accomplished by transition to non-fossil, renewable energy resources. However, here we concern ourselves with eliminating the remaining carbon dioxide emissions. Given the continuing and highly desirable worldwide economic growth, we expect this to be a large fraction of the total required emission reductions.

A source of carbon dioxide that is particularly difficult to manage is the transportation sector. A transition to electric or hydrogen fueled vehicles is in principle possible but would take a long time to accomplish. Even though it has been proposed [6], it does not appear to be economically viable to collect the carbon dioxide of a vehicle directly at the source. The mass flows would be prohibitively large. Generally, even stationary, small sources would be difficult to deal with. A unit mass of fuel results in roughly three mass units of gaseous CO_2 that would need to be temporarily stored at the source and later shipped to a disposal site. The mass of the stored material would be more than doubled once more, if one were to store the CO_2 absorbed onto some substrate, like CaO. Capturing CO_2 on board of an airplane is simply not possible because of the mass involved; in a car it would be prohibitively expensive; and even in a home it would not be practicable, as it would require a huge infrastructure for removal and transport of CO_2 to a disposal site.

Distributed carbon dioxide sources account for approximately half of the total emissions. While it may not be necessary to address them initially, for carbon management to be successful in the long term, they cannot be ignored.

Carbon dioxide capture from the atmosphere, in principle, can deal with any source, large or small. Indeed, the appeal of biomass for sequestration and of credits for growing trees is based on the very same premise. Since photosynthesis takes the carbon it needs from the air, it can compensate for any emission, and ideally it can be done at the disposal site eliminating the need for long distance surface transportation.

Thus, it is our objective to explore the feasibility of CO_2 capture from air. We would like to find out whether it is physically possible, whether it could be done at acceptable cost,and whether the scale of such an operation would be acceptable. We will show in the following that CO_2 capture is physically and economically feasible, and that the scale of operation is actually small compared to other renewable options that are considered as possible replacements for fossil energy.

13.3 APPROACH

Carbon dioxide capture from air is certainly possible. Plants during photosynthesis routinely accomplish this task. Chemical processes also can capture CO_2. A classic chemistry experiment is to bubble air through a calcium hydroxide solution and to remove the air's CO_2 in this fashion. Other means work as well and have been used in the past in industrial processes to generate CO_2 free air. However, in capturing CO_2 one is very much constrained by economic considerations. One can hardly spent any effort in handling the air as any cost is amplified by the dilution ratio, which is roughly one part in three thousand.

It is not economically possible to perform significant amount of work on the air, which means one cannot heat or cool it, compress it or expand it. It would be possible to move the air mechanically but only at speeds that are easily achieved by natural flows as well. Thus, one is virtually forced into considering physical or chemical adsorption from natural airflow passing over some recyclable sorbent [1, 2]. Once the CO_2 has been taken out of the air, the down stream processing deals with volumes and masses that are of the same order of magnitude as the CO_2 itself and is therefore not subject to the large amplification factor that results from the dilute nature of CO_2 in air.

To get an appreciation for the scales, let us measure the CO_2 content of air in energy units. At 365 ppm of CO_2 in the air, a cubic meter (or 40 moles of air) contains 0.015 moles of CO_2. If this CO_2 were extracted from the air to compensate for an equivalent CO_2 emission by a gasoline engine somewhere else, we could relate the amount of CO_2 in a cubic meter of air with the heat released in the combustion of gasoline resulting in the emission of the same 0.015 moles of CO_2. This heat of combustion amounts to 10,000 J. Thus removing the CO_2 from one cubic meter of air and disposing of it opens the door for generating 10,000 J of heat from gasoline anywhere in the world. Combined, these two actions are carbon neutral.

This approach to a net zero carbon economy works, because CO_2 in the air is not harmful and the natural amount in the air is large compared to the amounts human activities add on short time scales. Current annual world emissions from human activities equal 1% of the total CO_2 in the air. Since mixing times are far shorter than a year, one can use the air as a conveyer that moves CO_2 from its source to its sink. As long as the total amount in transit is small compared to the air's CO_2 content, moving CO_2 in this fashion to the sink would not unduly distort atmospheric CO_2 concentrations. Locally, mixing is very fast and therefore local CO_2 depletion or enrichment is not likely to pose a problem either. If this were not the case, emissions from power plants would cause large local deviations. In the same fashion as CO_2 enriched air mixes rapidly with ambient air to maintain constant levels of CO_2, air depleted in CO_2 will also mix rapidly and return to ambient conditions. It is, however, this mixing rate which sets the limit of how closely one could space CO_2 extraction units [2, 4].

We note that the CO_2 content of a volume of air, as measured by the heat of combustion its removal could compensate for, is far larger than the kinetic energy the same volume of air would have assuming reasonable wind velocity. At 10 m/s, which is a wind stronger than is usually assumed to prevail in windmill operations [7], the kinetic energy of a cubic meter of air is 60 J, which should be compared to 10,000 J for the heat of combustion that would generate the CO_2 content of a cubic meter of air. A windmill that operates by extracting kinetic energy from natural airflow needs to be two orders of magnitude larger than a CO_2 collector that captures CO_2 to compensate for the emissions from a diesel engine that generates the same amount of electricity. Since windmills appear economically viable, this suggests that the capturing apparatus should not be too expensive to build.

One can pursue this line of reasoning a little further by looking at the same data in a slightly different fashion. Windmills are rated by energy flux per unit area. In effect the wind carries with it a flow of kinetic energy, a part of which a windmill transfers into electric energy. Thus a windmill at wind speed of 10 m/s would face an energy flux of 600 W/m^2. The equivalent CO_2 flux through the same area corresponds to 100,000 W/m^2. Thus an air "filtration" system could extract CO_2 from a stream that represents power generation of 100,000 W for every square meter of airflow. By this measure, CO_2 is far more concentrated than the kinetic energy harnessed by the windmill.

By invoking a measure of power per unit area we can also compare the efficacy of our approach to collecting solar energy. Peak fluxes of solar energy on the ground are around 1,000 W/m^2. Average fluxes in desert climates accounting for weather and day and night are around 200 W/m^2. Photovoltaic panels can capture maybe 25% of this flux. Under conditions of intensive agriculture, biomass growth can capture maybe 1.5% of this flux, and thus would rate at roughly 3 W/m^2 [8]. Typical unmanaged forest growth would fall far short of capturing even that much carbon equivalent.

The purpose of this discussion is to establish an estimate of a system's size necessary to collect CO_2 generated by an energy source of a given size. If one could maintain a flow of 3 m/s through some filter system, and collect half the CO_2 that passes through it, then the system would collect per square meter the CO_2 output from 15 kW of primary energy. This is more than the per capita primary energy consumption in the US, which is approximately10 kW. The size of a CO_2 collection system would thus have to be less than 1m$_2$ per person. Covering the same energy demand with wind-generated electricity instead would require an area at least a hundred times larger.

Even before having defined specific filters and sorbent materials, this discussion already suggests that the cost of CO_2 collection is not prohibitively high. Prior to any specific designs, let us assume that in collecting CO_2 from a natural airflow, one needs equipment that is different from but similar in size and cost to what one would use for a windmill harvesting wind energy from the same cross sectional flow area. We furthermore assume that both systems have the same capture efficiency. If the cost of a windmill operating on wind speeds of 6m/s is 5¢/kWh, then the equally sized CO_2 collector will collect 100 kg of CO_2 for every kWh its windmill partner collects. Thus, according to this simple comparison, the capture process should add 50¢ to the cost of a ton of CO_2. Of course this estimate is very crude, but even if the actually implementation were 5 times more expensive, the basic argument would not be affected.

However, the cost of contacting the air and scrubbing out the CO_2 is not the only cost one needs to consider in extracting CO_2 from air. For most sorbents that could have captured the CO_2 one needs to recover the sorbent and release the CO_2 in a concentrated stream ready for disposal. These process steps are likely to be far more expensive than the capture itself. Processing a ton of material tends to be measured in dollars not cents. Using cement manufacture to set the scale for this cost, Gilberto Rozenchan arrived at a price on the order of $10 to $15 per ton of CO_2. Even at this price, the approach would have great promise, in that it would allow capturing the CO_2 from gasoline for 9¢ to 14¢ per gallon of gasoline. For comparison, the cost of the crude oil ($30/barrel) going into the generation of one ton of CO_2 amounts to $80.

Thus, we need to develop a technology that would allow the capture of CO_2 from natural or man-made airflows that would enable us to recycle the sorbent and create a concentrated stream of CO_2. In the following section we shall discuss options for such processes.

13.4 TECHNOLOGY

A collector capturing CO_2 from a natural airflow is akin to a windmill. In one case one extracts CO_2 out of the airflow, in the other case one extracts kinetic energy. However, one should not pursue this analogy too far. A modern windmill has an aerodynamic design that maximizes momentum flow from the air to the airfoil. Unlike momentum that can be transported independently of mass flow, material flows are intimately tied to mass flow and thus require drastically different designs. The first task in developing CO_2 capture from air would be to define an optimal design. Candidates include filter banks standing in the airflow like snow fences, designs that resembles leaves on a tree, or systems akin to cooling towers that actively move the air.

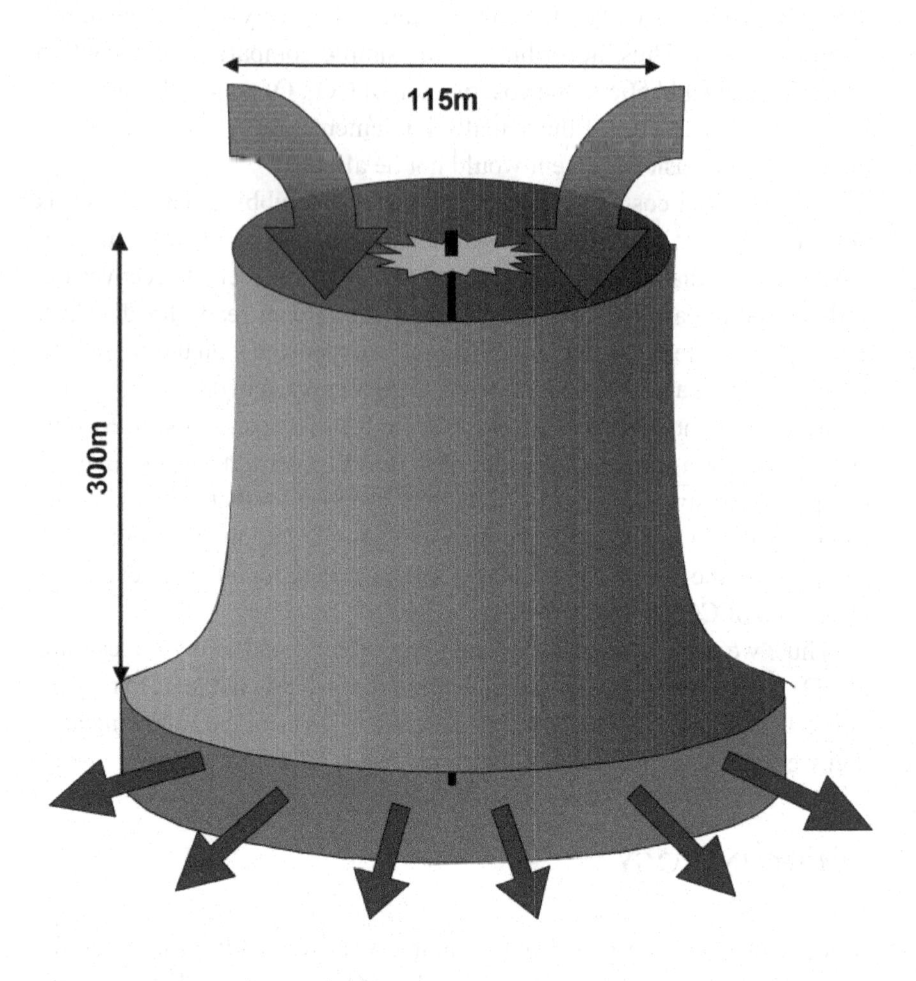

FIGURE 1: Sketch of a convection tower that could either provide electricity or CO_2 capture. Water pumped to the top cools the air, which causes a downdraft inside the tower. The tower has a 10,000 m^2 opening. Cooling the air to the degree possible in a desert climate would cause in the absence of obstructions a downdraft in excess of 15 m/s generating a flow of nearly 15 km^3 of air per day through the tower. The air leaving at the bottom could drive wind turbines or flow over CO_2 absorbers. Based on the volumes of air flowing and the potential energy of the cold air generated at the top of the tower, the tower could generate 3 to 4 MW of electricity after pumping water to the top. The same airflow would carry 9,500 tons of CO_2 per day through the tower. This CO_2 flow equals the output of a 360 MW power plant.

To illustrate this with an example: some years ago, a wind energy technology was suggested that could operate in a dry climate. Inside a large tower, water is pumped to the top, where it cools the air by evaporation. The cold air, being denser, would cause a downdraft inside the convection tower. The potential energy of the air falling down is eight times larger than the potential energy of the water that has to be pumped up. The air flows through the lightweight tower structure and escapes at the bottom where its kinetic energy is harnessed by a number of wind turbines. This effort had grown from preliminary designs to a consortium that was planning on building such a tower in the Negev desert [9]. For such a tower to be economically viable it would cost maybe $3,000 per kWe. Such a cost does not appear unreasonable. Nevertheless in the end, these towers were not built. However, our point here is only to show how much more efficient such a tower would be at extracting CO_2 rather than kinetic energy.

Figure 1 shows a simple design of a convective tower that would generate 3 to 4 MW of electricity. It also passes 9,500 tons/day of CO_2 through itself, which corresponds to the CO_2 output of conventional 360 MW coal fired power plant. The CO_2 flux is also equivalent to the CO_2 output of the vehicle fleet of a city of 700,000 people, indicating the usefulness of the concept for dealing with emissions from the transportation sector. The first comparison to the coal-fired power plant reiterates our earlier observation. The cost of the collection tower, even if it exceeded the $9 million implied by a cost of $3,000/kW for its electricity generating cousin, would still be extremely cheap compared to the cost of the coal fired power plant, which would be approximately $300 to $400 million. Thus, the cost of the collection tower would be dwarfed by the cost of the corresponding power plant. While we are not advocating this specific design for CO_2 capture, it shows once again that the physical structure required to capture the CO_2 is not going to drive the cost of the process. There are a number of different design options, and further work will have to tell which ones are most advantageous.

If we make the assumption that sorbents can be found, which are chemically reactive and have equilibrium partial pressures of CO_2 over them that are substantially lower than ambient partial pressures, then one can estimate what sizes of filters will be needed to collect a substantial fraction of the CO_2 in the air passing through. As a simple proof of principle we consider a slurry and solution of $Ca(OH)_2$. For such slurry the rate of reaction is reasonably fast, and the partial pressure of CO_2 would substantially lower than ambient partial pressure in air. We have performed a more detailed analysis elsewhere [2] and only note here that the diffusion of CO_2 through air is for many designs the rate limiting step. We found that surfaces with square millimeter orifices and passage length of a few centimeters would remove most of the CO_2 from laminar airflow. Thin absorbing threads like in conventional air filters may be even more advantageous and would allow for lighter structures. Details of such designs would have to await the choice of sorbent, e.g. whether it is liquid or solid, as such details would have direct influence on the specific choice.

The choice of sorbent needs to be carefully considered. Calcium hydroxide is an obvious candidate for a sorbent, but it is likely that there are better choices. For one, the binding energy of the carbonation reaction of calcium hydroxide or calcium oxide is far larger than would be required on thermodynamic grounds. The free energy penalty of concentrating CO_2 to 1 bar is RT log P/P_0. Where P_0 is 1 bar and P is the ambient partial pressure of CO_2. At ambient temperatures this number is approximately 20 kJ/mole. In the case of CaO, the binding energy is 180 kJ. For $Ca(OH)_2$ the penalty at 120 kJ is somewhat lower, but it would be very difficult to avoid a transitional step which makes lime from lime stone before the lime is slaked.

A good sorbent should not escape in large quantities from the capturing system, and it should be environmentally benign. It needs to be either extremely cheap, or can withstand many recycle loops.

13.5 RESULTS

Results of our dimensional analysis suggest that the collection of CO_2 directly from air is feasible. Collecting CO_2 from air is far more efficient than collecting wind energy. We emphasize that we can make this statement without having determined an optimal collection system or having settled on an optimal choice of sorbents. Even looking at the most simple implementations suggests that the cost of the effort is tolerable. Our simple analysis suggests, that filter systems using alkaline solutions of $Ca(OH)_2$, or sodium or potassium hydroxide could easily capture CO_2 from air. The major cost of any such process is in the recovery of the sorbent. A preliminary analysis assuming $Ca(OH)_2$ as a possible sorbent suggests, that the cost will be on the order of $10 to $15 per ton of CO_2 and that the additional CO_2 generated in the process of collection is substantially less than the amount of CO_2 captured. In any event, one would design the sorbent recovery system so that it would capture its own CO_2. Since this process would be a large operation at a good disposal site, it is a prime candidate for on-site capture. The energy penalty for this approach is about 200 kJ for every 700 kJ of heat of combustion from gasoline. Per gallon of gasoline one would need 3 cents worth of coal to accomplish the CO_2 recovery from lime. Other sorbents, with better chemical kinetics and lower binding energies could substantially improve the cost of the overall process design.

We have also looked at the overall scale of the collection effort. As mentioned earlier, the cross sectional area needed in the US is slightly less than 1 m^2 per person. However, one could not combine all these units in a single location, as they would tend to interfere with each other. Units down wind from other units could not capture the nominal value of CO_2 as they would be processing air already depleted in CO_2.

What limits the amount of CO_2 one can remove from airflow in one location is the rate of turbulent diffusion from higher altitude to the ground. A recent study [4] suggests that the overall rate of uptake is indeed what one would expect from turbulent diffusion coefficients that are on the order of $10m^2/s$.

Even a worldwide collection system does not have to be extremely large. Per person the cross sectional area facing the wind would have to be about $0.12m^2$. The area would increase $0.65m^2$ per person if the world's per capita energy consumption would reach the current US per capita consumption. At present rate, 380,000 collection units eaching taking up 100 m × 100 m in land area could collect all the CO_2 emissions from human activities. One would need one such unit – roughly two football fields – for every 16,000 people. These units could share the land with other activities. For example each unit could consist of 5 vertical subunits 19 m wide by 19 m tall. The 380,000 units would have to be spread out over an area at least 530 km by 530 km of which they would occupy 1.4%.

13.6 BENEFITS

The method of extracting carbon dioxide from the air we outlined above could operate on a scale large enough to deal with all the carbon dioxide emitted in the world. The only limit to the use of this approach would be from other technologies that for specific emissions may be more cost-effective. One advantage of extraction from air is that it would be possible to sequester more CO_2 than is generated, thereby reducing the total CO_2 load of the atmosphere.

Quite likely, in the long term one would limit extraction from air to the capture of CO_2 from distributed sources and to excess CO_2 already in the air. If included into the design from the start, the collection of CO_2 at a concentrated source is always cheaper than first letting the CO_2 dilute in the air and recapture it later. There are, however, additional issues to consider. One is the cost of transporting CO_2. Transport by the air comes free, and typical cost estimates for long distance transport of CO_2 are around $10/ton. At that cost, a careful economic analysis would be required to decide whether in a given case atmospheric convection would not allow for a cheaper solution to the problem. The cost of carbon capture could well be comparable to the cost of shipping carbon. Furthermore, extraction from air would open up resource sites for carbon disposal, which are simply too far away from all sources to compete by any other means. This additional effect may well compensate for a slightly higher cost in capture relative to transport to a more nearby sink.

Consider some examples: Disposal in the deep ocean would only be feasible from a platform at the disposal site. In this case, CO_2 capture on site, may well be cheaper than CO_2 shipping from distant harbors. Secondly, just like some of the best oil reservoirs turn out to be in remote locations, some of the best underground deposits for CO_2 may also be in isolated locations. Again, it may be easier to serve such sites by extracting CO_2 from air rather than shipping it over long distances. Finally, many sites for successful mineral sequestration would again be in remote sites, as for example in Alaska or the Canadian Northwest.

For mineral carbonate disposal, remoteness would facilitate mining. The overall area affected by mining is not very large, but mining in or near populated areas is problematic. On the other hand, sites in remote locations would not be useful, unless the CO_2 is directly taken from the air.

As we mentioned in the introduction, a major advantage of carbon capture from the air is that it does not require abandonment of existing infrastructure. Extraction from the air could be introduced in parallel to other methods that sequester carbon dioxide directly captured at the source. It would allow the removal of CO_2 virtually immediately and it could be grown rapidly over the course of the next few decades. The cost of the process is independent of the amount of consumption. While on-site capture becomes more and more expensive as one is trying to drive emissions to zero, net-zero emissions obtained by matching extraction from air to the output of some plant, does not incur such increases in cost. Indeed one could aim for 80%, 100% or even $1_20\%$ capture without substantively changing the cost structure. By having capture exceed emissions, one could actually aim for reducing CO_2 in the air.

How fast such a method will be introduced depends on many variables. If we assume that the overall cost of the process is $15 per ton of CO_2 and if we further assume that roughly half of this cost is in capital investment, then the elimination of 22 billion tons of CO_2 would represent an annual cost of $330 billion worldwide. The capital cost involved would be on the order of 1.6 trillion, which is a huge number, but it is again not so large as to be prohibitive. If one were to aim at an implementation in the course of a decade, the total worldwide capital investment would be comparable to the current discussion on tax cuts in the US alone. New industries like the electronic industry have shown that investments on this scale can indeed be made in a matter of decades. Whether or not it will be done depends on the perceived urgency of the problem.

13.7 FUTURE ACTIVITIES

To move from a simple dimensional analysis to a full development of the technology, a number of R&D issues will need to be addressed. One is the modeling and understanding of the airflow in order to define the maximum level of CO_2 that can be removed at any given site without untoward side effects. Preliminary studies suggest the feasibility of the approach in this regard [4].

Secondly, one needs to choose between various designs for contacting natural airflows. The situation is right now wide open and somewhat reminiscent of the early days in windmill design. Many vastly different designs competed with each other until finally a handful of particularly elegant and simple solutions took over.

Thirdly one needs to find a good sorbent. Currently the only sorbent that is environmentally acceptable and guaranteed to work is $Ca(OH)_2$. Other possibilities will need to be explored.

We are planning the analysis of several process implementations for the extraction of CO_2 from air. A successful process design, combined with any of the methods proposed for carbon dioxide disposal would be a major step toward solving the greenhouse gas problem and toward establishing a net zero carbon economy that would not have to abandon the vast fossil energy resources that could fuel economic prosperity for generations.

BIBLIOGRAPHY

1. Lackner, K.S., H.-J. Ziock, and P. Grimes. Carbon Dioxide Extraction from Air: Is it an Option? in Proceedings of the $_2$4th International Conference on Coal Utilization & Fuel Systems. 1999. Clearwater, Florida.
2. Lackner, K.S., P. Grimes, and H.-J. Ziock, The Case for Carbon Dioxide Extraction from Air. SourceBook--The Energy Industry's Journal of Issues, 1999. 57(9): p. 6-10.
3. Lackner, K.S., P. Grimes, and H.-J. Ziock, Carbon Dioxide Extraction from Air? 1999, Los Alamos National Laboratory: Los Alamos, NM.
4. Elliott, S., et al., Compensation of Atmospheric CO_2 Buildup through Engineered Chemical Sinkage. Geophysical Research Letters, $_2$001. $_2$8(7): p. 1$_2$35-1$_2$38.
5. Climate Change 1994, Radiative Forcing of Climate Change and an Evaluation of the IPCC IS9$_2$ Emission Scenario., ed. J.T. Houghton, et al. 1995, Cambridge: Cambridge University Press.
6. Seifritz, W., Partial and total reduction of CO_2 Emissions of Automobiles Using CO_2 Traps. Int. J. Hydrogen Energy, 1993. 18: p. $_2$43--$_2$51.
7. Gipe, P., Wind Energy Comes of Age. 1995, New York: John Wiley & Sons.
8. Ranney, J.W. and J.H. Cushman, Energy From Biomass, in The Energy Source Book, R. Howes and A. Fainberg, Editors. 1991, American Institute of Physics: New York.
9. Bishop, J.E., Wind Tower May Yield Cheap Power, in Wall Street Journal. June 9, 1993.

AUTHOR NOTES

CHAPTER 2

Acknowledgments

The authors thank the Indian Institute of Technology, Kharagpur, India, where this work was carried out.

CHAPTER 3

Acknowledgments

This project is financially supported by Key program of National Natural Science Foundation of China (50736001), the National Natural Science Foundation of China (51106018), the High-tech Research and Development Program of China (2006AA09A209-5), the Major State Basic Research Development Program of China (2009CB219507), the China Postdoctoral Science Foundation (2011M500553), the Scientific Research Foundation for Doctors of Liaoning Province (20111026) and the Fundamental Research Funds for the Central Universities of China.

CHAPTER 4

Acknowledgments

Special thanks are due to L.H. Spangler, L.M., Dobeck from Montana State University at Bozeman and S. Mitra from Brookhaven National Laboratory at Upton NY, for their assistance in preparing and carrying out the experiments at the ZERT facility. Support was provided by the U.S. Department of Energy, under Contract No. DE-AC02-98CH10886.

CHAPTER 6

Competing Interests

We have no competing interests with any organization to publish the manuscript 'On the potential for CO2 mineral storage in continental flood basalts – PHREEQC batch- and 1D diffusion-reaction simulations'.

Authors' Contributions

VTH Pham has made substantial contributions to conception and designs the manuscript. Her contributions are to acquisition of data, analysis and interpretation of data. PA did revising and has given final approval of the version to be published. HH involved in drafting the manuscript, writing methodology part and revising of the final version. All authors read and approved the final manuscript.

Acknowledgments

We highly appreciated constructive comments and suggestions from the reviewers. This work has been funded by SSC-Ramore (Subsurface storage of carbon dioxide - risk assessment, monitoring and remediation) project and (partially) by SUCCESS centre for CO2 storage under grant 193825/S60 from Research Council of Norway (RCN). SUCCESS is a consortium with partners from industry and science, hosted by Christian Michelsen Research as.

CHAPTER 7

Competing Interests

The authors declare that they have no competing interests.

Authors' Contributions

SAC is the primary author. She designed and directed the experiments, WWM modeled the experiments, and SCT conducted the experiments. All authors have read and approved the final manuscript.

Acknowledgments

This document was prepared as an account of work sponsored by an agency of the United States government. Neither the United States government nor Lawrence Livermore National Security, LLC, nor any of their employees makes any warranty, expressed or implied, or assumes any legal liability or responsibility for the accuracy, completeness, or usefulness of any information, apparatus, product, or process disclosed, or represents that its use would not infringe privately owned rights. Reference herein to any specific commercial product, process, or service by trade name, trademark, manufacturer, or otherwise does not necessarily constitute or imply its endorsement, recommendation, or favoring by the United States government or Lawrence Livermore National Security, LLC. The views and opinions of authors expressed herein do not necessarily state or reflect those of the United States government or Lawrence Livermore National Security, LLC, and shall not be used for advertising or product endorsement purposes.

This work performed under the auspices of the U.S. Department of Energy by Lawrence Livermore National Laboratory under Contract DE-AC52-07NA27344.

We thank three anonymous reviewers for their comments which significantly improved the manuscript. We acknowledge funding from and data provided by the Joint Industry Project (a consortium of BP, Statoil and Sonatrach) and the U.S. Department of Energy to investigate the importance of geochemical alteration at the In Salah CO2 storage project. We also thank Bill Ralph for his contributions to experiments early in the project and Mike Singleton and Pihong Zhao for chemical analyses, and Phil Ringrose for interest in geochemistry.

CHAPTER 8

Funding

This work was supported by National Institutes of Health grant GM083113 (to ETH) and by the Agriculture and Food Research Initiative Competitive Grants Program Grant no. 2010-65110-20488 from the USDA National Institute of Food and Agriculture. The funders had no role in study design, data collection and analysis, decision to publish, or preparation of the manuscript.

Competing Interests

The authors have declared that no competing interests exist.

Acknowledgments

We acknowledge Jeff F. Miller, Laura Weyrich and Xuqing Zhang for critical review of this manuscript and all members of the Harvill lab for support and helpful discussion. Additionally, we thank Alexia Karanikas and the Eunice Kennedy National Insitute of Child Health and Human Development (NICHD) Collaborative Pediatric Critical Care Research Network (CPCCRN) for B. pertussis strain CHOC 0012. The USDA is an equal opportunity provider and employer.

Authors' Contributions

Conceived and designed the experiments: SEH ML TN ETH. Performed the experiments: SEH ML TN DN. Analyzed the data: SEH ML TN DN. Contributed reagents/materials/analysis tools: TN ETH. Wrote the paper: SEH TN ETH.

CHAPTER 9

Funding

This work was supported by the Smithsonian Marine Science Network and a National Science Foundation Biocomplexity award (DEB-9981535). The funders had no role in study design, data collection and analysis, decision to publish, or preparation of the manuscript.

Competing Interests

The authors have declared that no competing interests exist.

Acknowledgments

We thank the Smithsonian Marine Science Network and the staff of the Carrie Bow Cay Research Station. We thank Dr Stephen Crooks, Dr Emily Pidgeon and other members of the Blue Carbon Working Group and acknowledge the support of the South East Queensland Climate Adaptation Research Initiative.

Authors' Contributions

Conceived and designed the experiments: CEL RWR ICF. Performed the experiments: CEL RWR ICF. Analyzed the data: CEL. Contributed reagents/materials/analysis tools: CEL RWR ICF. Wrote the paper: CEL RWR ICF.

CHAPTER 10

Funding

Lei Cheng was supported by a fellowship from the USDA-ARS Plant Science Research Unit (Raleigh, NC) and was in part by an USDA grant to SH (2009-35101-05351). The funders had no role in study design, data collection and analysis, decision to publish, or preparation of the manuscript.

Competing Interests

The authors have declared that no competing interests exist.

Acknowledgments

We thank Walter Pursley and Erin Silva of USDA-ARS Plant Science Research Unit for field assistance and maintaining CO_2 and O_3 facilities. We are grateful to Guillermo Ramirez and Lisa Lentz of Soil Science Department of North Carolina State University for their help in nitrogen analyses. We are also grateful to Dr. Yiqi Luo and two anonymous reviewers for their constructive suggestions.

Authors' Contributions

Conceived and designed the experiments: KOB FLB ELF SH LC CT. Performed the experiments: LC FLB CT. Analyzed the data: LC SH. Contributed reagents/materials/analysis tools: HDS TWR JLD. Wrote the paper: LC SH.

CHAPTER 11

Acknowledgments

The authors gratefully acknowledge Canay Ataoz librarian of the Eastern Mediterranean University for her endless support and kindness.

CHAPTER 12

Acknowledgments

This work has been supported by FEDER, ACCIO, and the Government of Catalonia (Funding TECRD12-1-0010).

INDEX

B

C

D

E

F

G

Q

R

S

T

U

V

W

X

Z

Printed in the United States
by Baker & Taylor Publisher Services